Titles in This Series

Volume

3 **Rémi Vaillancourt and Andrei L. Smirnov, Editors**
Asymptotic methods in mechanics
1993

2 **Philip D. Loewen**
Optimal control via nonsmooth analysis
1993

1 **M. Ram Murty, Editor**
Theta functions
1993

Asymptotic Methods in Mechanics

Volume 3

CRM
PROCEEDINGS &
LECTURE NOTES

Centre de Recherches Mathématiques
Université de Montréal

Asymptotic Methods in Mechanics

Rémi Vaillancourt
Andrei L. Smirnov
Editors

American Mathematical Society
Providence, Rhode Island USA

The production of this volume was s ur la Formation de Chercheurs et l'Aide à la Recherc l Sciences and Engineering Research Council of Canad

1991 *Mathematics Subject Classificat* 5B40, 73K15.

Library of Congress Cataloging-in-Publication Data

Asymptotic methods in mechanics/Rémi Vaillancourt, Andrei L. Smirnov.
p. cm. — (CRM proceedings & lecture notes, ISSN 1065-8580; v. 3)
Includes bibliographical references.
ISBN 0-8218-6993-0
1. Asymptotic expansions—Congresses. 2. Differential equations—Asymptotic theory—Congresses. 3. Mechanics, Applied—Congresses. I. Vaillancourt, Rémi, 1934– . II. Smirnov, Andrei L., 1956– .
QA295.A89 1993 93-41470
620.1—dc20 CIP

∞ The paper used in this book is acid-free and falls within the guidelines
established to ensure permanence and durability.
♻ Printed on recycled paper.

This volume was typeset using $\mathcal{A}\mathcal{M}\mathcal{S}$-TeX,
the American Mathematical Society's TeX macro system,
and submitted to the American Mathematical Society in camera-ready
form by the Centre de Recherches Mathématiques.

10 9 8 7 6 5 4 3 2 1 98 97 96 95 94 93

“To our understanding wives,
Luba and Jeanne”

Contents

Foreword

These Proceedings contain papers on mechanics (mostly for thin structures) essentially using the methods of asymptotic integration and asymptotic approaches in the choice of the mathematical models.

Only some of the problems arising in practice have exact solutions. Thus, one is usually forced to limit oneself to approximate solutions, among which one can distinguish the analytical and the numerical solutions.

In the beginning of this century, the analytical solutions took the main place, but lately, especially during the last decades, due to the advent of powerful computers and well-developed numerical methods, in particular, the finite element methods, the situation changed dramatically. Nowadays, most problems can be solved by means of numerical methods. Consequently, the approximate analytical solutions have been unjustly underestimated.

The analytical and numerical solutions have to complement each other. The approximate analytical solutions, among which the asymptotic solutions occupy a central place, clarify qualitatively the solution behaviour and provide a first approximation within a reasonable error. Note that the accuracy of the asymptotic solutions increases as the difficulties in the construction of the numerical solution increase. This is the case, for example, as the plate or shell relative thickness decreases.

These Proceedings consist of two parts.

The first part is a survey of the asymptotic methods applied to the analysis of thin-wall structures. It contains only the main results and a vast bibliography, to which the reader may refer for a detailed study.

The properties of the asymptotic expansions, including the asymptotic expansions of solutions with a parameter, the Laplace method, and the stationary phase and saddle point methods are also discussed.

In this survey, asymptotic expansions of both the singularly perturbed differential equations and the corresponding boundary value problems are analyzed. Both ordinary and partial differential equations are examined. For the linear equations, the WKB approximations and their generalizations to the case of turning points or transition lines (caustics) are considered and Vishik–Lyusternik's method is described.

Then the application of the asymptotic integration methods to free vibrations and buckling of thin shells is considered. Methods developed by Gol′denveizer, Maslov, and Wasow are discussed and numerous references to works containing applications of these methods to problems of shell theory are given.

On the other hand, the one-dimensional beam equation and two-dimensional shells equations themselves are the results of the reductions of three-dimensional problems of the theory of elasticity to one- or two-dimensional problems in the narrow zones. In this review the asymptotic algorithm of such reduction is briefly discussed.

In the last part of the survey, the singular degeneracy of nonlinear boundary value problems is considered. The algorithm of Vasil′eva–Butuzov and the method of Lyapunov–Schmidt are described.

The second part of these Proceedings contains original papers on mechanics in which an asymptotic approach is used.

The main group of papers [**19, 58, 59, 129, 188, 240, 241**] is devoted to the application of asymptotic methods to thin shell analysis.

In [**240**], the nonlinear axisymmetric deformation of a shell of revolution is considered. The effect of the load intensity on the asymptotic solution is examined.

In [**129**], the buckling of a convex shell of revolution under nonaxisymmetric loading is studied. The buckling mode localized in neighbourhoods of the edges is constructed.

In [**58, 59**], low-frequency free vibrations of a cylindrical shell with a slanted edge and connected shells are considered. The asymptotic solutions, which take into account the influence of the boundary conditions, are obtained for the free frequencies and vibration modes.

In [**189**], the acoustic problem for the vibrations of two concentric cylindrical shells in air is examined. The ratio of the shell radii for which the vibration mode is localized near the excitation zone is found.

Papers [**19**] and [**241**] contain a review of the authors' work on the asymptotic integration of thin shell equations. These papers were presented as reports at the First European Congress of Mathematics in July of 1992 in Paris.

Similar to paper [**189**], the papers [**9**], [**20**], [**47**], [**48**], [**152**] and [**192**] are concerned with the equations describing the behaviour of three-dimensional continuous media.

In [**9**] the perturbation method is used for the construction of an approximate solution to an axisymmetric electromagnetic problem which occurs in the diagnosis of material defects by nondestructive eddy current testing.

In [**20, 47, 48**], [**152**] and [**192**], one- and two-dimensional models for thin structures are used.

In [**47**] and [**48**], a coupled problem for thin flexible profiles moving in an ideal incompressible flow is considered. The case of two flexible profiles moving in an ideal incompressible flow, the interconnection of which is provided by the flow and the vortex traces behind the back edges of the profiles and the case of a flexible airfoil encountering a gust in an incompressible flow, are analyzed.

Papers [**20**] and [**152**] are concerned with the determination of the temperature deformations of metallic mirrors. In these papers the real structures are modeled by means of a system of one-dimensional plasto-elastic beams ([**152**]) or two-dimensional elastic thin plates ([**20**]), respectively. Numerical studies reveal the good quality of the proposed models.

In paper [**192**], the dynamic buckling of a rotating anisotropic shaft is considered. The asymptotic method is used to study the stability of a thin anisotropic shaft with a crack. This problem is very important for an early diagnosis is of a crack.

The general and main aim of these Proceedings demonstrate the importance of asymptotic methods and simplified asymptotic models and their possible applications to practical problems of mechanics.

Petr E. Tovstik

Preface

These Proceedings are the result of the scientific cooperation of researchers from the Laboratory of Applied Mechanics of the Institute of Mathematics and Mechanics at St. Petersburg State University, the Departments of Theoretical and Applied Mechanics and the Department of Hydroelasticity of the Faculty of Mathematics and Mechanics of the same University, the Department of Acoustics of the St. Petersburg Institute of Cinema Engineering, the Department of Applied Mathematics of the Riga Technical University and the Department of Mathematics of the University of Ottawa.

The contents are papers on mechanics (mostly on mechanics of thin structures) in which, basically, the problems are formulated by means of an asymptotic approach and the solutions are obtained by the methods of asymptotic integration.

Initially, it was planned to limit the proceedings to original papers containing new results. These papers are presented in the second part of this collection. However, as the papers were being collected, it was realized that, for the convenience of the reader, it would be useful to include a survey on the asymptotic methods in mechanics which are used in these papers. This survey, which comprises the first part of this book, may also serve as an introduction to the field.

A special feature of the collection is the vast bibliography on the applications of asymptotic methods to problems of mechanics, including Russian literature on asymptotic methods, which is generally translated in English, but still not widely known, despite its high quality. The editors made every effort to include the number of pages, in parentheses, for the monographs listed in the bibliography to the Survey, because many of the Russian books may not be easily available to the Western reader and it was thought that this extra piece of information would be useful.

The book is intended for researchers and engineers working in the analysis and construction of thin-walled structures and continuous media, and applied mathematicians, who are interested in asymptotic methods in problems of mechanics. The book would also be useful to advanced undergraduate and postgraduate students in Mathematics, Physics and Engineering.

This work was supported in part by the Natural Sciences and Engineering Research Council of Canada under grant A7691, the Centre de recherches mathématiques of the Université de Montréal, the Ministry of Science, Higher Education and Technical Policy of Russia under the theme #01910056114, and the Institute of Mathematics and Mechanics of St. Petersburg State University.

The opportunity given to the first editor to work in Professor J. A. Goldak's Laboratory in Mechanical Engineering, at Carleton University, through the Scientific Exchange Program between Carleton University and St. Petersburg State

University under the direction of Professor D. R. F. Taylor, has been very helpful in the final stage of preparation of the book.

The editors would like to underline the special role of Professor P. E. Tovstik, who initiated the study of asymptotic methods applied to problems of solid mechanics at St. Petersburg (then Leningrad) State University more than 25 years ago and who lectured to more than half of the contributors.

The editors did the major translation and edition work for the book. Our friend, Dr. A. A. Kolyshkin, while visiting Ottawa, did the formidable task of inserting all the corrections needed in the first revisions of the text. Thanks are due Mr. N. Filippov (St. Petersburg State University) for the excellent bitmap drawings, which were used as templates, by the editors, for the final postcript figures. A very special thank is due to Mrs. V. Sergeeva (St. Petersburg State University) who typeset the text in $\mathcal{AMS}$-TEX.

Andrei L. Smirnov and Rémi Vaillancourt

Part 1

A Survey

Asymptotic Methods in Mechanics with Applications to Thin Shells and Plates

Svetlana M. Bauer, Sergei B. Filippov,
Andrei L. Smirnov, and Petr E. Tovstik

Introduction

Engineering and applied mechanics problems are often solved by analytical or numerical methods. Since in the applications known exact analytical solutions are rather rare, in most cases one resorts to approximate solutions. Among approximate analytical solutions, asymptotic methods based on expansions in powers of small or large parameters occupy a central place.

Asymptotic expansions are often adapted to the problems to be solved. The aim of the present survey is to describe classes of asymptotic expansions which occur in the approximate evaluation of integrals and in the solution of algebraic and transcendental equations. A special attention will be devoted to asymptotic solutions of boundary value problems for ordinary and partial differential equations. In considering these problems, we shall discuss only the so-called singular degeneracy characterized by the existence of a small parameter in the highest order derivatives. This degeneracy usually provides divergent asymptotic series.

In Chapter 1, we define asymptotic expansion and give expansions for the evaluation of integrals and the solution of algebraic equations.

Chapter 2 is concerned mainly with homogeneous systems of linear first order differential equations with a small parameter in the derivative,

$$\mu \frac{dy}{dx} = A(x, \mu) y, \qquad \mu > 0,$$

and methods for constructing the asymptotic expansions of its solutions as $\mu \to 0$. The asymptotic integration of equations with turning points is considered in detail.

In Chapter 3, we consider asymptotic methods for the solution of linear one-dimensional singular problems which degenerate to lower-order problems. We focus our attention on the regular Vishik–Lyusternik degeneracy. Nonregular degeneracy is also discussed. Cases where the system of equations has integrals with different indices of variation are also considered. Finally, applications to problems of shell vibrations and buckling and other problems of solid mechanics are analyzed.

In Chapter 4, since asymptotic methods for the integration of partial differential equations depends on the particular problems considered, we limit our discussion to those methods which are applied to problems of solid mechanics, especially to

1991 *Mathematics Subject Classification.* Primary: 34E; Secondary: 35B40, 73K15.

Key words and phrases. Asymptotic methods, singular perturbations, shell buckling and vibration.

This is the final form of the paper.

problems of the linear theory of thin shells and plates. We shall not consider any of the methods used in problems of viscous fluids.

In Chapter 5, we consider singular perturbation problems which have applications to nonlinear problems of the theory of shells. Vasil′eva's method for nonlinear ordinary differential equations and Lyapunov–Schmidt's method for constructing small solutions of boundary value problems for nonlinear partial differential equations.

Some of the problems of asymptotic integration are not discussed in this review, in particular, perturbation methods and asymptotic expansions for the solution of ordinary differential equations with right-hand sides which depend regularly upon a small parameter (see for example the monographs by N. N. Bogolubov and Yu. A. Mitropolsky [**25, 26**], K. V. Kholshevnikov [**92**], I. G. Malkin [**122**], N. N. Moiseev [**135**] and A. H. Nayfeh [**139, 140**]) and the integration of equations with rapidly changing coefficients (see for example the paper by I. V. Andrianov *et al.* [**8**] and the monographs by N. S. Bakhvalov and G. P. Panasenko [**16**] and E. Sanchez-Palencia [**185**]).

While discussing the different classes of asymptotic expansions, we formulate the main results and provide references, where the problems are discussed in greater detail. Since the literature on this subject is very vast, our list of references is necessarily incomplete. We prefer to refer to monographs and handbooks whenever possible.

The authors' interests in the theory of thin plates and thin shells explains why the literature in this area is discussed in greater detail.

CHAPTER 1

Asymptotic Expansions

In this chapter we define asymptotic expansion and give expansions for the evaluation of definite integrals and the solution of algebraic equations. We introduce also the index of variation of a function.

1.1. Function estimations

We recall the definitions of the symbols O, o and $\sim$. Let the functions $f(z)$ and $g(z)$ be defined on a set S of the complex numbers, $\mathbb{C}$, or the real numbers, $\mathbb{R}$, and let a be a point of accumulation of points of S. In this section we assume that $z \in S$.

NOTATION 1.1. We write

$$f(z) = O\left(g(z)\right), \qquad \text{as } z \longrightarrow a,$$

if there exists a constant C and a neighbourhood U of a, such that

$$|f(z)| \leq C|g(z)|, \qquad \forall z \in U.$$

We write

$$f(z) = O\left(g(z)\right)$$

if there exists a constant C such that the inequality

$$|f(z)| \leq C|g(z)|$$

holds for all z in S.

We write

$$f(z) = o\left(g(z)\right), \qquad \text{as } z \longrightarrow a,$$

if

$$\lim_{z \to a} \frac{f(z)}{g(z)} = 0. \tag{1.1.1}$$

It follows that $f = O(g)$ means that the order of the function f is not larger than the order of the function g, and $f = o(g)$ means that the order of f is less than the order of g, as $z \longrightarrow a$.

If the orders of f and g as $z \longrightarrow a$ are equal, we write

$$f \sim g, \qquad \text{as } z \longrightarrow a.$$

In this case $f = O(g)$ and $g = O(f)$ hold simultaneously, as $z \longrightarrow a$. Many authors use the symbol $\sim$ only if the functions f and g are equivalent, i.e.

$$\lim_{z \to a} \frac{f(z)}{g(z)} = 1. \tag{1.1.2}$$

In the sequel we shall use $f \sim g$ even if equality (1.1.2) does not hold with the constant 1.

We list some examples in which $S = \mathbb{R}$, the set of real numbers:

$$
\begin{aligned}
\ln x &= o(x^\alpha), && \text{as } x \to \infty, \text{ if } \alpha > 0;\\
\sin x &= O(1), && \forall x \in \mathbb{R};\\
x^\alpha &= o(e^x), && \text{as } x \to \infty, \text{ if } \alpha > 0;\\
x^{\alpha_1} &= o(x^{\alpha_2}), && \text{as } x \to 0, \text{ if } \alpha_1 > \alpha_2 > 0;\\
\sin x &\sim x, && \text{as } x \to 0;\\
\sin x - x &\sim x^3, && \text{as } x \to 0.
\end{aligned}
\tag{1.1.3}
$$

Operations on the symbols O and o and a large number of examples may be found, for example, in the monographs by N. G. de Bruijn [**28**], A. Erdélyi [**46**], M. F. Fedoryuk [**51, 53**] and F. W. J. Olver [**147**].

1.2. Asymptotic series

1.2.1. Asymptotic sequences and expansions. Consider a sequence of functions

$$\varphi_n(z), \qquad n = 0, 1, 2, \ldots, \tag{1.2.1}$$

defined on S and let a be a point of accumulation of S.

DEFINITION 1.1. The sequence (1.2.1) is said to be *asymptotic* as $z \to a$, if for any natural number $n \geq 0$,

$$\varphi_{n+1}(z) = o\left(\varphi_n(z)\right), \qquad \text{as } z \to a. \tag{1.2.2}$$

We list a few examples of asymptotic sequences, where $S = \mathbb{R}$:

$$
\begin{aligned}
&(z-a)^n, && \text{as } z \to a,\\
&z^{-n}, && \text{as } z \to \infty,\\
&e^{\lambda_n z}, && \text{as } z \to \infty, && \text{if } \Re(\lambda_{n+1} - \lambda_n) < 0,\\
&\Phi(z)(z-a)^{\alpha_n}, && \text{as } z \to a, && \text{if } \alpha_{n+1} > \alpha_n,
\end{aligned}
\tag{1.2.3}
$$

where $\Phi(z)$ is an arbitrary function in S and $\Re\lambda$ denotes the real part of $\lambda \in \mathbb{C}$. The first two sequences of (1.2.3) are called *power sequences*.

DEFINITION 1.2. If the function $f(z)$ is defined on S, then the series

$$f(z) \simeq \sum_{n=0}^{\infty} a_n \varphi_n(z), \qquad \text{as } z \to a, \tag{1.2.4}$$

is called an *asymptotic expansion of $f(z)$ in the Poincaré sense* by means of the asymptotic sequence (1.2.1) if there are constants a_n such that for any integer $N \geq 0$

$$R_N(z) = f(z) - \sum_{n=0}^{N} a_n \varphi_n(z) = o\left(\varphi_N(z)\right), \qquad \text{as } z \to a. \tag{1.2.5}$$

Note that the series (1.2.4) may diverge. Relation (1.2.5) generates a sequence of approximate formulae:

(1.2.6)
$$f(z) \approx \sum_{n=0}^{N} a_n \varphi_n(z), \qquad N = 0, 1, 2, \dots,$$

the error of which, $R_N(z)$, is of the order of the first term omitted in series (1.2.4),

$$R_N(z) \sim \varphi_{N+1}(z).$$

A large number of asymptotic expansions for special functions may be found in the handbook [**1**] and in the books by M. F. Fedoryuk [**53**], E. Jahnke *et al.* [**83**], and F. W. J. Olver [**147**]. For example, the exponential integral has the asymptotic expansion [**1**]:

(1.2.7)
$$E(z) = \int_z^{\infty} t^{-1} e^{-t} dt \simeq e^{-z} \sum_{n=0}^{\infty} \frac{(-1)^n n!}{z^{n+1}}, \qquad \text{as } z \to \infty,$$

where the series on the right-hand side is divergent for all z, but the error $R_N(z)$ (see (1.2.5)) is not larger than the first term omitted.

1.2.2. Properties of asymptotic series. Properties of asymptotic series and operations on them are discussed in the books [**28, 46, 51, 53, 147**] quoted previously. We mention only the main operations.

The asymptotic expansion (1.2.4) of a function $f(z)$ by a sequence (1.2.1) is unique, i.e. the coefficients a_n in (1.2.4) are determined uniquely by the limit

(1.2.8)
$$a_n = \lim_{z \to a} \frac{f(z) - \sum_{k=0}^{n-1} a_k \varphi_k(z)}{\varphi_n(z)}.$$

On the other hand, an asymptotic series does not determine a function $f(z)$ uniquely. Indeed, the functions $f(z) + f_1(z)$ and $f(z)$ have the same asymptotic expansion (1.2.4) if for each $n \geq 0$

(1.2.9)
$$f_1(z) = o\left(\varphi_n(z)\right), \qquad \text{as } z \to a.$$

For example, the functions $E(z)$ and $E(z) + e^{-2z}$ have the same expansion (1.2.7).

Therefore, in choosing the asymptotic sequence (1.2.1) for the asymptotic expansion of a function $f(z)$, it is necessary to take into account the behaviour of $f(z)$ as $z \to a$. Hence there are limitations on the functions $\varphi_n(z)$. For an improper choice of $\varphi_n(z)$, expansion (1.2.4) either does not exist or yields the trivial result $f(z) \simeq 0$. For example, if $\varphi_n(z) = z^{-n}$, $e^{-z} \simeq 0$ as $z \to \infty$, i.e. each $a_n = 0$ in (1.2.4).

Asymptotic expansions may be added and multiplied by functions, and, under specific assumptions, they can be differentiated and integrated. The asymptotic power series

(1.2.10)
$$f(z) \simeq \sum_{n=0}^{\infty} a_n z^n, \qquad g(z) \simeq \sum_{n=0}^{\infty} b_n z^n, \qquad \text{as } z \to 0,$$

can be multiplied, $f(z)g(z)$, divided, $f(z)/g(z)$, if $b_0 \neq 0$ and substituted one into another, $f\left[g(z)\right]$, if $b_0 = 0$, and the resulting asymptotic expansions are obtained by the rules for convergent MacLaurin series.

A convergent power series of a complex variable z,

$$\sum_{n=0}^{\infty} a_n z^n, \tag{1.2.11}$$

on an open disk $S = \{z : |z| < R\}$ determines an *analytic function*, $f(z)$, on S and its asymptotic expansion as $z \to 0$ coincides with (1.2.11). It is not difficult to find series (1.2.11) which diverge for all $z \neq 0$ (for example, if $a_n = n!$). It is interesting to answer the following question: Under which conditions on the coefficients a_n, is the series (1.2.11) the asymptotic expansion of a function $f(z)$, analytic in a sector

$$S = \{z : |z| < R, \qquad \alpha_1 < \arg z < \alpha_2\}, \tag{1.2.12}$$

as $z \to 0$? The answer is unexpected: no conditions need be imposed either on the coefficients a_n nor on the angle (α_1, α_2) (see also A. Erdélyi [**46**]).

We note the main difference between convergent series and divergent asymptotic series from the computational point of view. If the series (1.2.4) is convergent for some z_0, then

$$R_N(z_0) \to 0, \qquad \text{as } N \to \infty, \tag{1.2.13}$$

i.e. the value of $f(z_0)$ may be found, in principle, to any precision by taking N large enough. On the other hand, if the asymptotic series (1.2.4) is divergent, then the error $R_N(z_0)$ of the series truncated after the $(N+1)$-st term reaches a minimum for some $N = N(z_0)$. In this case, however, the error decreases as $z_0 \to a$.

For simplicity, in this work we shall use the term *analytic function* for a real analytic function of a real variable and a complex analytic function of a complex variable, unless the context is not clear. The term *holomorphic function* which, in this work, is equivalent to complex analytic function will not be used.

In very easy cases, for example for special functions, the complete asymptotic series is already constructed. In other cases, in particular, for the asymptotic integration of differential equations, because of the numerous calculations, we are forced to restrict ourselves to one or two terms of the series (see the following sections).

In the following Sections 1.3–1.5, we consider the evaluation of integrals with large parameters.

1.3. Laplace's method

1.3.1. Asymptotic expansions of the simplest integrals. Let us consider the integral

$$F(\lambda) = \int_a^b \varphi(x)\, e^{\lambda h(x)}\, dx, \qquad -\infty \le a < b \le \infty, \tag{1.3.1}$$

which contains the large parameter $\lambda > 0$. We look for the asymptotic expansion of $F(\lambda)$ as $\lambda \to \infty$. A vast literature is devoted to this question, for example the books [**28, 46, 50, 51, 53, 147**] and the handbooks [**1, 75, 83, 174**]. We mention, in particular, the three volumes of the monograph by E. Ya. Riekstynsh [**174**], in which integrals of type (1.3.1) and integrals of more general types are studied. Below we mention only the simplest results for $\varphi(x)$ and $h(x)$ analytic on $a < x < b$ and $h(x)$ real.

CASE A. If $h(x) = -x$ and $\varphi(x)$ has a convergent or asymptotic power expansion

$$\varphi(x) = \sum_{n=0}^{\infty} a_n x^{n+\alpha}, \qquad \alpha > -1, \tag{1.3.2}$$

in a neighbourhood of $x = 0$, then the integral

$$F(\lambda) = \int_0^b \varphi(x)\, e^{-\lambda x}\, dx \tag{1.3.3}$$

has the asymptotic expansion

$$F(\lambda) \simeq \sum_{n=0}^{\infty} \frac{\Gamma(n+\alpha+1)}{\lambda^{n+\alpha+1}}\, a_n, \qquad \text{as } \lambda \to \infty, \tag{1.3.4}$$

where $\Gamma(z)$ is Euler's Gamma function. Indeed a term by term integration in (1.3.3) yields

$$F(\lambda) = \sum_{n=0}^{\infty} a_n F_n(\lambda), \qquad F_n(\lambda) = \int_0^b x^{n+\alpha}\, e^{-\lambda x}\, dx.$$

In the case $c > b > 0$, it may be shown, by integration by parts, that

$$\int_b^c x^{n+\alpha}\, e^{-\lambda x}\, dx = o(\lambda^N),$$

where N is an arbitrary integer. The last relation is valid also for $c = \infty$. Therefore

$$F_n(\lambda) = \int_0^{\infty} x^{n+\alpha}\, e^{-\lambda x}\, dx - \int_b^{\infty} x^{n+\alpha}\, e^{-\lambda x}\, dx = \int_0^{\infty} x^{n+\alpha}\, e^{-\lambda x}\, dx + o(\lambda^N).$$

Since

$$\int_0^{\infty} x^{n+\alpha}\, e^{-\lambda x}\, dx = \frac{1}{\lambda^{n+\alpha+1}} \int_0^{\infty} z^{n+\alpha}\, e^{-z}\, dz = \frac{\Gamma(n+\alpha+1)}{\lambda^{n+\alpha+1}},$$

then we obtain formula (1.3.4).

REMARK 1.1. Whether the series (1.3.4) diverges or converges depends upon the coefficients a_n. But even if this series converges, its sum $\tilde{F}(\lambda)$ is not necessarily equals to $F(\lambda)$. For example, if we apply (1.3.4), with $\alpha = 0$, to the integral

$$F(\lambda) = \int_0^{\pi/2} e^{-\lambda x} \sin x\, dx, \tag{1.3.5}$$

we get

$$F(\lambda) \simeq \sum_{k=0}^{\infty} \frac{(-1)^k}{\lambda^{2k+2}} = \frac{1}{1+\lambda^2} = \tilde{F}(\lambda). \tag{1.3.6}$$

The exact value of the integral,

$$F(\lambda) = \frac{1}{1+\lambda^2}\left(1 - \lambda\, e^{-\pi\lambda/2}\right), \tag{1.3.7}$$

differs from the value of the function $\tilde{F}(\lambda)$ to which the series (1.3.6) converges, because, in this and the following cases, in the series (1.3.4) or (1.3.6) only the contribution of a neighbourhood of the point where $e^{\lambda h(x)}$ attains its maximum is taken into account in the integral while the exponentially small terms are neglected.

CASE B. If, in (1.3.1), we take

$$\varphi(x) = \sum_{n=0}^{\infty} a_n x^n, \qquad h(x) = -x^2, \qquad a < 0,\, b > 0, \tag{1.3.8}$$

then the integral has the expansion

$$\int_a^b \varphi(x)\, e^{-\lambda x^2}\, dx \simeq \sum_{n=0}^{\infty} \frac{\Gamma(n+1/2)}{\lambda^{n+1/2}}\, a_{2n}, \qquad \text{as } \lambda \to \infty. \tag{1.3.9}$$

This series represents the contribution of the integral in a neighbourhood of $x = 0$ where $e^{-\lambda x^2}$ attains its maximum.

1.3.2. Asymptotic expansions of more complex integrals. Let $h(x)$ be analytic on $a \le x \le b$ and suppose that h attains its maximum at $x = c$, and $h(x) < h(c)$ if $x \ne c$. We consider two particular cases.

CASE A. Let $c = a$ and $h'(a) < 0$. If, in (1.3.1), we change the variable of integration x to $u = h(a) - h(x)$, then the integral reduces to (1.3.3) with the asymptotic expansion

$$F(\lambda) \simeq e^{\lambda h(a)} \sum_{n=0}^{\infty} n!\, a_n \lambda^{-n-1}, \qquad \text{as } \lambda \to \infty. \tag{1.3.10}$$

From the above change of variable we obtain $x(u)$ as an analytic function in a neighbourhood of the point $x = a$. The coefficients a_n in (1.3.10) are determined from the expansion

$$\varphi\left[x(u)\right] x'(u) = \sum_{n=0}^{\infty} a_n u^n. \tag{1.3.11}$$

CASE B. Now let $a < c < b$ and $h''(c) < 0$. If we substitute the variable $h(c) - h(x) = u^2$ into (1.3.1), then the integral is transformed into (1.3.8) and

$$F(\lambda) \simeq e^{\lambda h(c)} \sum_{n=0}^{\infty} \frac{\Gamma(n+1/2)}{\lambda^{n+1/2}}\, a_{2n}, \qquad \text{as } \lambda \to \infty, \tag{1.3.12}$$

where the coefficients a_n are again determined from (1.3.11). In particular, if we keep only the first term of the series (1.3.12), we have

$$F(\lambda) = e^{\lambda h(c)} \left[\varphi(c) \sqrt{-\frac{2\pi}{\lambda h''(c)}} + O\left(\lambda^{-3/2}\right) \right], \qquad \text{as } \lambda \to \infty. \tag{1.3.13}$$

We can similarly consider cases where higher order derivatives are zero at $x = c$.

1.4. Oscillatory integrals. The stationary phase method

1.4.1. Asymptotic expansions of the integral of an oscillating function. We consider the integral

$$F(\lambda) = \int_a^b \varphi(x)\, e^{i\lambda S(x)}\, dx, \qquad -\infty \le a < b \le \infty, \tag{1.4.1}$$

where $\lambda > 0$ is a large parameter and $S(x)$ is a real function. At first, let $[a, b]$ be a finite interval in which

$$\varphi(x) \in C^{N+1}, \qquad S(x) \in C^{N+2}, \qquad S'(x) \neq 0, \tag{1.4.2}$$

i.e. $\varphi(x)$ and $S(x)$ have $N+1$ and $N+2$ continuous derivatives, respectively. Then

$$F(\lambda) = \sum_{n=0}^{N} \frac{1}{(i\lambda)^{n+1}} \left\{ e^{i\lambda S(x)} \left[\frac{1}{S'(x)} \frac{d}{dx} \right]^n \left[\frac{\varphi(x)}{S'(x)} \right] \right\} \Bigg|_{x=a}^{x=b} + o\left(\lambda^{-N-1}\right), \quad \text{as } \lambda \to \infty. \tag{1.4.3}$$

Formula (1.4.3) is obtained by integration by parts and use of the Riemann–Lebesgue lemma (see N. G. de Bruijn [**28**] and F. W. J. Olver [**147**]) which states that

$$\int_a^b \varphi(x)\, e^{i\lambda x}\, dx = o(1), \qquad \text{as } \lambda \to \infty, \tag{1.4.4}$$

for piecewise continuous functions $\varphi(x)$.

Under assumptions (1.4.2) it follows that the asymptotic expansion (1.4.3) of $F(\lambda)$ depends only upon the values of the functions $\varphi(x)$ and $S(x)$ and their derivatives at the end points of the interval of integration. If $\varphi(x)$ and $S(x)$ or their derivatives have a finite number of points of discontinuity, the interval of integration, $[a, b]$, can be split into subintervals in such a way that the points of discontinuity coincide with the end points of the subintervals, and formula (1.4.3) can be applied to each subinterval.

1.4.2. Stationary point. A point $x = c$, in which $S'(c) = 0$, is called a *stationary* point. If there exist stationary points, then expansion (1.4.3) is not valid anymore. We consider integral (1.4.1) with only one stationary point, $x = c$, and suppose that

$$\varphi(x),\ S(x) \in C^{\infty}, \qquad S''(c) \neq 0, \qquad a < c < b, \tag{1.4.5}$$

and

$$\varphi^{(n)}(a) = \varphi^{(n)}(b) = 0, \qquad n = 0, 1, 2, \ldots \tag{1.4.6}$$

Conditions (1.4.6) ensure that the integration picks up the contribution of a neighbourhood of the stationary point and excludes the contribution of the end points $x = a$ and $x = b$. Thus we obtain

$$F(\lambda) = e^{i\lambda S(c) + (\pi i/4) S_0} \sum_{n=0}^{\infty} \frac{\Gamma(n + 1/2)}{\lambda^{n+1/2}}\, a_{2n}\, e^{(n\pi i/4) S_0}, \tag{1.4.7}$$

where

$$a_m = \left\{ \left[S_1(x, c) \frac{d}{dx} \right]^m [S_1(x, c) \varphi(x)] \right\} \Bigg|_{x=c} \tag{1.4.8}$$

and

$$S_1(x, c) = \frac{\sqrt{2 S_0 [S(x) - S(c)]}}{S'(x)}, \qquad S_0 = \operatorname{sgn}\left[S''(c)\right],$$

where

$$\operatorname{sgn} t = \begin{cases} +1, & \text{if } t > 0, \\ -1, & \text{if } t < 0. \end{cases}$$

1.4.3. Neutralizer. The neutralizers which were introduced by Van der Korput (see A. Erdélyi [**46**]) allow one to express an integrand as the sum of two functions, each of which, together with its derivatives, is zero at one end of the interval of integration. Thus we have the following definition.

DEFINITION 1.3. A *neutralizer* is a smooth function, $\nu(x) \in C^\infty$, which satisfies the following conditions:

$$\nu(a) = 1, \qquad \nu(b) = 0, \qquad \nu^{(n)}(a) = \nu^{(n)}(b) = 0, \qquad n = 1, 2, \dots \tag{1.4.9}$$

The function (see [**46**])

$$\nu(x) = \left[\int_a^b \exp\left(-\frac{1}{t-a} - \frac{1}{b-t}\right) dt\right]^{-1} \int_x^b \exp\left(-\frac{1}{t-a} - \frac{1}{b-t}\right) dt \tag{1.4.10}$$

is an example of neutralizer.

By splitting the interval (a, b) into two parts and representing $\varphi(x)$ in the form

$$\varphi(x) = \varphi(x)\nu(x) + \varphi(x)\,(1 - \nu(x)) \tag{1.4.11}$$

we can achieve that only the neighbourhood of one of the end points of the interval of integration contributes to the integral. To evaluate this contribution in the case where the functions $\varphi(x)$ and $S(x)$ have singularities at $x = 0$ one can use Erdélyi's lemma (see M. F. Fedoryuk [**51**]), which we quote without proof.

LEMMA 1.1 (ERDÉLYI). *Let*

$$F(\lambda) = \int_0^b x^{\beta-1}\varphi(x)\, e^{i\lambda x^\alpha}\, dx. \tag{1.4.12}$$

If

$$\alpha \geq 1, \qquad \beta > 0, \qquad \varphi(x) \simeq \sum_{n=0}^{\infty} a_n x^n, \qquad \varphi^{(n)}(b) = 0, \qquad n = 0, 1, 2, \dots,$$

then

$$F(\lambda) \simeq \sum_{n=0}^{\infty} e^{\pi i(n+\beta)/(2\alpha)}\Gamma\left(\frac{n+\beta}{\alpha}\right)\frac{a_n}{\alpha}\,\lambda^{-(n+\beta)/\alpha}, \qquad \text{as } \lambda \to \infty. \tag{1.4.13}$$

Therefore only the neighbourhoods of the end points, a and b, of the interval of integration, the points of discontinuity of $\varphi(x)$ and $S(x)$ and the stationary points of $S(x)$ contribute to the integral (1.3.1). In neighbourhoods of any other points the oscillation of $\exp(i\lambda S(x))$ annihilates any contribution to the integral.

For example, the asymptotic expansion (1.4.3) of the integral

$$F(\lambda) = \int_{-\infty}^{+\infty} \frac{e^{i\lambda x}}{1+x^2}\, dx = \pi\, e^{-\lambda} \tag{1.4.14}$$

contains only zero terms because there are no points of discontinuity nor stationary points and the contributions of any neighbourhoods of the ends points are equal to zero because $\varphi^{(n)} \to 0$ as $x \to \pm\infty$.

Consider a smooth periodic function $\varphi(x) \in C^\infty$, $\varphi(x+2\pi) = \varphi(x)$. The Fourier coefficients of $\varphi(x)$ are

$$F(n) = \frac{1}{2\pi}\int_0^{2\pi} \varphi(x)\, e^{inx}\, dx, \qquad n \in \mathbb{Z}. \tag{1.4.15}$$

Because $\varphi(x)$ is periodic, the contributions of the end points of the interval of integration cancel out and we get the estimate

$$F(n) = o(n^{-N}), \qquad \text{as } n \to \infty,$$

for any $N > 0$.

1.5. The saddle point method

1.5.1. Curvilinear integrals in the complex plane. We seek the asymptotic expansion of the integral

$$F(\lambda) = \int_\gamma \varphi(z)\, e^{\lambda h(z)} dz, \qquad \lambda > 0, \tag{1.5.1}$$

as $\lambda \to \infty$, where $\varphi(z)$ and $h(z)$ are analytic in a domain S of the complex plane $\mathbb{C}$ and γ is a curve in S, by means of the saddle point method. This method comprises the following two steps:

(1) a deformation of the contour γ into a more convenient contour γ_0 for asymptotic estimations and
(2) the evaluation of the integral along the deformed contour by Laplace's method (see Section 1.3).

We provide a short description of this method; for a full description, see, for example, the previously quoted books [**28, 46, 51, 53, 147**]. As $\lambda \to \infty$, the maximum absolute value of the integrand is attained at the points z, in which the function $\Re h(z)$ is maximum. Since the integrand is an analytic function, the value of the integral is invariant under any deformation of γ in S which keeps the endpoints fixed. Let us suppose that among these contours there exists a contour $\gamma_0 \subset S$ such that the value of the minimax

$$\min_\gamma \max_{z\in\gamma} \Re h(z)$$

is attained. We assume also that a deformation of the contour γ into γ_0 inside the domain S is possible. Then

$$F(\lambda) = \int_{\gamma_0} \varphi(z)\, e^{\lambda h(z)} dz. \tag{1.5.2}$$

The asymptotic expansion for integral (1.5.2) may be obtained by Laplace's method (see Section 1.3). For simplicity, let $z = c \in \gamma_0$ be the only point where $\Re h(z)$ is maximum.

If $z = c$ coincides with an endpoint of γ_0 and $h'(c) \neq 0$, the expansion $F(\lambda)$ can be found by formula (1.3.10).

1.5.2. Saddle points. If $z = c$ is an interior point of γ_0, then it follows from the minimax property of the contour γ_0 that c is a *stationary point* of the function $f(x,y) = \Re h(z)$, where $z = x + iy$, since, at this point, $\frac{\partial f}{\partial x} = \frac{\partial f}{\partial y} = 0$ and it follows from the Cauchy–Riemann equations that $h'(c) = 0$.

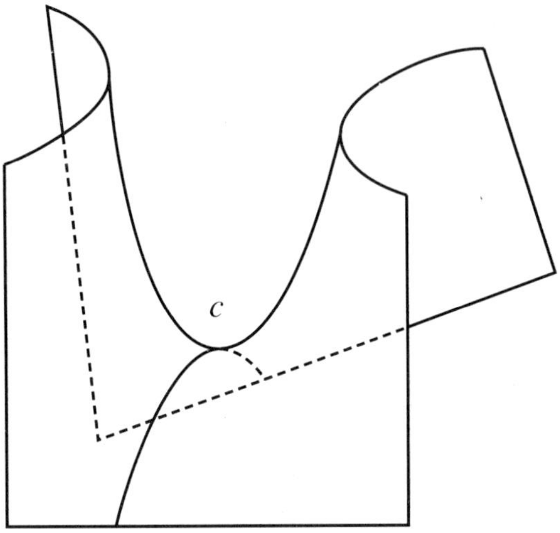

FIGURE 1. A simple saddle point.

DEFINITION 1.4. Suppose that the value of the minimax

$$\min_{\gamma} \max_{z\in\gamma} \Re h(z)$$

is attained on the contour $\gamma_0 \subset S$. An interior stationary point $z = c$ of γ_0 is called a *saddle point*, and the value $\Re h(c)$ is called the *amplitude* of the saddle point. If $h''(c) \neq 0$, the saddle point is said to be *simple*.

A simple saddle point c is shown in Fig. 1.

The asymptotic expansion of the contribution of a neighbourhood of a simple saddle point to integral (1.5.1) is given by formula (1.3.12).

The two lines $\Im\,(h(z) - h(c)) = 0$, along which the function $f(x, y)$ changes fastest, intersect at right angle through this point. The one of these lines along which the function $f(x, y)$ decreases away from the point $z = c$, is called *the line of steepest descent.*

The contour γ_0 may be deformed in such a way that, in a neighbourhood of the point $z = c$, it coincides with the line of steepest descent. Then, in this neighbourhood, $\Im h(z) = \text{const}$ and it is therefore convenient to use Laplace's method (see (1.3.12)) for the estimation of integral (1.5.2).

There are no general algorithms to construct the curve γ_0. For the specific integral (1.5.1), it is first necessary to find the saddle points and draw the lines of steepest descent through them. Then one has to try to deform the contour γ so that it partly coincides with the lines of steepest descent and perhaps other curves along which the integrals may be neglected for being asymptotically small.

The saddle point method allows a generalization to integrals of type (1.5.1) for functions of several variables (see M. F. Fedoryuk [**51, 53**]).

The following Section 1.6 illustrates an application of the saddle point method.

1.6. Asymptotic properties of Airy's functions

1.6.1. Representation of Airy's functions by contour integrals. Airy's functions can be used for the asymptotic integration of differential equations with

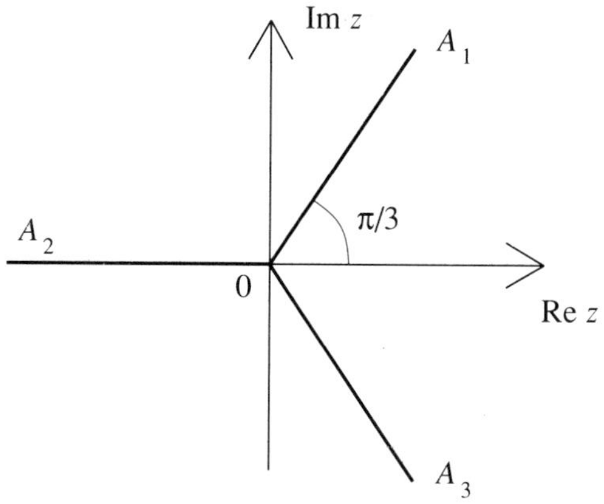

FIGURE 2. Contours of integration A_1OA_3 and $A_2OA_1 + A_2OA_3$.

a parameter and turning points (see Subsection 2.3.4).

The solutions of Airy's equation,

$$\frac{d^2w}{d\eta^2} - \eta w = 0, \tag{1.6.1}$$

are entire functions of the complex argument η. Applying the Laplace transformation we can represent the solutions $w_k(\eta)$ in the form of contour integrals

$$w_k(\eta) = C_k \int_{\gamma_k} e^{-\eta z + z^3/3}\, dz, \tag{1.6.2}$$

where the contours γ_k approach infinity along rays OA_n with $\arg z = \frac{\pi}{3}(-1+2n)$, $n = 1, 2, 3$ (see Fig. 2), and along which the integrand in (1.6.2) goes to zero as $z \to \infty$.

We obtain the standard Airy functions, $\mathrm{Ai}(\eta)$ and $\mathrm{Bi}(\eta)$, which are real for real η, if, in (1.6.2), we assume

$$\begin{aligned} &\mathrm{Ai}(\eta) = w_1(\eta), \qquad C_1 = \frac{i}{2\pi}, \qquad \gamma_1 = A_1OA_3, \\ &\mathrm{Bi}(\eta) = w_2(\eta), \qquad C_2 = \frac{1}{2\pi}, \qquad \gamma_2 = A_2OA_1 + A_2OA_3. \end{aligned} \tag{1.6.3}$$

Taylor's expansions of Airy's functions are obtained from (1.6.2) by using the expansion

$$e^{-\eta z} = \sum_{k=0}^{\infty} \frac{(-\eta z)^k}{k!} \tag{1.6.4}$$

and integrating term by term with respect to z along the rays OA_n (see [**1**]); thus

$$\mathrm{Ai}(\eta) = a_1 f_1(\eta) - a_2 f_2(\eta), \qquad \mathrm{Bi}(\eta) = \sqrt{3}\left[a_1 f_1(\eta) + a_2 f_2(\eta)\right], \tag{1.6.5}$$

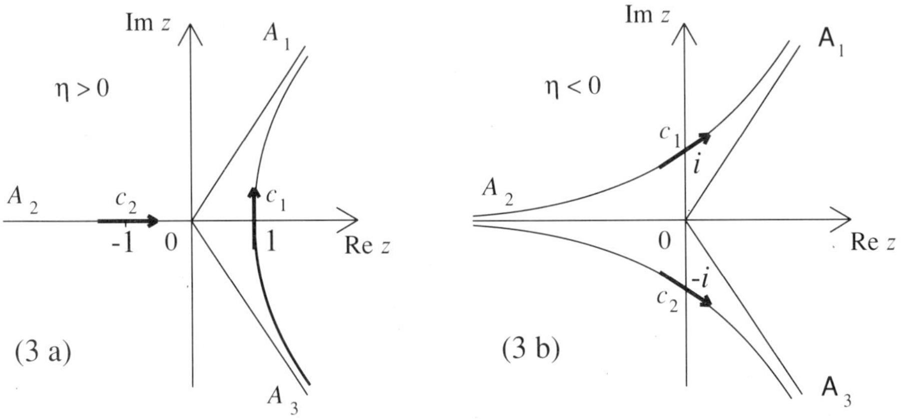

FIGURE 3. The saddle points and the lines of steepest descent for Airy's functions.

where

$$f_1 = \sum_{k=0}^{\infty} b_k \eta^{3k}, \qquad b_0 = 1, \qquad b_k = \frac{b_{k-1}}{3k(3k-1)}, \qquad a_1 = 3^{-2/3}\Gamma\left(\frac{2}{3}\right),$$

$$f_2 = \sum_{k=0}^{\infty} d_k \eta^{3k+1}, \qquad d_0 = 1, \qquad d_k = \frac{d_{k-1}}{3k(3k+1)}, \qquad a_2 = 3^{-1/3}\Gamma\left(\frac{1}{3}\right).$$

1.6.2. Asymptotic expansions of Airy's functions. To obtain asymptotic expansions of Airy's functions as $\eta \to \infty$ we use the saddle point method. If $\eta > 0$, we substitute $z = \sqrt{\eta}\, z_1$ in (1.6.2) so that formula (1.6.2) becomes

$$w_k(\eta) = C_k \sqrt{\eta} \int_{\gamma_k} e^{\lambda h_1(z_1)}\, dz_1, \qquad h_1(z_1) = \frac{z_1^3}{3} - z_1, \qquad \lambda = \eta^{3/2}. \tag{1.6.6}$$

Similarly if $\eta < 0$, substituting $z = \sqrt{-\eta}\, z_1$, we transform (1.6.2) in the form

$$w_k(\eta) = C_k \sqrt{-\eta} \int_{\gamma_k} e^{\lambda h_2(z_1)}\, dz_1, \qquad h_2(z_1) = \frac{z_1^3}{3} + z_1, \qquad \lambda = |\eta|^{3/2}. \tag{1.6.7}$$

The saddle point method can be conveniently applied to (1.6.6) and (1.6.7). If $\eta > 0$ the saddle points are $c_{1,2} = \pm 1$, and if $\eta < 0$ they are $c_{1,2} = \pm i$. Since the lines of steepest descent, shown in Fig. 3, have the rays OA_n as asymptotes, then the asymptotic expansions of the integrals (1.6.6) and (1.6.7) consist only of the contributions in neighbourhoods of the turning points c_1 and c_2.

For $\eta > 0$ and $\eta < 0$, we denote these contributions, respectively, by I_k^+ and I_k^-, where $k = 1, 2$. Then, by (1.6.3),

$$\begin{aligned} &\mathrm{Ai}(\eta) \simeq -I_1^+, \qquad &&\mathrm{Bi}(\eta) \simeq 2I_2^+, \qquad &&\text{as } \eta \to +\infty, \\ &\mathrm{Ai}(\eta) \simeq I_2^- - I_1^-, \qquad &&\mathrm{Bi}(\eta) \simeq I_1^- + I_2^-, \qquad &&\text{as } \eta \to -\infty. \end{aligned} \tag{1.6.8}$$

For this purpose, the contours γ_k are deformed in the following two ways.

If $\eta > 0$, we have

$$\gamma_1 = A_1 c_1 A_3, \qquad \gamma_2 = A_2 c_2 c_1 A_1 + A_2 c_2 c_1 A_3.$$

The first contour, γ_1, consists entirely of the line of steepest descent through the saddle point c_1. The second contour, γ_2, goes twice through the saddle point c_2; this is why the contribution of I_2^+ in (1.6.8) is doubled. Moreover, since the amplitude of the saddle point c_2 is larger than that of c_1, the contribution of c_1 is neglected.

If $\eta < 0$, we have

$$\gamma_1 = A_1c_1A_2 + A_2c_2A_3, \qquad \gamma_2 = A_2c_1A_1 + A_2c_2A_3.$$

The first contour consists of two lines of steepest descent, the ends of which go to infinity, as shown in Fig. 3 where the arrows indicate the directions along which the integration is taken in the evaluation of the contributions $I_k^{\pm}$. If these directions are opposite to the directions of the contours γ_k, then a minus sign appears in formula (1.6.8). After the evaluation of the contributions $I_k^{\pm}$ as $\eta \to \infty$, we get (see [**1**]):

$$\begin{aligned}
\mathrm{Ai}(\eta) &\simeq \frac{1}{2\sqrt{\pi}}\eta^{-1/4}e^{-\zeta}\sum_{k=0}^{\infty}\frac{(-1)^k c_k}{\zeta^k}, \qquad |\arg\eta| < \pi,\\
\mathrm{Bi}(\eta) &\simeq \frac{1}{\sqrt{\pi}}\eta^{-1/4}e^{\zeta}\sum_{k=0}^{\infty}\frac{(-1)^k c_k}{\zeta^k}, \qquad |\arg\eta| < \frac{\pi}{3},\\
\mathrm{Ai}(-\eta) &\simeq \frac{1}{\sqrt{\pi}}\eta^{-1/4}\sin\left(\zeta+\frac{\pi}{4}\right)\left[\sum_{k=0}^{\infty}\frac{(-1)^k c_{2k}}{\zeta^{2k}}\right.\\
&\qquad \left. - \cos\left(\zeta+\frac{\pi}{4}\right)\sum_{k=0}^{\infty}\frac{(-1)^k c_{2k+1}}{\zeta^{2k+1}}\right], \qquad |\arg\eta| < \frac{2\pi}{3},\\
\mathrm{Bi}(-\eta) &\simeq \frac{1}{\sqrt{\pi}}\eta^{-1/4}\cos\left(\zeta+\frac{\pi}{4}\right)\left[\sum_{k=0}^{\infty}\frac{(-1)^k c_{2k}}{\zeta^{2k}}\right.\\
&\qquad \left. + \sin\left(\zeta+\frac{\pi}{4}\right)\sum_{k=0}^{\infty}\frac{(-1)^k c_{2k+1}}{\zeta^{2k+1}}\right], \qquad |\arg\eta| < \frac{2\pi}{3},
\end{aligned} \tag{1.6.9}$$

where

$$\zeta = \frac{2}{3}\eta^{3/2}, \qquad c_k = \frac{\Gamma(3k+1/2)}{54^k k!\Gamma(k+1/2)}.$$

The series in (1.6.9) are divergent and they are valid even for complex η. The inequalities in (1.6.9) determine the sectors in neighbourhoods of the point $\eta = \infty$, in which these expansions are asymptotic.

The graphs of the functions $\mathrm{Ai}(\eta)$ and $\mathrm{Bi}(\eta)$, for the real η, are shown in Fig. 4.

1.7. The index of variation of a function

1.7.1. The notion of the index of variation. The solution to an ordinary differential equation which contains a parameter $\lambda > 0$ is a function, $F(x, \lambda)$, of two variables. For an asymptotic integration of such an equation and a qualitative study of its solution as $\lambda \to \infty$, it is necessary to have a method to describe the rate of change of F with respect to x, i.e. $\partial F/\partial x$, as $\lambda \to \infty$. In [**66, 67**] the index of variation of a function has been introduced for the solution of the equations of the theory of thin shells, and found to be very convenient.

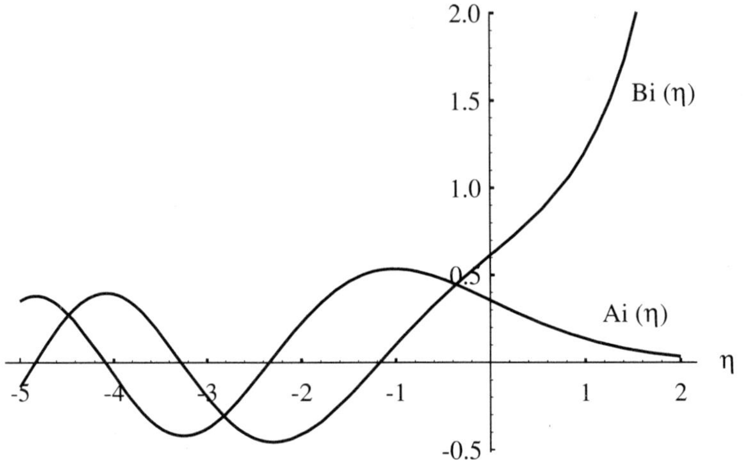

FIGURE 4. Airy's functions of a real argument.

DEFINITION 1.5. The *index of variation* of a function $F(x, \lambda)$ as $\lambda \to \infty$ is the number t such that

$$\frac{\partial F}{\partial x} \sim \lambda^t F, \qquad \text{as } \lambda \to \infty, \tag{1.7.1}$$

where the symbol $\sim$ means that the two functions are of the same order (see Section 1.1). For $t > 0$ the function F is of fast variation and for $t < 0$ it is of slow variation.

As an example we consider the functions

$$F(x, \lambda) = g(x)\, e^z \tag{1.7.2}$$

and

$$F(x, \lambda) = g(x) \sin z, \tag{1.7.3}$$

where

$$z = \lambda^t f(x), \qquad \text{as } \lambda \to \infty, \qquad \text{for all } t \geq 0 \qquad \text{and } f'(x) \neq 0. \tag{1.7.4}$$

The index of variation of the functions (1.7.2) and (1.7.3) is equal to t. In fact, for the function (1.7.3) the estimate (1.7.1) is not valid in a neighbourhood of the points where $\sin z$ or $\cos z$ are equal to zero. But by introducing the index of variation this fact can be ignored. For the function e^{iz}, this defect in definition (1.7.1) does not exist.

The index of variation may differ in different areas of variation of x. As an example we consider the equation

$$\frac{\partial^2 F}{\partial x^2} - \lambda q(x) F = 0, \tag{1.7.5}$$

where the coefficient, $q(x)$, is analytic. If $q(x) \neq 0$, then the index of variation of the solution $F(x, \lambda)$ of equation (1.7.5) is $t = 1/2$.

If now $q(x_0) = 0$ and $q'(x_0) = a \neq 0$, then $t = 1/3$ if $x - x_0 \sim \lambda^{-1/3}$. To prove this we transform x to η by $x = x_0 + (\lambda a)^{-1/3}\eta$. Then equation (1.7.5) becomes

$$\frac{\partial^2 F}{\partial \eta^2} - (\eta + R)F = 0, \qquad R = O\left(\lambda^{-1/3}\eta^2\right). \tag{1.7.6}$$

Taking $R = 0$, we obtain Airy's equation (1.6.1) without the parameter λ. Since $\partial F/\partial \eta \sim F$, then $\partial F/\partial x \sim \lambda^{1/3} F$.

The *characteristic length of the deformation profile*, $l \sim \lambda^{-t}$, is connected with the index of variation t (this concept is used to describe the field of deformation of solids).

1.7.2. General and partial indices of variation. The index of variation is also introduced for functions of several variables. We consider, for example, the function $F(x, y, \lambda)$. Its *general index of variation* is the value of the parameter t such that

$$\max\left\{\left|\frac{\partial F}{\partial x}\right|, \left|\frac{\partial F}{\partial y}\right|\right\} \sim \lambda^t F, \qquad \text{as } \lambda \to \infty. \tag{1.7.7}$$

If the derivatives of F have different orders in different directions of the (x, y)-plane, then a *partial index of variation* $t_1 < t$ is also introduced. If, for example, $|\partial F/\partial x| \ll |\partial F/\partial y|$, then

$$\frac{\partial F}{\partial x} \sim \lambda^{t_1} F, \qquad \text{as } \lambda \to \infty. \tag{1.7.8}$$

1.7.3. The index of intensity. To compare the orders of several functions we introduce their *indices of intensity*. If two given functions, $F_1(x, \lambda)$ and $F_2(x, \lambda)$, can be represented in the form

$$F_k = g_k(x, \lambda) H_k(x, \lambda), \qquad k = 1, 2, \tag{1.7.9}$$

and

$$g_k \sim \lambda^{p_k}, \qquad k = 1, 2, \qquad H_1 \sim H_2, \qquad \text{as } \lambda \to \infty, \tag{1.7.10}$$

then the quantities p_1 and p_2 are called the indices of intensity of the functions F_1 and F_2, respectively. If, for example, $F_k = g_k \sin(z + \alpha_k)$ is of type (1.7.3), then $H_1 \sim H_2 \sim 1$, and the indices of intensity coincide with the orders of the functions F_k, i.e.

$$F_k \sim \lambda^{p_k}. \tag{1.7.11}$$

Formula (1.7.11) is valid if $H_1 \sim H_2 \sim 1$. In other cases, for example, for functions F of type (1.7.2) the relation (1.7.11) is not valid. In these cases, we represent F in the form of a product (1.7.9) and find p_k from (1.7.10).

If a function F has index of variation $t > 0$ and index of intensity p, then, by (1.7.1), the index of variation p_1 of its derivative $\partial F/\partial x$ is equal to $p_1 = p + t$.

However the index of variation cannot be introduced for every function. It will be shown in Chapter 3 that solutions of differential equations with a parameter may be represented as a sum of several functions with different indices of variation.

1.8. Newton polygon

We consider a linear ordinary differential equation of the n-th order with constant coefficients, $a_k(\mu)$, which depend upon a positive parameter, $\mu > 0$,

$$\sum_{k=0}^{n} a_k(\mu)\frac{d^{n-k}z}{dt^{n-k}} = 0. \tag{1.8.1}$$

By setting $z = e^{xt}$, we obtain the following algebraic equation of degree n (see [**76, 146, 243**]):

$$\sum_{k=0}^{n} a_k(\mu)x^{n-k} = 0. \tag{1.8.2}$$

If the coefficients a_k have the form

$$a_k(\mu) = \sum_{j=0}^{m_k} a_{kj}\mu^{\alpha_{kj}}, \tag{1.8.3}$$

where

$$a_{kj} \neq 0, \qquad \alpha_{k,j+1} > \alpha_{kj}, \qquad 0 \leq m_k \leq \infty, \qquad k = 0, 1, \dots, n,$$

then

$$a_k(\mu) \sim \mu^{\alpha_{k0}}, \qquad \text{as } \mu \to 0.$$

For $a_p(\mu) \equiv 0$ we assume that $m_p = 0$ and $\alpha_{p0} = \infty$.

If $a_n(\mu) \neq 0$ and $a_0(\mu) \neq 0$, then the asymptotic expansions of the roots, x_q, of equation (1.8.2) are

$$x_q \simeq \sum_{j=0}^{\infty} x_{qj}\mu^{\beta_{qj}}, \qquad \text{as } \mu \to 0, \tag{1.8.4}$$

where $\beta_{q,j+1} > \beta_{qj}$ and $q = 1, \dots, n$.

As an example, we consider the equation

$$\mu^2x^5 + x^4 - 2x^3 + x^2 - \mu = 0. \tag{1.8.5}$$

We find the first term in expansion (1.8.4). If we omit the index q and substitute $x = x_0\mu^{\beta_0}$ in (1.8.5), then we get

$$\mu^{2+5\beta_0}x_0^5 + \mu^{4\beta_0}x_0^4 - 2\mu^{3\beta_0}x_0^3 + \mu^{2\beta_0}x_0^2 - \mu = 0. \tag{1.8.6}$$

Now we choose the main terms in (1.8.6) and determine the values of β_0 for which not less than two main terms have the same order. For $\beta_0 < 0$, the first two terms are the main ones. The value $\beta_0 = -2$ is obtained by equating their orders, $2 + 5\beta_0 = 4\beta_0$, and $x_0 = -1$ is found from the equation $x_0^5 + x_0^4 = 0$.

To establish the main terms and to determine the values of the parameter β_0 it is convenient to apply the so-called *Newton polygon* algorithm. This algorithm may be found in the papers by I. F. Obrazstov, B. V. Nerubailo and I. V. Andrianov [**146**] and V. A. Trenogin [**242**] and is described below.

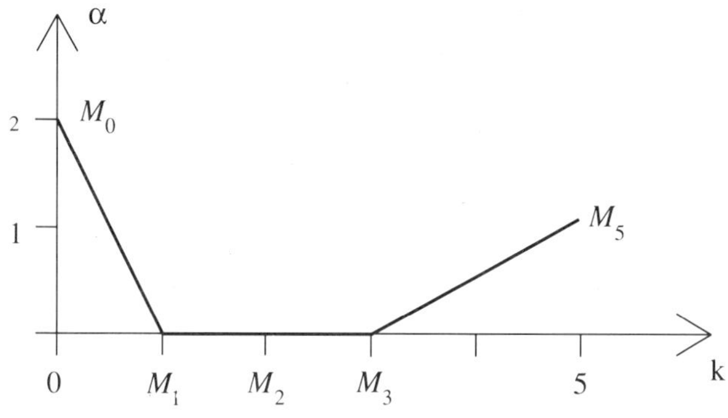

FIGURE 5. Newton polygon for equation (1.8.5).

In the (k, α)-plane, we plot $n+1$ points, $\{k, \alpha_{k0}\}$, $k = 0, \ldots n$, with integer abscissae and plot the segment M_0M_1 (see Fig. 5).

The tangent of the angle between a segment and the k-axis is equal to the value β_0 for which the orders of the first and the second terms coincide; it is easy to check that the terms with higher order in μ correspond to the points above the straight line passing through M_0 and M_1. In the case considered here, all the other points lie above the straight line M_0M_1.

If we draw the segment M_0M_2, we find the value β_0 for which the first and the third terms have the same order. But in our case, these terms are not the main terms, since the order in μ of the second term for such value of β_0 is less (the point M_1 is below the line M_0M_2). Since we are interested only in the main terms, we connect the points M_k by segments in such a way that any point M_l which does not belong to these segments lies above the plotted broken line.

To construct this broken line we rotate the ray plotted from the point $M_0^* = M_0$ vertically down in the counter-clockwise direction. We denote by M_1^* the first of the points M_k which is touched by this ray. Then we rotate the ray plotted from the point M_1^* vertically down in the same direction until it touches M_2^*. This process is continued until the last ray touches the point $M_s^* = M_{n+1}$.

The broken line, L, connecting the points M_i^* is called the *Newton polygon.* In fact, it is the lower part of the boundary curve of the smallest convex set containing the points M_i.

The Newton polygon displays the following properties of the main terms of (1.8.4):

(1) The tangent of the angle between a segment $M_i^*M_{i+1}^*$ and the k-axis gives the order β_{q0} of the root x_q.
(2) The length of the projection of this segment on the k-axis is equal to the number of roots x_q of this order.
(3) The number of points M_k through which the segment passes is equal to the number of terms of the equation used for the evaluation of x_{q0}.

Newton polygon for equation (1.8.5) is shown in Fig. 5. It contains three parts with downward, constant and upward slopes, respectively, and gives

$$\beta_{10} = -2, \qquad \beta_{20} = \beta_{30} = 0, \qquad \beta_{40} = \beta_{50} = \frac{1}{2}.$$

To evaluate the next term of the series (1.8.4) we substitute $x_q = \mu^{\beta_{q0}}(x_{q0}+\Delta_q)$ into (1.8.2). Then we get an algebraic equation for Δ_q similar to (1.8.2), but we seek only the roots Δ_q which converge to zero as $\mu \to 0$. The number of such roots is equal to the multiplicity of the root x_{q0}.

The approximate values of the roots of (1.8.5) are

$$x_1 = -\mu^{-2} - 2, \qquad x_{2,3} = 1 \pm \mu^{1/2}, \qquad x_{4,5} = \pm\mu^{1/2} + \mu.$$

CHAPTER 2

Singular Perturbation of Linear Differential Equations

In this chapter, we consider the asymptotic integration of linear ordinary differential equations of the n-th order and first-order systems with variable coefficients and powers of a small parameter, μ, multiplying the derivatives, as $\mu \to 0$. Fundamental sets of solutions are constructed with and without turning points in Sections 2.3 and 2.2 respectively. In Section 2.4, these systems will be used in the solution of boundary eigenvalue problems. In Section 2.5, special attention is given to the construction of localized eigenfunctions. In Section 2.6, two-dimensional shell theories are considered. In Section 2.7, general formulae are used for the solution of boundary value problems for free vibrations and buckling of thin shells, when the problems can be reduced to one-dimensional problems.

In this survey, we do not discuss perturbation methods and asymptotic expansions of ordinary differential equations with a right-hand side which depends regularly on a small parameter [**25, 92, 122, 135, 139, 140**] and integration of equations with rapidly varying coefficients [**8, 16, 185**].

2.1. Introduction

2.1.1. The main assumption on the matrix $A(x, \mu)$. As mentioned above, this chapter is concerned mainly with homogeneous systems of n linear first-order differential equations with a small parameter multiplying the derivative,

$$\mu \frac{dy}{dx} = A(x, \mu) y, \qquad \mu > 0, \tag{2.1.1}$$

and methods for constructing the asymptotic expansions of their solutions as $\mu \to 0$. Here y is an n-vector and A is an $n \times n$-matrix. We make the following assumption on A (see [**52, 73, 117, 135, 247, 266**]).

ASSUMPTION 2.1. The $n \times n$ matrix A of system (2.1.1) can be represented in the form

$$A(x, \mu) \simeq \sum_{k=0}^{\infty} \mu^k A_k(x), \qquad \text{as } \mu \to 0, \tag{2.1.2}$$

where the coefficients, $A_k(x)$, are analytic on a real or complex domain S.

We also mention more general systems of the type [**135, 246, 266**]

$$\mu^h \frac{dy}{dx} = A(x, \mu) y, \qquad h > 1, \tag{2.1.3}$$

under the same assumptions on the matrix A as above. The positive integer h is called the *rank* of the system.

Many authors use the large parameter $\lambda = \mu^{-1}$ instead of the small parameter μ; in this case, systems (2.1.1) and (2.1.3) become

$$\frac{dy}{dx} = A^*(x,\lambda)y, \qquad A^* \simeq \sum_{k=0}^{\infty} \lambda^{h-k} A_k(x), \qquad \text{as } \lambda \to \infty, \tag{2.1.4}$$

respectively. In the sequel, λ will be replaced by μ^{-1} in quoted results.

2.1.2. The problems. We raise the following questions about the systems (2.1.1) and (2.1.3):

- How to construct the formal (see 2.1.3) asymptotic solution?
- Under which conditions a formal asymptotic solution is an asymptotic expansion of the solution?
- How to have an asymptotic splitting of the system, i.e. how to reduce the system to several subsystems of lower order?
- How to obtain the solutions of boundary value problems and eigenvalue problems for systems (2.1.1) and (2.1.3)?

2.1.3. Formal asymptotic solutions. We shall need the following definition.

DEFINITION 2.1. An asymptotic series which transforms the system under consideration into an identity is called a *formal asymptotic solution.*

We introduce this terminology independently of whether a formal solution is an asymptotic expansion of the solution or not.

Under certain assumptions, the formal asymptotic solution of system (2.1.3) has the form [**135, 246, 266**]

$$y(x,\mu) \simeq \sum_{k=0}^{\infty} U_k(x)\mu^k \exp\left\{\int_{x_0}^{x} \sum_{k=0}^{h-1} \mu^{k-h}\lambda^{(k)}(\tilde{x})\, d\tilde{x}\right\}, \tag{2.1.5}$$

where $U_k(x)$ and $\lambda^{(k)}(x)$ are analytic functions of x. In particular, formula (2.1.5) is valid if the characteristic equation

$$\det[A_0(x) - \lambda^{(0)} I] = 0 \tag{2.1.6}$$

has simple roots, $\lambda_p = \lambda_p^{(0)}(x)$, $p = 1, \dots, n$, for all x. In [**246**] cases, where formulae like (2.1.5) are also valid for multiple roots, are studied. In these cases, it is necessary that the roots of (2.1.6) be identically multiplied by x and the structure of the Jordan canonical form of the matrix $A_0(x)$ does not change with x. Other limitations may also be necessary on the matrices $A_k(x)$.

2.1.4. Turning points. We define turning points.

DEFINITION 2.2. Points $x = x_*$, where the multiplicity of the roots of equation (2.1.6) or the structure of the Jordan canonical form of the matrix $A_0(x)$ changes, are called *turning points* or *transition points.*

In neighbourhoods of turning points, formula (2.1.5) is invalid because some $U_k(x) \to \infty$ as $x \to x_*$. The general problem of constructing a uniform asymptotic solution over the entire domain S when turning points exist, is open. Some particular results are discussed below.

2.2. Solution of system (2.1.1) without turning points

2.2.1. Formal asymptotic solution in the case of simple roots. We consider system (2.1.1) under assumption 2.1 on the matrix $A(x,\mu)$. If $\lambda_p(x)$ is a simple root of the characteristic equation (2.1.6), which is a polynomial equation of degree n in $\lambda^{(0)}$, then the following formal asymptotic solution of system (2.1.1) corresponding to this root is the vector-valued function, in $\mathbb{R}^n$ or $\mathbb{C}^n$,

$$y^{(p)}(x,\mu) \simeq \sum_{k=0}^{\infty} U_k^{(p)}(x)\mu^k \exp\left\{\frac{1}{\mu}\int_{x_0}^{x} \lambda_p(\tilde{x})\,d\tilde{x}\right\}, \qquad \mu > 0. \tag{2.2.1}$$

The vector-valued functions $U_k^{(p)}(x)$ are determined recursively after substitution of (2.2.1) into (2.1.1). We consider in greater detail the construction of the first term in the series (2.2.1). We write the equations

$$\begin{aligned} (A_0 - \lambda_p I)U_0^{(p)} &= 0, \\ (A_0 - \lambda_p I)U_1^{(p)} + A_1 U_0^{(p)} &= \frac{dU_0^{(p)}}{dx}, \end{aligned} \tag{2.2.2}$$

obtained as a result of equating the coefficients of μ^0 and μ^1. The first of these equations provides

$$U_0^{(p)} = \varphi_0(x)V^{(p)}(x),$$

where $\varphi_0(x)$ is a scalar function and $V^{(p)}(x)$ is an eigenvector of A_0, i.e.

$$[A_0(x) - \lambda_p(x)I]\,V^{(p)}(x) = 0.$$

The function $\varphi_0(x)$ is obtained from the next approximation. The second equation in (2.2.2) is considered as a system of linear nonhomogeneous equations in the unknown vector $U_1^{(p)}$. The determinant of this system is equal to zero and its compatibility condition has the form

$$W^{(p)T} A_1 U_0^{(p)} = W^{(p)T} \frac{dU_0^{(p)}}{dx},$$

where $W^{(p)}(x)$ is an eigenvector of $A_0^T(x)$:

$$\left[A_0^T(x) - \lambda_p(x)I\right] W^{(p)}(x) = 0,$$

and A_0^T is the *transpose* of A_0. From the compatibility condition we derive the following first-order ordinary differential equation for the function $\varphi_0(x)$,

$$\frac{d\varphi_0(x)}{dx} = b_0(x)\varphi_0(x), \qquad b_0 = \frac{W^T(A_1 V - dV/dx)}{W^T V},$$

which may be integrated by quadratures.

Then we get

$$U_1^{(p)} = \varphi_1(x)V^{(p)}(x) + U_1^{(p)*},$$

where $U_1^{(p)*}$ is a particular solution of the second equation in (2.2.2), and $\varphi_1(x)$ is a scalar function, determined from the next higher approximation.

We remark that the functions $\lambda_p(x)$ and $U_k^{(p)}(x)$ are analytic on S.

2.2.2 Fundamental set of solutions. If, for each $x \in S$, all the roots $\lambda_p(x)$, $p = 1, \dots, n$, of equation (2.1.6) are simple, then formula (2.2.1) provides a fundamental matrix $\widetilde{Y}(x, \mu)$ of the formal asymptotic solutions. This matrix consists of the n columns (2.2.1). Is there a matrix $Y(x, \mu)$ of the exact solutions of system (2.2.1), for which

$$Y(x,\mu) \simeq \widetilde{Y}(x,\mu), \qquad \text{as } \mu \to 0, \tag{2.2.3}$$

i.e. is the matrix $\widetilde{Y}$ an asymptotic expansion? Relation (2.2.3) assumes that the corresponding matrix elements of Y and $\widetilde{Y}$ satisfy the estimate (1.2.5) of Chapter 1 for all $x \in S$. This question has been studied, for example, in [**52, 54, 187, 266**] and its answer depends first of all on the domain S.

If for some $x = x_0$ the eigenvalues $\lambda_p(x_0)$ are simple, then there exists a neighbourhood of x_0, where the asymptotic relation (2.2.3) is valid (see [**266**]).

A sufficient condition to have (2.2.3) in the absence of multiple roots, λ_p, is that the differences

$$\Re\left(\lambda_p(x) - \lambda_j(x)\right), \qquad p, j = 1, \dots, n,$$

do not change sign for all $x \in S$ (see [**52**]), i.e. for any pair of roots $\lambda_p(x)$ and $\lambda_j(x)$ one of the following relations,

$$\Re(\lambda_p(x) - \lambda_j(x)) \geq 0, \qquad \Re(\lambda_p(x) - \lambda_j(x)) \leq 0, \tag{2.2.4}$$

is satisfied for all $x \in S$. In this case, the asymptotic relation (2.2.3) is uniform in x in any closed subset $S_1 \subset S$.

In [**187**], S is a real interval, $x_1 \leq x \leq x_2$, and conditions (2.2.4) on the roots $\lambda_p(x)$ are replaced with conditions on their integrals, i.e. for any pair of roots $\lambda_p(x)$ and $\lambda_j(x)$ one of the following inequalities,

$$\Re \int_{x_1}^{x} (\lambda_p - \lambda_j)\, d\tilde{x} \geq 0, \qquad \Re \int_{x_2}^{x} (\lambda_p - \lambda_j)\, d\tilde{x} \leq 0, \tag{2.2.5}$$

has to be satisfied.

2.2.3. Linear differential equations of the n-th order. We consider a linear differential equation of the n-th order:

$$\sum_{k=0}^{n} \mu^k a_k(x,\mu) \frac{d^k y}{dx^k} = 0, \tag{2.2.6}$$

where

$$a_k(x,\mu) \simeq \sum_{j=0}^{\infty} \mu^j a_{kj}(x), \qquad \mu \to 0, \qquad a_{n0}(x) \neq 0, \qquad \forall x \in S,$$

and the coefficients $a_{jk}(x)$ are analytic in S. By introducing the variables

$$y_j = \mu^j \frac{d^j y}{dx^j}, \qquad j = 0, 1, \dots, n-1,$$

we transform (2.2.6) into the first-order system (2.1.1); hence all the statements which are valid for system (2.1.1) are also valid for equation (2.2.6). Here the characteristic equation (2.1.6) has the form

$$P_0(\lambda) := \sum_{k=0}^{n} a_{k0}(x)\lambda^k = 0. \tag{2.2.7}$$

If (2.2.7) has no multiple roots, the fundamental set of solutions (2.2.6) is given by the same formulae (2.2.1), but in this case $y^{(p)}$ and $U_k^{(p)}$ are scalar functions of x.

The following remark will simplify later statements on the order of a system.

REMARK 2.1. To determine the *order*, m, *of a system of linear ordinary differential equations*

$$\sum_{j=1}^{n} a_{ij}\left(x, \frac{d}{dx}\right) u_j = 0, \qquad i = 1, \ldots, n,$$

where $a_{ij}(x,p)$ are polynomials in p, we evaluate the symbolic determinant

$$\det[a_{ij}(x,p)] = a_0(x)p^m + a_1(x)p^{m-1} + \cdots,$$

and, if $a_0(x) \neq 0$, then m is the order of this system.

2.2.4. Singular degeneracy. We consider a linear differential equation of order $m+n$,

$$\sum_{k=1}^{n} \mu^k a_{k+m}(x,\mu)\frac{d^{k+m}y}{dx^{k+m}} + \sum_{k=0}^{m} a_k(x,\mu)\frac{d^k y}{dx^k} = 0, \tag{2.2.8}$$

which, for $\mu = 0$, degenerates into an equation of order m,

$$\sum_{k=0}^{m} a_k(x,0)\frac{d^k y}{dx^k} = 0. \tag{2.2.9}$$

We assume that the coefficients $a_k(x,\mu)$ satisfy (2.2.6), with $a_{m+n,0}(x) \neq 0$, $\forall x \in S$. Multiplying each term of (2.2.8) by μ^m, we obtain (2.2.6). The characteristic equation of type (2.2.7) corresponding to (2.2.8) is

$$\sum_{k=0}^{n} a_{k+m,0}(x)\lambda^{k+m} = 0. \tag{2.2.10}$$

This equation has the root $\lambda \equiv 0$ of multiplicity m. If $a_{m,0}(x) \neq 0$, $\forall x \in S$, and the nonzero roots, $\lambda_p(x)$, $p = 1, \ldots, n$, of equation (2.2.10) are simple for all $x \in S$, then equation (2.2.8) has n formal asymptotic solutions of type (2.2.1) with index of variation $t = 1$ (see Section 1.7) and m solutions with index of variation $t = 0$. Hence

$$y^{(q)}(x,\mu) \simeq \sum_{k=0}^{\infty} \mu^k U_k^{(q)}(x), \qquad q = n+1, \ldots, n+m, \qquad \text{as } \mu \to 0, \tag{2.2.11}$$

where $U_0^{(q)}(x)$ are linearly independent solutions of the degenerate equation (2.2.9).

The formal series (2.2.1) and (2.2.11) are asymptotic expansions of the exact solutions of equation (2.2.6) as $\mu \to 0$ under the same assumptions (2.2.4) or (2.2.5) on the roots of (2.2.7) (including the root $\lambda \equiv 0$).

2.2.5. A particular case of singular degeneracy. We consider the following particular case of equation (2.2.8) which often arises in mechanical problems,

$$\mu^n Ny + Ly = 0, \tag{2.2.12}$$

where the differential operators N and L are

$$Ny = \sum_{k=0}^{m+n} a_k(x)\frac{d^k y}{dx^k}, \qquad Ly = \sum_{k=0}^{m} b_k(x)\frac{d^k y}{dx^k}, \tag{2.2.13}$$

and the coefficients $a_k(x)$ and $b_k(x)$ are assumed to be analytic. The characteristic equation (2.2.10), in this case, is the binomial equation

$$a_{m+n}(x)\lambda^{m+n} + b_m(x)\lambda^m = 0. \tag{2.2.14}$$

As for (2.2.6), we assume that $a_{m+n}(x) \neq 0$, $\forall x \in S$. If $b_m(x) \neq 0$, $\forall x \in S$, equation (2.2.12) has m solutions of type (2.2.11) and n solutions of type (2.2.1). The case where $b_m(x)$ is equal to zero at some (turning) points is discussed in Section 2.3.

The Orr–Sommerfeld equation describing the stability of the laminar flow of a viscous liquid [**116, 167, 265**] is of type (2.2.12) with $m = n = 2$. The equations for the axisymmetric ($m = 2$, $n = 4$) and cyclic symmetric ($m = n = 4$) deformations (and also for the vibrations and stability) of thin shells of revolution are of the same type (see Chapter 3 and [**64, 66, 67, 73, 180, 193, 266, 268**]).

2.2.6. Second-order equations. A particular case of equation (2.2.8) is the second-order equation

$$\mu^2\frac{d^2y_1}{dx^2} + \mu p_1(x,\mu)\frac{dy_1}{dx} + q_1(x,\mu)y_1 = 0, \tag{2.2.15}$$

which, after the substitution

$$y_1 = y\, e^{-(2\mu)^{-1}\int p_1 dx}, \tag{2.2.16}$$

becomes

$$\mu^2\frac{d^2y}{dx^2} - q(x,\mu)y = 0. \tag{2.2.17}$$

Numerous studies are devoted to equations (2.2.15) and (2.2.17) (see, for example, [**28, 46, 52, 54, 61, 80, 135, 139, 140, 143, 147, 247, 266**]). Many mechanical and physical problems can be transformed into these equations. The asymptotic solutions of equations (2.2.15) or (2.2.17) of type (2.2.1) are called *WKB-approximations*, and the methods of construction of such solutions are called *WKB-methods*. This name comes from the first letters of the authors' names: Wentzel, Kramers and Brillouin, who first applied this method to problems of quantum mechanics. Later the term "WKB-method" was also used to mean a "method of asymptotic integration" for a wider class of problems [**124, 126**].

The asymptotic solution of equation (2.2.17), in the case $q(x,\mu) \equiv q(x)$, has the form (2.2.1), i.e.

$$\lambda_p^2 - q(x) = 0, \qquad p = 1, 2,$$

and

$$
\begin{aligned}
U_0^{(p)} &= [q(x)]^{-1/4}, \\
U_{k+1}^{(p)} &= -\frac{2}{\sqrt{\lambda_p}} \int_{x_0}^{x} U_k^{(p)\prime\prime} \sqrt{\lambda_p}\, d\tilde{x}, \qquad k = 0, 1, \ldots
\end{aligned} \tag{2.2.18}
$$

It is clear from (2.2.18) that the neighbourhoods of the turning points $x = x_*$ (where $q(x_*) = 0$) need a special consideration because $U_k^{(p)}(x_*) = \infty$.

2.2.7. The lifting method to a space of higher dimension. The easiest way to determine the unknown coefficients of the series (2.2.1) is by substituting it into the corresponding equations and equating the coefficients of the same power of μ. In [**117**] another approach to these problems, to be discussed in Sections 2.3–2.7 below, is proposed. Instead of the function $y(x, \mu)$, we seek a function $Y(x, \eta_p, \mu)$, of the variables $x, \eta_1, \ldots, \eta_n$, such that

$$
y(x, \mu) = Y\left(x, \frac{1}{\mu} \int_{x_0}^{x} \lambda_p(\tilde{x})\, d\tilde{x}, \mu\right), \qquad p = 1, \ldots, n. \tag{2.2.19}
$$

As $\mu \to 0$, the problem of constructing $Y(x, \eta_p, \mu)$ is a regular perturbed problem. This method, sometimes called the *lifting method to a space of higher dimension*, is closely related to the *multi-scale method* considered by A. H. Nayfeh and others [**37, 139, 140**].

The two-scale method is convenient for the construction of solutions of equations with rapidly varying coefficients. We consider the system of equations

$$
\frac{dz}{dx} = A(x, \eta) z, \qquad \eta = \frac{x}{\mu}, \tag{2.2.20}
$$

where $\mu > 0$ is a small parameter, z is an n-dimensional vector and the matrix A is analytic in both arguments and 1-periodic in η,

$$
A(x, \eta + 1) = A(x, \eta).
$$

We seek the solution of system (2.2.20) in the form

$$
z(x, \mu) = u(x, \eta, \mu), \tag{2.2.21}
$$

where $u(y, \eta, \mu)$ is an auxiliary function depending on y and η. This function satisfies the partial differential equation

$$
\frac{\partial u}{\partial y} + \frac{1}{\mu} \frac{\partial u}{\partial \eta} = A(y, \eta) u. \tag{2.2.22}
$$

We seek the solutions to this equation in the form of formal series in powers of μ,

$$
u(y, \eta, \mu) = \sum_{n=0}^{\infty} u_n(y, \eta) \mu^n, \tag{2.2.23}
$$

where $u(y, \eta + 1) = u(y, \eta)$.

Substitution of (2.2.23) into (2.2.22) produces the sequence of equations

$$
\frac{\partial u_0}{\partial \eta} = 0, \qquad \frac{\partial u_{k+1}}{\partial \eta} = A u_k - \frac{\partial u_k}{\partial y}, \qquad k = 0, 1, \ldots \tag{2.2.24}
$$

We denote the average value of the 1-periodic function F by

$$\langle F\rangle = \int_0^1 F(\eta)d\eta$$

and represent u_k as the sum of a slowly varying function u_k^0 and a rapidly varying function with zero average value u_k^*,

$$u_k(y,\eta) = u_k^0(y) + u_k^*(y,\eta), \qquad \langle u_k^*(y,\cdot)\rangle = 0. \tag{2.2.25}$$

Then equations (2.2.24) gives

$$\begin{aligned} \frac{\partial u_k^0}{\partial y} &= \langle A\rangle u_k^0 + \langle Au_k^*\rangle, \qquad k = 0,1,\dots, \\ \frac{\partial u_{k+1}^*}{\partial \eta} &= A(u_k^0 + u_k^*) - \langle A(u_k^0 + u_k^*)\rangle - \frac{\partial u_k^*}{\partial y}. \end{aligned} \tag{2.2.26}$$

The first equation in (2.2.26) does not contain the rapid variable η and one can find a solution which satisfies given initial or boundary conditions.

In particular, in the zeroth approximation, we have

$$\frac{du_0^0}{dy} = \langle A\rangle u_0^0, \qquad u_0^* = 0, \tag{2.2.27}$$

which corresponds to averaging the coefficients of system (2.2.20).

The second equation in (2.2.26) has only one periodic solution in η which satisfies condition (2.2.25).

As an example, we consider the second-order equation

$$\frac{d}{dx}\left(p(x,\eta)\frac{dw}{dx}\right) + q(x,\eta)w = 0, \tag{2.2.28}$$

where

$$p(x,\eta+1) = p(x,\eta) > 0, \qquad q(x,\eta+1) = q(x,\eta),$$

and find the averaged equation in the zeroth approximation. Introducing the auxiliary unknown functions

$$z_1 = p\frac{dw}{dx}, \qquad z_2 = w,$$

we transform equation (2.2.28) to a system of two equations of the first order of type (2.2.20) with the matrix A in the form

$$A = \begin{pmatrix} 0 & q \\ p^{-1} & 0 \end{pmatrix}.$$

Similar to (2.2.27), averaging by η and then coming back to the initial unknown function w, we get the averaged equation

$$\frac{d}{dx}\left(\langle p^{-1}\rangle^{-1}\frac{dw_0}{dx}\right) + \langle q\rangle w_0 = 0,$$

which does not coincide with the equation obtained by direct averaging of the coefficients of (2.2.20).

2.3. Construction of solutions with turning points

2.3.1. The Stokes multiplier method. We consider the system (2.1.1) with turning points in the domain, S, of integration. As it was mentioned above, in a neighbourhood of a turning point, x_*, the exponential representation (2.2.1) is invalid since some of the functions $U_k^{(p)}(x)$ approach infinity as $x \to x_*$. However, in a neighbourhood of any non-turning point, x_0, formulae (2.2.1) give an asymptotic expansion of a fundamental set of solutions for (2.1.1) (see [**266**]).

There exist two methods for constructing the asymptotic expansion of a fundamental set of solution for (2.1.1).

The first of these does not describe the solutions in neighbourhoods of the turning points but construct solutions in some subdomain $\widetilde{S} \subset S$, not containing the turning points (see [**52, 61, 80, 266**]). We give some detail. The asymptotic expansion of any solution $y(x, \mu)$ of (2.1.1) may be represented by a linear combination of the formal series (2.2.1),

$$y(x, \mu) \simeq \sum_{p=1}^{n} C_p(\mu) y^{(p)}(x, \mu), \qquad \text{as } \mu \to 0. \tag{2.3.1}$$

The problem now is that the coefficients $C_p(\mu)$ depend not only on the chosen solution $y(x, \mu)$, but also on the subdomain $\widetilde{S}$. The coefficients C_p are called *Stokes' multipliers*, and the lines across which Stokes' multipliers change by a jump are called *Stokes' lines.* Stokes' lines which originate from a turning point, $x = x_*$, are defined by the relation [**52**]

$$Re \int_{x_*}^{x} [\lambda_p(\tilde{x}) - \lambda_j(\tilde{x})]\, d\tilde{x} = 0, \qquad \lambda_p(x_*) = \lambda_j(x_*), \tag{2.3.2}$$

where $\lambda_p(x)$ are the roots of equation (2.1.6).

For example, for Airy's equation $\mu^2 y'' - xy = 0$, the rays $\varphi = \frac{2k\pi}{3}$, $k = 0, 1, 2$, emanating from the origin are Stokes' lines. Examples of asymptotic expansions of many special functions of type (2.3.1) can be found in handbooks, say [**1, 75, 83**], where the coefficients C_p depend on the sector in which x varies.

In the general case, the construction of Stokes' lines and Stokes' multipliers is an unsolved problem. The second-order equation (2.2.17) has been studied in sufficient detail (see [**52, 61, 80**]), but for higher-order equations only particular results are known (see, for example, [**52, 221, 245, 266**]).

2.3.2. The standard function method. We discuss in greater detail a second method for the construction of asymptotic expansions, i.e. the standard function method.

The first step towards the asymptotic integration of system (2.1.1) of n equations is to split it into several smaller systems.

Let the roots $\lambda_k(x)$ of the characteristic equation (2.1.6) form two distinct sets

$$\{\lambda_1(x), \dots, \lambda_p(x)\}, \qquad \{\lambda_{p+1}(x), \dots, \lambda_{p+q}(x)\},$$

where $p + q = n$, i.e.

$$\lambda_i(x) \neq \lambda_{p+j}(x), \qquad x \in S, \qquad 1 \le i \le p, \qquad 1 \le j \le q. \tag{2.3.3}$$

Then (see [**54, 266**]) there exists a linear nonsingular transformation of the dependent variables

$$y(x,\mu) \simeq \widetilde{P}(x,\mu)z(x,\mu), \qquad \widetilde{P}(x,\mu) \simeq \sum_{k=0}^{\infty} P_k(x)\mu^k, \tag{2.3.4}$$

such that the vector function z satisfies a differential equation,

$$\mu \frac{dz}{dx} = B(x,\mu)z, \tag{2.3.5}$$

of the same type as (2.1.1), but with a block-diagonal matrix B,

$$B(x,\mu) = \begin{bmatrix} B^{(1)}(x,\mu) & 0 \\ 0 & B^{(2)}(x,\mu) \end{bmatrix},$$

with

$$B^{(i)}(x,\mu) \simeq \sum_{k=0}^{\infty} B_k^{(i)}(x)\mu^k, \qquad i = 1,2, \tag{2.3.6}$$

where $B^{(1)}$ and $B^{(2)}$ are $p \times p$ and $q \times q$ matrices, respectively. The eigenvalues of $B_0^{(1)}(x)$ and $B_0^{(2)}(x)$ are $\lambda_1(x), \ldots, \lambda_p(x)$ and $\lambda_{p+1}(x), \ldots, \lambda_{p+q}(x)$, respectively. The matrix elements of $P_k(x)$ and $B_k^{(i)}(x)$ are analytic on S and the matrix $P_0(x)$ is nonsingular on S.

The expansions of $\widetilde{P}$ and $B^{(i)}$, of (2.3.4) and (2.3.6), with respect to μ are, in general, formally divergent series. It is shown, in [**54, 266**], that under some extra conditions there exists an analytic change of variables, $y = P(x,\mu)z$, which transforms the matrix A into a block-diagonal form and such that the series $\widetilde{P}(x,\mu)$ is the asymptotic expansion of $P(x,\mu)$ as $\mu \to 0$.

If all the roots, $\lambda_j(x)$, of the characteristic equation (2.1.6) are simple, then by splitting the system several times we can reduce it to n equations of the first order,

$$\mu\frac{dz_j}{dx} = b^{(j)}(x,\mu)z_j, \qquad b^{(j)} \simeq \sum_{k=0}^{\infty} b_k^{(j)}(x)\mu^k, \qquad b_0^{(j)}(x) = \lambda_j(x), \qquad j = 1,\ldots,n, \tag{2.3.7}$$

which can be integrated by quadratures. In fact, this is another way to obtain series (2.2.1).

2.3.3. A single turning point. Now we consider a domain S, which contains only one turning point, x_*. System (2.1.1) can be split into subsystems in such a way that, for one of these, equation (2.1.6) has only one root $\lambda(x_*)$ at $x = x_*$, and this root is of multiplicity p. Without loss of generality, we may assume that $x_* = 0$.

We suppose that equation (2.1.6), i.e.

$$\det\left(A_0(0) - \lambda I\right) = 0, \tag{2.3.8}$$

has only the root zero, of multiplicity p. The case $\lambda(0) \neq 0$ can be reduced to the case under consideration by the following change of dependent variable,

$$y_1 = y e^{\mu^{-1}\int_0^x \varphi(x)dx}, \qquad \varphi(0) = -\lambda(0). \tag{2.3.9}$$

The asymptotic integration of this new system is discussed in [**266**]. Furthermore, we assume that this system is transformed into a single differential equation of the n-th order (we assume $p = n$) of type (2.2.6), the characteristic equation (2.2.7) of which, for $x = 0$, has the root zero of multiplicity n.

In the standard function method, we consider, together with equation (2.2.6), a simpler *standard* equation and require that the asymptotic solutions of equation (2.2.6) be expressed by means of the solutions of the standard equation (and perhaps their derivatives) and be analytic in x on S.

The standard equations are studied, for example, in [**6, 52, 116, 166, 194–196, 217, 221, 245, 264–266**].

In the general case, it is still unknown how to choose the simplest standard equation. The order of the standard equation may be equal to, or less than, n. The latter case occurs, in particular, for equations of type (2.2.8) with $a_m(0) = 0$. In this case, only some of the solutions of equation (2.2.8) are expressed by means of the solutions of the standard equation (see below).

2.3.4. Particular cases. Now we come to particular cases. We start with the second-order equation (2.2.17):

$$\mu^2 \frac{d^2 y}{dx^2} - q(x,\mu) y = 0, \qquad q(x,\mu) = \sum_{k=0}^{\infty} q_k(x) \mu^k. \tag{2.3.10}$$

We assume that $q_k(x)$ are analytic on the domain S and that S contains the origin $x = 0$. Let $q_0(0) = 0$ and assume that $q_0(x) = 0$ has no other root in S. If $q_0'(0) \neq 0$ the turning point $x = 0$ is called *simple*. Let $q_0' > 0$. As standard equation we use Airy's equation (1.6.1)

$$\frac{d^2 v}{d\eta^2} - \eta v = 0. \tag{2.3.11}$$

The asymptotic expansions for the solutions of (2.3.10) may be represented, according to R. E. Langer [**106**], in the form

$$y(x,\mu) = a^{(0)}(x,\mu) v\left(\eta(x,\mu)\right) + \mu^{1/3} a^{(1)}(x,\mu) \frac{dv}{d\eta}, \tag{2.3.12}$$

where

$$\begin{gathered} \eta(x,\mu) = \mu^{-2/3} \xi(x), \qquad \xi(x) = \left[\frac{3}{2} \int_0^x \sqrt{q_0(\tilde{x})}\, d\tilde{x} \right]^{2/3} \\ a^{(j)}(x,\mu) \simeq \sum_{k=0}^{\infty} a_k^{(j)}(x) \mu^k, \qquad j = 0, 1. \end{gathered} \tag{2.3.13}$$

The function $\xi(x)$ and the coefficients $a_k^{(j)}(x)$ are determined by substituting them in (2.3.10) and by equating the coefficients of $\mu^k v$ and $\mu^k \frac{dv}{d\eta}$. This function and these coefficients are analytic on S, including the point $x = 0$. The first two coefficients in (2.3.13) are

$$a_0^{(0)} = \frac{1}{\sqrt{\xi'}} \cos\nu, \qquad a_0^{(1)} = \frac{1}{\sqrt{\xi \xi'}} \sin\nu, \qquad \nu = \int_0^x \frac{q_1(x)}{2\xi' \sqrt{\xi}}\, dx. \tag{2.3.14}$$

All the coefficients $a_k^{(j)}(x)$ are expressed by means of $q_i(x)$ in the form of quadratures.

The expansion (2.3.12) is valid uniformly on all S, including any neighbourhood of the origin $x = 0$. If $|\eta| \gg 1$, one can use the asymptotic expansions of Airy's function (1.6.9) and express (2.3.12) by the functions (2.2.1).

Suppose, for example, that S contains the real interval $-1 \le x \le 1$, the function $q(x, \mu)$ is real for real x and $q_1(x) \equiv 0$. We consider the exponentially decreasing solution of equation (2.3.10) for $x > 0$. We assume in (2.3.12) that $v(\eta) = \mathrm{Ai}(\eta)$ and write the first terms of the asymptotic series as $\mu \to 0$:

$$y(x,\mu) = \frac{1}{\sqrt{\xi'}} \mathrm{Ai}(\eta)\,[1 + O(\mu)] + \mathrm{Ai}'(\eta)\, O\left(\mu^{4/3}\right), \qquad -1 \le x \le 1,$$

$$(2.3.15)\quad y(x,\mu) = \frac{\mu^{1/6}}{\sqrt{4\pi q_0}} \exp\left(-\frac{1}{\mu}\int_0^x \sqrt{q_0(\tilde{x})}\, d\tilde{x}\right)[1 + O(\mu)], \qquad x > 0,$$

$$y(x,\mu) = \frac{\mu^{1/6}}{\sqrt{-\pi q_0}} \left\{\cos\left[\frac{1}{\mu}\int_0^x \sqrt{-q_0(\tilde{x})}\, d\tilde{x} - \frac{\pi}{4}\right] + O(\mu)\right\}, \qquad x < 0.$$

These formulae are uniformly valid for $-1 \le x \le 1$, $x > 0$ and $x < 0$, respectively.

The representation

$$(2.3.16)\qquad y(x,\mu) = a(x,\mu) v\left(\eta(x,\mu)\right),$$

where a and η are series in powers of $\mu^{1/3}$, has been used, instead of (2.3.12), by T. M. Cherry [**34**].

A similar expansion in fractional powers of μ is used in the asymptotic integration of partial differential equations in two independent variables (see (4.2.8) and also [**12**]).

The case of two simple turning points is studied in [**52, 73, 91, 107**]. A first way of constructing the solutions in this case is to use the standard differential equation for a parabolic cylinder:

$$(2.3.17)\qquad \frac{d^2 v}{d\eta^2} + \left(u - \eta^2\right) v = 0.$$

Another way is to construct solutions in neighbourhoods of the turning points separately and then to piece them together (see [**73**]).

The fourth-order equation ($m = n = 2$) of type (2.2.12) with $a_4 = 1$, $b_2(0) = 0$ and $b_2'(0) \ne 0$ is studied in [**116, 166, 167, 264, 265**]. The characteristic equation (2.2.14), i.e.

$$\lambda^4 + b_2(x)\lambda^2 = 0,$$

has the root $\lambda = 0$ of multiplicity four for $x = 0$.

2.3.5. An equation of order six. We consider in detail the integration of the equation of order six ($m = 2$, $n = 4$) of type (2.2.12), which describes the axisymmetric vibrations of shells of revolution. In this equation (see [**73**])

$$(2.3.18)\qquad \mu^4 = \frac{h^2}{12R^2(1-\nu^2)}, \qquad \mu > 0, \qquad b_2 = \frac{1}{R_2^2(x)} - \Lambda, \qquad a_6 = 1,$$

h is the shell thickness, which is assumed to be small as compared with its characteristic size R, ν is Poisson's ratio and Λ is a parameter proportional to the square of the natural frequency (see (2.6.15)). Moreover, the shell radius of curvature,

$R_2(x)$, is equal to the distance from the shell surface to its axis of symmetry measured along the shell normal, x is the arc length of the generator and $y(x)$ is the deflection (normal displacement).

Cylindrical and spherical shells, for which $R_2(s) = \text{const}$, have no turning points so that equation (2.2.12) transforms into an equation with constant coefficients [**213**]. For shells of other forms, there are turning points in the frequency interval $\Lambda^{(1)} \leq \Lambda \leq \Lambda^{(2)}$, where

(2.3.19)

$$\Lambda^{(1)} = \min_{x\in S}\left\{\frac{1}{R_2^2(x)}\right\}, \qquad \Lambda^{(2)} = \max_{x\in S}\left\{\frac{1}{R_2^2(x)}\right\}, \qquad S = \{x : x_1 \leq x \leq x_2\},$$

since there exists an x_* such that $b_2(x_*) = 0$, and $\lambda = 0$ is a root of multiplicity six of equation (2.2.14).

The asymptotic expansions of solution (2.2.12) for $b_2'(x_*) \neq 0$ in a small neighbourhood of the turning point x_* were obtained in three stages.

In the first stage [**6**], these expansions were constructed for a conical shell by using the standard equation

$$-\mu^4\left(\frac{d^6v}{d\eta^6} + \gamma_0\frac{d^5v}{d\eta^5} + \gamma_1\frac{d^4v}{d\eta^4}\right) + \eta\frac{d^2v}{d\eta^2} + (2+\gamma_0\eta)\frac{dv}{d\eta} + (\gamma_2 + \gamma_1\eta)v = 0, \tag{2.3.20}$$

where the constants γ_i depend on the form of the shell and its natural frequency.

In the second stage [**214**], it was shown that for an arbitrary shell of revolution one may assume, in (2.3.20), that $\gamma_1 = \gamma_2 = 0$.

In the third stage [**217**], five sequences of standard functions, $v_k^{(i)}(\eta)$, independent of the shell parameters, were introduced as solutions of the following five equations:

$$\begin{gathered}\frac{d^5v_k^{(i)}}{d\eta^5} - \eta\frac{dv_k^{(i)}}{d\eta} + (k-1)v_k^{(i)} = \delta_{i5}\delta_{k0}\frac{2\eta^{k-1}}{(k-1)!},\\ v_k^{(i)} = \frac{dv_{k-1}^{(i)}}{d\eta}, \qquad i = 1,\dots,5,\end{gathered} \tag{2.3.21}$$

where δ_{ij} is Kronecker's symbol ($\delta_{ii} = 1$, $\delta_{ij} = 0$, $i \neq j$).

2.3.6. Piecing asymptotics together. In a neighbourhood of $x = x_*$ five of the integrals of equation (2.2.12) have index of variation $t = 4/5$ and their asymptotic expansions may be written in the form (see [**73**])

$$y^{(i)}(x,\mu) \simeq \sum_{k=0}^{\infty}\varepsilon^k C_k(x)v_k^{(i)}(\eta) + \varepsilon\delta_{i5}y^*(x,\mu), \qquad i = 1,\dots,5, \tag{2.3.22}$$

where

$$\varepsilon = \mu^{4/5} > 0, \qquad \eta = \frac{1}{\varepsilon}\xi(x), \qquad \xi = \left(\frac{5}{4}\int_{x_*}^{x} b_2^{1/4}\,d\tilde{x}\right)^{4/5}, \qquad C_0 = \frac{1}{\sqrt{B}}(\xi')^{-3/2}.$$

Here the functions $\xi(x)$, $C_k(x)$ and $y^*(x,\mu)$ are real analytic and $B(x)$ is the distance between the axis of symmetry and the current point on the shell.

To construct the sixth integral we consider equation (2.2.12) for $\mu = 0$,

$$Ly_0 = b_2(x)\frac{d^2y_0}{dx^2} + b_1(x)\frac{dy_0}{dx} + b_0(x)y_0 = 0. \tag{2.3.23}$$

This equation has a regular singular point, $x = x_*$, at which $b_1(x_*) = 2b_2'(x_*)$. If we look for the solution of equation (2.3.23) in the form

$$y_0 = (x - x_*)^p \sum_{k=0}^{\infty} y_k (x - x_*)^k,$$

then, for the *characteristic index* p, we get the equation

$$b_2'(x_*)p(p-1) + b_1(x_*)p = 0,$$

from which we obtain $p_5 = -1$ and $p_6 = 0$. Since the characteristic indices differ by an integer, one solution, $y_0^{(6)}(x)$, is regular at $x = x_*$, and another, $y_5^{(0)}(x)$, has a singularity of the following type:

$$y_0^{(5)} = \frac{1}{x - x_*} + \sum_{k=0}^{\infty} y_{5k}(x - x_*)^k + \gamma y_0^{(6)}(x)\ln(x - x_*). \tag{2.3.24}$$

The regular solution $y_0^{(6)}(x)$ generates the asymptotic solution

$$y^{(6)}(x, \mu) \simeq \sum_{k=0}^{\infty} \mu^{4k} y_k^{(6)}(x), \qquad \text{as } \mu \to 0, \tag{2.3.25}$$

of equation (2.2.12)

We now return to expansions (2.3.22). They are asymptotic expansions as $\varepsilon\eta \to 0$, i.e. as $x \to x_*$. If $x - x_* \sim 1$, then $\eta \sim \varepsilon^{-1}$. Considering the asymptotics of the functions $v_k^{(i)}(\eta)$ as $\eta \to \pm\infty$, we can conclude (see [**73**]) that, for $i = 1, 2, 3$, the series (2.3.22) are asymptotic for all $x \in S$. For $i = 4$ and 5, the series (2.3.22) are not asymptotic as $x - x_* \sim 1$ since all the terms of these series have the same order as $\varepsilon \to 0$. To construct asymptotic expansions for these integrals in the whole domain S, one has to solve the "unpleasant" problem of piecing together the asymptotics (2.3.22) with simpler asymptotics of the types (2.2.1) and (2.3.25), which are valid for $x_1 \le x < x_*$ and $x_* < x \le x_2$. At present, the problem of constructing asymptotic expansions for the solutions of equation (2.2.12) which are uniformly valid for the entire domain $x_1 \le x \le x_2$ is still unsolved.

2.3.7. Other results. We briefly mention other results. Nonaxisymmetric vibrations of shells of revolution are considered in [**73, 222**]. In this case, unlike in the case of axisymmetric vibrations, $m = n = 4$ in (2.2.12) and the expansions (2.3.22) are valid under the same assumptions; however there are three solutions of type (2.3.25).

In [**221**], Tovstik considers the asymptotic integration of system (2.2.12) for arbitrary m and n under the assumptions that $b_m(x) = x$ and $b_{m-1}(0)$ is not an integer nor zero.

The *double* turning point at the origin, i.e. $b_2(0) = b_2'(0) = 0$ and $b_2''(0) \neq 0$, is considered in [**219**] for axisymmetric vibrations of shells of revolution ($m = 2$, $n = 4$). Axisymmetric vibrations of shells of revolution with two simple turning points in the interval of integration are considered by G. N. Chernyshev [**33**].

2.4. Boundary value problems and Sturm–Liouville problems

2.4.1. Boundary value problems. Let the domain of integration S be the real interval $[x_1, x_2]$. We consider the boundary value problem which consists of equation (2.2.6) together with the boundary conditions

$$\frac{d^k y}{dx^k} = q_k^{(j)} \qquad \text{for } x = x_j,\ k = 0, 1, \dots, n_j - 1,\ j = 1, 2,\ n_1 + n_2 = n. \tag{2.4.1}$$

Moreover, we assume that the coefficients $a_k(x, \mu)$ in (2.2.6) are real.

2.4.2. Different types of integrals. The behaviour of the integrals of equation (2.2.6) depends upon the roots $\lambda_p(x)$ of equation (2.2.7). In a neighbourhood of any non-turning point, x, the asymptotic solution (2.2.1) is

- exponentially increasing if $\Re\lambda_p(x) > 0$,
- exponentially decreasing if $\Re\lambda_p(x) < 0$ and
- oscillating if, for all x, $\Re\lambda_p(x) = 0$ and $\Im\lambda_p(x) \neq 0$.

If $\lambda_p(x) \equiv 0$, a case which can occur for equations (2.2.8) and (2.2.12), the integral (2.2.11) is *slowly varying.*

If the following conditions:

$$\Re\lambda_p(x) < 0, \qquad \Re\int_{x_1}^{x} \lambda_p(\tilde{x})d\tilde{x} < 0, \qquad x_1 < x \leq x_2, \tag{2.4.2}$$

are satisfied, the integral is called the *integral of the edge effect* or the *integral of the boundary layer* near the left end, $x = x_1$. This integral decreases exponentially away from the edge $x = x_1$ and becomes exponentially small for all $x > x_1$. In solving the boundary value problem with an error of the order $e^{-c/\mu}$, where $c > 0$, we may take the value of this integral to be equal to zero at the right end, $x = x_2$. One introduces, in a similar way, the integral of the edge effect at the right end, $x = x_2$.

There may exist integrals which do not belong to any of the above types. We note that some of these integrals may oscillate on a part S_0 of the interval S. Under the above assumptions, only the turning points, or the end points x_1 and x_2, may be boundary points of S_0.

2.4.3. A problem with a small parameter. We consider the Sturm–Liouville problem with a small parameter μ and unknown eigenvalues Λ,

$$\begin{gathered}\mu^2 \frac{d}{dx}\left[r(x)\frac{dy}{dx}\right] + p(x, \Lambda)y = 0,\\ y(x_1) = y(x_2) = 0, \qquad r(x) > 0, \qquad x \in S.\end{gathered} \tag{2.4.3}$$

The problem is to find the asymptotics of the eigenfunctions $y(x)$ and eigenvalues Λ as $\mu \to 0$. To find the asymptotics of the n-th eigenvalue, $\lambda_n \sim n^2$ as $n \to \infty$, the traditional formulation of the Sturm–Liouville problem,

$$\frac{d}{dx}\left[r(x)\frac{dy}{dx}\right] + [\lambda p_1(x) - p_2(x)]\,y = 0, \tag{2.4.4}$$

reduces to (2.4.3). For that purpose we introduce a fixed $\mu \sim n^{-1}$ and assume that $\Lambda = \mu^2\lambda$. The solution of problem (2.4.3) is considered, for example, in [**42, 52, 147**]. We mention the main results.

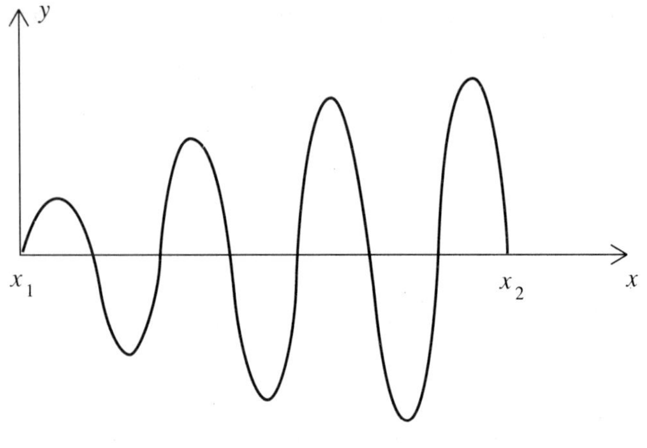

FIGURE 6. Oscillating eigenfunction.

If Λ is such that $p(x,\Lambda) < 0$ for every $x \in S$, then problem (2.4.3) has no nontrivial solutions, i.e. Λ is not an eigenvalue.

Let now $p(x,\Lambda) > 0$ for every $x \in S$. Then equation (2.4.3) has two oscillating integrals U and V. Their real linear combination,

$$y(x,\mu) = U\cos\varphi + V\sin\varphi,$$

which is equal to zero for $x = x_1$, has the expansion

$$U(x,\mu) \simeq \sum_{k=0}^{\infty}(-1)^k a_{2k+1}(x)\mu^{2k+1}, \qquad V(x,\mu) \simeq \sum_{k=0}^{\infty}(-1)^k a_{2k}(x)\mu^{2k}, \tag{2.4.5}$$

where

$$\varphi(x,\mu) = \frac{1}{\mu}\int_{x_1}^{x}\sqrt{\frac{p}{r}}d\tilde{x}, \qquad a_0 = (pr)^{-1/4},$$

and

$$a_{k+1} = \frac{a_0}{2}\int_{x_1}^{x} a_0(ra_k')'d\tilde{x}, \qquad k = 0,1,\dots$$

The condition $y(x_2) = 0$ yields the following equation for the evaluation of Λ:

$$\varphi_2(\Lambda) := \varphi(x_2,\mu) = n\pi - \arctan\frac{U(x_2,\mu)}{V(x_2,\mu)}, \tag{2.4.6}$$

where n is an integer. The second term on the right-hand side of (2.4.6) is of the order μ, and, in the first approximation, may be neglected. Because $\partial\varphi_2/\partial\Lambda \sim \mu^{-1}$, the distance between the neighbouring roots of equation (2.4.6) is of the order μ.

The eigenfunctions oscillate on the entire interval (x_1, x_2) (see Fig. 6)

2.4.4. Simple turning point at the left edge. Let

$$p(x_1,\Lambda) = 0, \qquad p_x'(x_1,\Lambda) > 0$$

and

$$p(x,\Lambda) > 0, \qquad x > x_1,$$

i.e. suppose that $x_* = x_1$ is a simple turning point. Applying the asymptotic representations (2.3.12) and (2.3.15) we have the following equation for Λ:

$$\varphi_2(\Lambda) = \pi\left(n - \frac{1}{12}\right) + O(\mu), \tag{2.4.7}$$

where $\varphi_2(\Lambda)$ is the same as in (2.4.6).

2.4.5. Interior simple turning point. Now let the simple turning point x_*, i.e. $p_x(x_*, \Lambda) = 0$ and $p'_x(x_*, \Lambda) > 0$, be situated inside the open interval $x_1 < x_* < x_2$ and suppose that $p(x, \Lambda) > 0$ for $x > x_*$ and $p(x, \Lambda) < 0$ for $x < x_*$.

The solutions of equation (2.4.3) oscillate for $x > x_*$ and are exponentially increasing or decreasing for $x < x_*$. As a standard function in (2.3.12) we use Airy's function $\mathrm{Ai}(\eta)$ which provides the exponentially decreasing solution for $x < x_*$. Neglecting the left boundary condition, at $x = x_1$, with an error of the order $e^{-c/\mu}$, and satisfying the right condition, at $x = x_2$, we get the equation for Λ in the form $\mathrm{Ai}(\eta(x_2)) = 0$, from which, by (2.3.13), we obtain

$$\varphi_2(\Lambda) = \frac{2}{3}|\eta_n|^{3/2} + O(\mu), \qquad n = 1, 2, \ldots, \tag{2.4.8}$$

where

$$\varphi_2(\Lambda) = \frac{1}{\mu}\int_{x_*}^{x_2} \sqrt{\frac{p}{r}}\, dx,$$

and η_n is the n-th zero of $\mathrm{Ai}(\eta)$. If we apply the asymptotic formula (1.6.9) for $n \gg 1$, then equation (2.4.8) may be rewritten in the form

$$\varphi_2(\Lambda) = \pi\left(n - \frac{1}{4}\right) + O(\mu), \tag{2.4.9}$$

which is convenient for a comparison with equations (2.4.6) and (2.4.7). Even for small n, the precision of equation (2.4.9) is acceptable. We list the first terms of $2/3|\eta_n|^{3/2}$ and (in parentheses) the corresponding values of $\pi(n - 1/4)$:

2.383 (2.356), 5.510 (5.498), 8.647 (8.639).

For an illustrative example, see Subsection 2.4.10 below. The eigenfunctions oscillate for $x_* < x \leq x_2$ and they decrease as x approaches the edge $x = x_1$ (see Fig. 7)

This and other cases related to the asymptotic solution of problem (2.4.3) are considered in [**42, 52, 80, 117, 147**]. The case of two turning points, where $p > 0$ along the segment joining them, with an oscillating solution for equation (2.4.3), is considered below in Section 2.5.

2.4.6. Boundary eigenvalue problem for $n > 2$. Now we consider boundary eigenvalue problems for $n > 2$. For system (2.1.1) we introduce the boundary conditions

$$B^{(j)}y = 0, \qquad \text{for } x = x_j, \qquad j = 1, 2, \tag{2.4.10}$$

where $B^{(j)}$ are $n \times n_j$ matrices. For equation (2.2.6) we assume that conditions (2.4.1) are homogeneous, i.e. $q_k^{(j)} = 0$. The coefficients of system (2.1.1) and equation (2.2.6) are assumed to depend upon the parameter Λ, i.e. $A_k = A_k(x, \Lambda)$ and $a_k = a_k(x, \Lambda)$. The boundary value problems for equations (2.2.8) and (2.2.12)

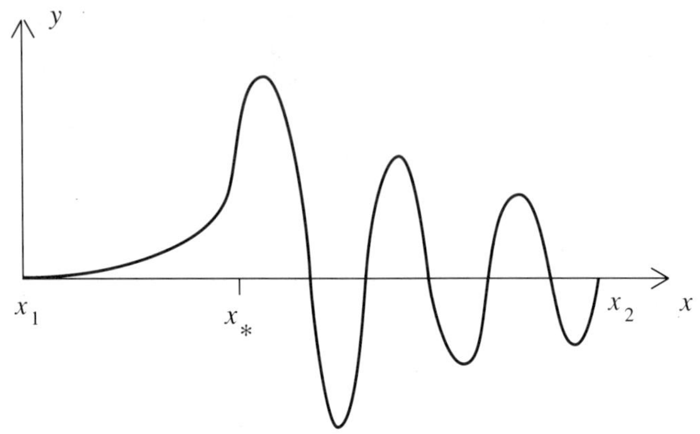

FIGURE 7. Eigenfunction with a turning point.

which, for $\mu = 0$, degenerate into equations of lower orders are considered in Chapter 3.

If we suppose that the asymptotic expansions, as $\mu \to 0$, for n linearly independent integrals of equation (2.1.1) are already constructed, then, to determine Λ, we use the equation

$$\Delta_n(\Lambda, \mu) = 0, \tag{2.4.11}$$

for the determinant of order n. This determinant is obtained by substituting a linear combination of the integrals into conditions (2.4.10) or (2.4.1). The study of equation (2.4.11) splits into particular cases, which depend upon the properties of the roots of equation (2.1.6).

We now suppose that there are no turning points and equation (2.1.1) has l_1 and l_2 edge effect integrals at the left and right ends of the interval $[x_1, x_2]$, respectively, and l_0 oscillating integrals ($l_0 + l_1 + l_2 = n$). (The authors of this survey never met in any application the cases where $l_1 > n_1$ or $l_2 > n_2$, i.e. the number of edge effect integrals is higher than the number of boundary conditions at the corresponding end.)

If $l_1 = n_1$ (the case $l_2 = n_2$ can be considered in a similar way), then, with an error of the order $e^{-c/\mu}$, equation (2.4.11) splits into the product of two equations

$$\Delta_{n_1}(\Lambda, \mu) = 0, \qquad \Delta_{n-n_1}(\Lambda, \mu) = 0. \tag{2.4.12}$$

The first is obtained by substituting a linear combination of the edge effect integrals at the edge $x = x_1$ into the boundary conditions (2.4.1) at this edge. A solution of the equation $\Delta_{n_1} = 0$, if any exists, corresponds to an eigenfunction which is exponentially decreasing away from the edge $x = x_1$. The eigenfunction corresponding to the root of the equation $\Delta_{n-n_1} = 0$, for $l_0 \neq 0$, can contain an oscillating component. In Section 2.7 we shall consider an example where $l_1 = n_1$.

2.4.7. A pair of pure imaginary roots. Let equation (2.1.6) have a pair of pure imaginary roots $\lambda_{1,2} = \pm ip(x, \Lambda)$ and $n_1 = l_1 + 1$ and $n_2 = l_2 + 1$. Then equation (2.4.11) can be reduced to

$$\begin{aligned} &D_1(\Lambda,\mu)\cos\varphi + D_2(\Lambda,\mu)\sin\varphi = 0, \\ &\varphi = \varphi(\Lambda,\mu) = \frac{1}{\mu}\int_{x_1}^{x_2} p(x,\Lambda)dx. \end{aligned} \tag{2.4.13}$$

where D_1 and D_2 are regular functions of μ (see also [**73**]). This equation determines a series of eigenvalues, which are at a distance of the order μ between each other, and the corresponding eigenfunctions contain an oscillating component. Equation (2.4.13) is similar to equation (2.4.6) but differs from equations (2.4.12). The roots of (2.4.12),

$$\Lambda = \Lambda_0 + \mu\Lambda_1 + \dots, \qquad \Delta_{n_1}(\Lambda_0, 0) = 0, \tag{2.4.14}$$

(if they exist) depend regularly upon μ; but this expansion is not valid for the roots of (2.4.13).

2.4.8. One turning point and imaginary roots. Suppose now that system (2.1.1) has one simple turning point, $x = x_*(\Lambda)$, such that, for $x > x_*$, equation (2.1.6) has a pair of pure imaginary roots $\lambda_{1,2} = \pm ip(x,\Lambda)$, and, for $x < x_*$, it has real roots $\lambda_1 < 0$ and $\lambda_2 > 0$. Moreover let equation (2.1.6) have $l_1 = n_1 - 1$ roots with $\Re\lambda < 0$ and $l_2 = n_2 - 1$ roots with $\Re\lambda > 0$. Under these assumptions, equation (2.4.11) splits into two equations. The first one is $\Delta_{n_1} = 0$ as in equation (2.4.12).

The second equation of (2.4.12) depends only upon the boundary conditions at the edge $x = x_2$ and, by the transformation (2.3.4), it becomes

$$D(\Lambda,\mu)\,\mathrm{Ai}(\eta(x_2)) + D^{(1)}(\Lambda,\mu)\frac{1}{\mu}\frac{d}{dx}\,\mathrm{Ai}[\eta(x_2)] = 0, \tag{2.4.15}$$

where, according to (2.3.13),

$$-\eta(x_2) = \left[\frac{3}{2\mu}\int_{x_*}^{x_2} p(x)dx\right]^{2/3} \equiv \left[\frac{3}{2}\varphi_2(\Lambda)\right]^{2/3}. \tag{2.4.16}$$

If $|\eta(x_2)| \gg 1$, by the asymptotic formula (6.9), equation (2.4.15) may be transformed into an equation of type (2.4.13) for $\varphi = \varphi_2(\Lambda)$ (see (2.4.8)). The last equation is similar to (2.4.13) and has a series of roots, Λ, which are at a distance of the order μ between each other, and the corresponding eigenfunctions are oscillating for $x_* < x \le x_2$.

For $x_2 - x_* \sim \mu^{2/3}$, $\eta(x_2) \sim 1$ and, for $D(\Lambda,\mu) \ne 0$, equation (2.4.15) is expressed as

$$\mathrm{Ai}\left(\eta(x_2)\right) + \mu^{1/3}D^*(\Lambda,\mu)\,\mathrm{Ai}'\left(\eta(x_2)\right) = 0. \tag{2.4.17}$$

This equation can be transformed into (2.4.8) with an error of the order $\mu^{1/3}$.

The one-dimensional problems for the free vibrations and buckling of shells to be considered later in Section 2.7 will illustrate the remarks made in the above Sections 2.4.1–2.4.8.

2.4.9. A beam of nonconstant thickness. In the present subsection, we consider the problem of determining the natural frequencies, ω, of a beam of nonconstant thickness with clamped edges. Thus we have the boundary value problem

$$\begin{gathered} \frac{d^2}{dx^2}\left[r(x)\frac{d^2y}{dx^2}\right] - q(x)\omega^2 y = 0, \\ y(0) = y'(0) = y(l) = y'(l) = 0, \\ r(x) = EI(x) > 0, \qquad q(x) = \rho S(x) > 0, \end{gathered} \tag{2.4.18}$$

where $y(x)$ is the deflection, $r(x)$ is the beam bending stiffness, E is Young's modulus, I is the moment of inertia of a cross-section, $q(x)$ is mass per unit length, ρ is the material density, $S(x)$, is the area of the cross-section and l is the length of the beam.

If we assume that $\omega^2 = \mu^{-4}$, then equation (2.4.18) is of the form (2.2.12), i.e.

$$\mu^4 \frac{d^2}{dx^2}\left[r(x)\frac{d^2y}{dx^2}\right] - q(x)y = 0. \tag{2.4.19}$$

Since the eigenvalue Λ is absent, we consider the unknown values of the small parameter μ for which equation (2.4.19) has nonzero solutions which satisfy the boundary conditions (2.4.18). The characteristic equation (2.2.7) is

$$r(x)\lambda^4 - q(x) = 0,$$

so that the assumptions under which equation (2.4.13) was derived are fulfilled. Then (2.4.13) has the form

$$\cos\left[\frac{1}{\mu}\int_0^l \left(\frac{q}{r}\right)^{1/4} dx\right] = O(\mu), \tag{2.4.20}$$

from which we find

$$\mu_n = \frac{1}{\pi(n+1/2)}\int_0^l \left(\frac{q}{r}\right)^{1/4} dx + O\left(\mu_n^3\right), \qquad n = 1, 2, \dots, \tag{2.4.21}$$

or

$$\omega_n = \pi^2\left(n + \frac{1}{2}\right)^2 \left[\int_0^l \left(\frac{q}{r}\right)^{1/4} dx\right]^{-2} + O(1), \qquad n = 1, 2, \dots \tag{2.4.22}$$

The precision of formulae (2.4.21) and (2.4.22) increases with n. For $n = 1$, the error in the application of formula (2.4.22) to a beam of constant cross-section ($p, q = \text{const}$) is 0.004. This error is caused by the fact that, in the derivation of formulae (2.4.20) and (2.4.13), we neglected the interference of the edges effects, i.e. the value of $e^{-c/\mu}$ as compared with 1 (for the beam $c = l(q/r)^{1/4}$).

2.4.10. A circular membrane. As an example with turning points we consider the vibrations of a circular membrane with a large wave number, m, in the circumferential direction:

$$T\left[\frac{d}{dr}\left(r\frac{dw}{dr}\right) - \frac{m^2 w}{r^2}\right] + \rho\omega^2 w = 0, \qquad w(R) = 0, \tag{2.4.23}$$

where $w(r)\cos m\varphi$ is the membrane deflection, T is the stretching force, ρ is mass per unit area, ω is the vibration frequency and R is the membrane radius. The exact solution of this problem is represented by means of the zeros, α_{mn}, of the Bessel functions J_m:

$$\omega_{mn} = \frac{1}{R}\sqrt{\frac{T}{\rho}}\,\alpha_{mn}, \qquad J_m(\alpha_{mn}) = 0, \qquad n = 1, 2, \dots \tag{2.4.24}$$

Here we consider the asymptotic expansion as $m \to \infty$. After the substitutions

$$\mu = \frac{1}{m}, \qquad w(x) = yx^{-1/2}, \qquad \omega = \frac{m}{R}\sqrt{\frac{T\Lambda}{\rho}}, \qquad r = Rx, \tag{2.4.25}$$

equation (2.4.23) is of the form (2.4.3), i.e.

$$\mu^2\frac{d^2y}{dx^2} + p(x, \Lambda)y = 0, \qquad p = \Lambda - \frac{1 - \mu^2/2}{x^2}, \qquad y(1) = 0. \tag{2.4.26}$$

The existence of a singular point at $x = 0$ does not prevent us from obtaining the asymptotic integration of this equation, since we get an exponentially decreasing solution $y(x)$. The term $\mu^2/2$ in the expression for p is neglected. For $\Lambda \le 1$, the boundary value problem (2.4.26) has no solution. If $\Lambda > 1$, the turning point $x_* = \Lambda^{-1/2}$ lies in the integration domain $0 \le x \le 1$, and we may use formula (2.4.8). Evaluating φ_2 in (2.4.8) approximately and assuming that x_* is close to 1, we find that

$$\Lambda_{mn} = \left[1 + \left(\frac{2}{m}\right)^{2/3}|\eta_n|\right] + O\left(\frac{1}{m}\right), \tag{2.4.27}$$

where η_n is the n-th root of $\mathrm{Ai}(\eta) = 0$. Comparing formulae (2.4.24) and (2.4.25) we get

$$\alpha_{mn} = m\Lambda_{mn}^{1/2}. \tag{2.4.28}$$

The asymptotic expansions of α_{mn} by negative powers of $m^{1/3}$ (the first two terms of which coincide with (2.4.28)) can be found in [**83**].

A generalization of this problem to an arbitrary convex membrane will be done in Section 4.3.

2.5. Localized solutions of ordinary differential equations

2.5.1. Localized solutions. In Section 2.4 we considered cases (see (2.4.12) and (2.4.15)), where the eigenfunctions were exponentially decreasing as x approached one of the endpoints of the interval $[x_1, x_2]$, and the equation for the eigenvalue Λ did not depend (to an error of the order $e^{-c/\mu}$) upon the boundary conditions at this end point. Now we consider cases where system (2.1.1), or equation (2.2.6), have a solution which is oscillating for $x_*^{(1)} < x < x_*^{(2)}$ ($x_1 < x_*^{(1)} < x_*^{(2)} < x_2$) and exponentially decreasing as x approaches the endpoints x_1 and x_2. This solution satisfies, with an error of the order $e^{-c/\mu}$, any homogeneous boundary conditions at $x = x_1$ and $x = x_2$. Such a solution is said to be a *localized solution.*

We assume that the coefficients of system (2.1.1) depend upon the parameter Λ, and intend to find the eigenvalues, Λ, for which a localized solution exists.

For example, some quantum mechanical problems for the one-dimensional Schrödinger equation, and problems in vibrations, dynamics and buckling of shells [**52,**

73, 80, 91, 107, 111, 238] can be reduced to the cases considered here. A rather complete study is found in [**52**]. We shall consider only the simplest cases.

2.5.2. A quantization condition and an example. We consider the second-order equation (2.4.3) and assume, as before, that $r(x) > 0$ on S. Let

$$p\left(x_*^{(k)}\right) = 0, \qquad p'\left(x_*^{(k)}\right) \neq 0, \qquad p(x) > 0, \qquad \text{for } x_*^{(1)} < x < x_*^{(2)}.$$

The necessary and sufficient condition for the existence of a localized solution is

$$\text{(2.5.1)} \qquad \frac{1}{\mu}\int_{x_*^{(1)}}^{x_*^{(2)}} f(x,\Lambda)dx = \frac{\pi}{2}(2n+1) + O(\mu), \qquad n = 0, 1, 2, \ldots,$$

where $f = \sqrt{p/r}$. This "quantization" condition may be obtained by means of several different methods:

(1) by using the standard equation (2.3.17) which describes the behaviour of the solution simultaneously in neighbourhoods of both turning points $x = x_*^{(k)}$ (see [**91, 107**]);
(2) by constructing the solutions (2.3.12) in neighbourhoods of each of the turning points $x_*^{(k)}$ and piecing together their asymptotics (2.3.15) on the interval $x_*^{(1)} < x < x_*^{(2)}$ (see [**73**]);
(3) by bypassing the turning points in the complex x-plane (see [**52, 80**]).

As an example, we consider the free vibrations of a spring laying on an elastic foundation,

$$\text{(2.5.2)} \qquad T\frac{d^2y}{dx^2} + \left[\omega^2\rho - c(x)\right]y = 0, \qquad y \to 0 \qquad \text{as } x \to \pm\infty,$$

where T is the spring stress, ρ is the linear density, ω is the unknown frequency and $c(x)$ is the variable stiffness of the foundation. Assuming that the function $c(x)$ has a minimum at $x = 0$, we look for the vibration modes localized near $x = 0$.

For $c(x) = c_0 + c_2x^2$, the exact solution of the problem [**84**] is

$$\text{(2.5.3)} \qquad \begin{aligned} y_n(x) &= D_n\left(z\sqrt{2}\right) = H_n(z)\,e^{-z^2/2}, \\ z &= \left(\frac{c_2}{T}\right)^{1/4} x, \qquad n = 0, 1, 2, \ldots, \\ \omega_n^2 &= \rho^{-1}\left[c_0 + \sqrt{Tc_2}\,(2n+1)\right], \end{aligned}$$

where $H_n(z)$ is Hermite's polynomial of degree n and $D_n(t)$ is the n-th parabolic cylinder function. The case $n = 6$ is shown in Fig. 8.

For a general function $c(x)$, formula (2.5.1) gives an approximate equation for ω_n in the form

$$\text{(2.5.4)} \qquad \int_{x_*^{(1)}}^{x_*^{(2)}} \sqrt{T^{-1}\left[\rho\omega_n^2 - c(x)\right]}\,dx = 2n+1, \qquad n = 0, 1, 2, \ldots,$$

the precision of which increases with n. Here $x_*^{(1)}$ and $x_*^{(2)}$ are the zeros of the expression under the radical in (2.5.4).

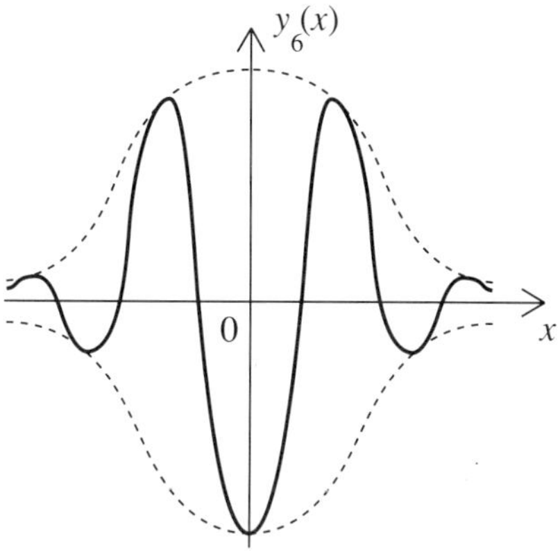

FIGURE 8. The function $y_6(x)$ of the parabolic cylinder.

2.5.3. Existence of a formal localized solution. Suppose now that the order of system (2.1.1) or equation (2.2.6) is greater than two and the characteristic equation (2.1.6) has exactly two imaginary roots for $x_*^{(1)} < x < x_*^{(2)}$,

$$\lambda_{1,2} = \pm i f(x), \qquad f(x) \sim \sqrt{(x_*^{(2)} - x)(x - x_*^{(1)})}, \qquad \text{as } x \to x_*^{(k)}, \tag{2.5.5}$$

which, for $x \notin \left[x_*^{(1)}, x_*^{(2)}\right]$, are real roots with opposite signs. The remaining roots of (2.1.6) have nonzero real parts. Then under the same condition (2.5.1), system (2.1.1) has a *formal* localized solution. This solution will be asymptotic provided the roots of (2.1.6) satisfy additional conditions of type (2.2.4) or (2.2.5).

2.5.4. A system of order greater than or equal to four. We mention one more case of localized solutions for systems of order greater than or equal to four (see [**73**]). Suppose that, for $x_*^{(1)} < x < x_*^{(2)}$, equation (2.1.6) has four pure imaginary roots of the form

$$\lambda_{1,2,3,4} = \pm i q(x) \pm i f(x), \qquad f(x) \sim \sqrt{(x_*^{(2)} - x)(x - x_*^{(1)})}, \qquad \text{as } x \to x_*^{(k)}, \tag{2.5.6}$$

where $q(x)$ is analytic. Suppose also that for $x \notin \left[x_*^{(1)}, x_*^{(2)}\right]$ these roots have nonzero real parts. Under the same assumptions on the remaining roots of equation (2.1.6), a necessary condition for the existence of localized solutions is again given by (2.5.1).

2.5.5. Lowest positive eigenvalue. The determination of the lowest positive eigenvalue, Λ, and the construction of the corresponding eigenfunction are of special interest in the analysis of vibration and buckling problems (see Chapter 3). It often occurs, in this case, that this eigenfunction is localized and the turning points $x_*^{(1)}$ and $x_*^{(2)}$ are close to each other $\left(x_*^{(2)} - x_*^{(1)} \sim \sqrt{\mu}\right)$; then the algorithm described below may be applied (see also [**111, 209, 238**]). This algorithm is a simplified one-dimensional version of the algorithm described by V. P. Maslov [**124**].

We consider the self-adjoint equation of order $2n$,

$$\sum_{k=0}^{n}(-i\mu)^{2k}\frac{d^k}{dx^k}\left[a_k(x)\frac{d^k y}{dx^k}\right]=0, \qquad x_1\le x\le x_2, \tag{2.5.7}$$

where the coefficients a_k are real analytic and depend linearly upon the parameter Λ,

$$a_k(x)=a_{1k}(x)-\Lambda a_{2k}(x), \qquad \Lambda>0. \tag{2.5.8}$$

We seek the eigenvalues, Λ, for which there exists a localized solution to equation (2.5.7) in a neighbourhood of the so-called *weakest point*, x_0 $(x_1<x_0<x_2)$. This solution is sought in the form of an asymptotic series,

$$\begin{aligned} y(x,\mu)&=\sum_{k=0}^{\infty}\mu^{k/2}y_k(\xi)\exp\left[i\left(\frac{1}{\sqrt{\mu}}p_0\xi+\frac{1}{2}a\xi^2\right)\right],\\ \xi&=\frac{1}{\sqrt{\mu}}(x-x_0),\\ \Lambda&=\Lambda_0+\mu\Lambda_1+\dots, \end{aligned} \tag{2.5.9}$$

where $y_k(\xi)$ are polynomials in ξ. Here p_0, x_0 and Λ_k are real and $\Im a>0$. This last condition ensures that the solution $y(x,\mu)$ in (2.5.9) is decreasing as $|x-x_0|$ increases.

Substituting $\lambda=ip$ into the characteristic equation (2.2.7) and solving for Λ, we get

$$\Lambda=\left[\sum_{k=0}^{n}a_{1k}(x)p^{2k}\right]\left[\sum_{k=0}^{n}a_{2k}(x)p^{2k}\right]^{-1}=:f(p,x). \tag{2.5.10}$$

We suppose that there exists only one point, (p_0,x_0), at which the positive minimum,

$$\Lambda_0=\min_{\substack{p,x\\ f>0}}\{f(p,x)\}=f(p_0,x_0), \tag{2.5.11}$$

is attained and

$$\begin{aligned} d^2f&=f^0_{pp}dp^2+2f^0_{px}dpdx+f^0_{xx}dx^2>0,\\ f^0_{px}&=\frac{\partial^2 f(p_0,x_0)}{\partial p_0\partial x_0},\dots \end{aligned} \tag{2.5.12}$$

We seek the positive minimum (2.5.11) over all $x\in[x_1,x_2]$ and $p_0\ge 0$. Then a solution of type (2.5.9) exists and

$$\begin{aligned} &y_0^{(m)}(\xi)=H_m(z), \qquad z=\sqrt{c}\,\xi, \qquad \Lambda^{(m)}=\Lambda_0+\mu\Lambda_1^{(m)}+O\left(\mu^2\right),\\ &a=\frac{ir-f^0_{px}}{f^0_{pp}}, \qquad c=\frac{r}{f^0_{pp}}, \qquad r=\sqrt{f^0_{pp}f^0_{xx}-\left(f^0_{px}\right)^2}>0,\\ &\Lambda_1^{(m)}=r\left(m+\frac{1}{2}\right), \qquad m=0,1,2,\dots, \end{aligned} \tag{2.5.13}$$

where $H_m(z)$ is Hermite's polynomial of degree m (in particular, $H_0(z) = 1$, $H_1(z) = z$, $H_2(z) = z^2 - 1/2$). Formulae (2.5.9) and (2.5.13) determine the eigenvalues, $\Lambda^{(m)}$, the lowest of which is obtained for $m = 0$.

For the eigenfunctions, we have two cases:

$$\text{(2.5.14)} \qquad \text{(A)} \quad p_0 = 0, \qquad \text{and} \qquad \text{(B)} \quad p_0 > 0.$$

In the case (A), $p_0 = 0$, the eigenfunctions have the form

$$\text{(2.5.15)} \qquad y^{(m)}(x,\mu) = \left[H_m(z) + O\left(\sqrt{\mu}\,\right)\right] e^{-z^2/2}, \qquad z = \sqrt{\frac{c}{\mu}}\,(x - x_0),$$

and the corresponding eigenvalues, $\Lambda^{(m)}$, are simple. The eigenfunctions have a similar behaviour for a pair of turning points, $x_*^{(1)}$ and $x_*^{(2)}$, if between them the characteristic equation (2.2.7) has a pair of pure imaginary roots (see (2.5.5)).

In the case (B), $p_0 > 0$, the function (2.5.9) is complex. Since the coefficients a_k in (2.5.7) are real, both the real and imaginary parts of (2.5.9) are solutions of equation (2.5.7). But it would be a mistake to conclude that the eigenvalues $\Lambda^{(m)}$ are double and an arbitrary linear combination of the real and imaginary parts of (2.5.9) yields the eigenfunction. In fact, the series (2.5.9) is not convergent but asymptotic and to the parameter $\Lambda^{(m)}$ there correspond two fixed real eigenfunctions,

(2.5.16)

$$y^{(m,j)}(x,\mu) = \left[H_m(z)\cos\Psi_j + O\left(\sqrt{\mu}\,\right)\right] e^{-\Im a(x-x_0)^2/(2\mu)},$$

$$\Psi_j = \frac{p_0 x}{\mu} + \frac{\Re a(x-x_0)^2}{2\mu} + \Theta_j, \qquad j = 1, 2, \qquad z = \sqrt{\frac{c}{\mu}}\,(x - x_0),$$

where the phases Θ_j are fixed ($0 \le \Theta_1,\ \Theta_2 < 2\pi$). The corresponding exact eigenvalues, $\Lambda^{(m,1)}$ and $\Lambda^{(m,2)}$, are distinct; however

$$\text{(2.5.17)} \qquad \Delta\Lambda = \Lambda^{(m,2)} - \Lambda^{(m,1)} = O\left(\mu^N\right)$$

for any N. Such eigenvalues are said to be *asymptotically double.* If equation (2.2.7) has a pair of pure imaginary roots (2.5.6) between the turning points, then the eigenvalues are also asymptotically double.

In Fig. 9, it is shown that, for $m = 0$, there is (a) one eigenfunction if $p_0 = 0$ and (b, c) two eigenfunctions if $p_0 > 0$.

2.6. Two-dimensional shell theories

Since, in this review, much of the attention is on thin shell theory, it is worth listing here the main points of the simplest two-dimensional variant of this theory. The equations of shell theory may be found in great detail, for example, in the monographs [**41, 66, 67, 77, 119, 143, 173, 261**].

2.6.1. Geometry of the neutral surface and its deformations. We introduce the system of orthogonal curvilinear coordinates α_1 and α_2 which coincide with the lines of curvature of the neutral surface, S, of the shell. Let a point M on S be determined by the radius vector $\mathbf{r} = \mathbf{r}(\alpha_1, \alpha_2)$.

The shell fills the volume

$$\text{(2.6.1)} \qquad (\alpha_1, \alpha_2) \in G, \qquad |z| \le \frac{h}{2},$$

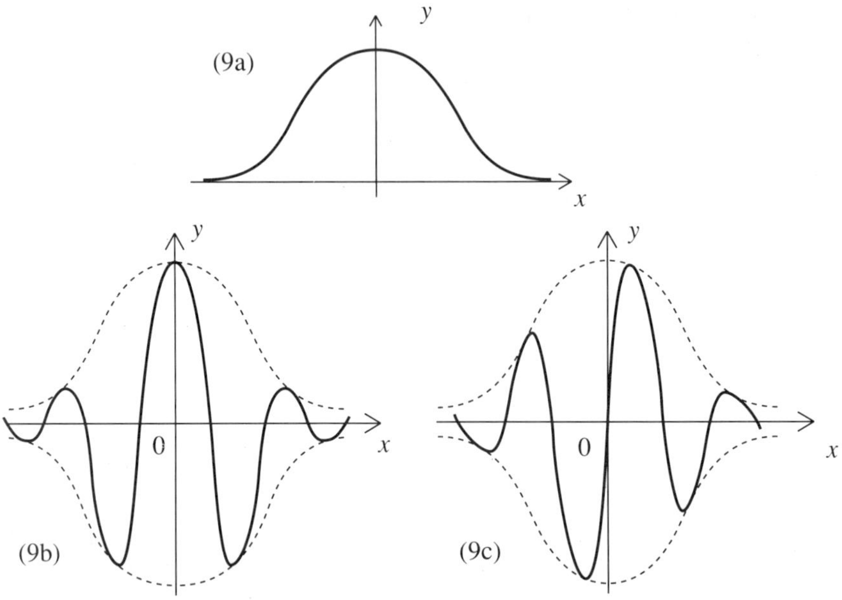

FIGURE 9. Different eigenfunctions: (a) – case of a simple eigenvalue; (b) and (c) – two eigenfunctions corresponding to a pair of asymptotically double eigenvalues.

where z is the distance along the normal to the neutral surface and h is the shell thickness. Let $\Gamma = \partial G$ be the boundary of the domain G. The shell is said to be thin if its relative thickness h/R is small, where R is the characterisic shell length.

We introduce a local orthogonal system of coordinates by means of the unit vectors $\mathbf{e}_1$, $\mathbf{e}_2$ and $\mathbf{n}$, where

$$\mathbf{e}_1 = \frac{1}{A_1}\frac{\partial \mathbf{r}}{\partial \alpha_1}, \qquad A_1 = \left|\frac{\partial \mathbf{r}}{\partial \alpha_1}\right|, \qquad (1 \leftrightarrow 2), \qquad \mathbf{n} = \mathbf{e}_1 \times \mathbf{e}_2. \tag{2.6.2}$$

Here and below, in this section, the notation $(1 \leftrightarrow 2)$ means that the formula with 1 and 2 interchanged is also valid.

The first and the second quadratic forms of the surface are

$$\begin{aligned} I &= ds^2 = A_1^2 d\alpha_1^2 + A_2^2 d\alpha_2^2, \\ II &= \frac{A_1^2}{R_1} d\alpha_1^2 + \frac{A_2^2}{R_2} d\alpha_2^2, \end{aligned} \tag{2.6.3}$$

where ds is arc length on the surface, A_1 and A_2 are Lamé's coefficients, R_1 and R_2 are the radii of curvature. We also use the notation $k_i = R_i^{-1}$.

We now describe the deformation of the neutral surface. Let u_1, u_2 and w be the projections of the displacements of a point M on the unit vectors $\mathbf{e}_1$, $\mathbf{e}_2$ and $\mathbf{n}$ before deformation. In the linear approximation, the tangential (membrane) surface deformations, ε_1, ε_2 and ω, are

(2.6.4)
$$\begin{aligned} \varepsilon_1 &= \frac{1}{A_1}\frac{\partial u_1}{\partial \alpha_1} + \frac{1}{A_1 A_2}\frac{\partial A_1}{\partial \alpha_2} u_2 - \frac{w}{R_1}, && (1 \leftrightarrow 2), \\ \omega_1 &= \frac{1}{A_1}\frac{\partial u_2}{\partial \alpha_1} - \frac{1}{A_1 A_2}\frac{\partial A_1}{\partial \alpha_2} u_1, && \omega = \omega_1 + \omega_2, \end{aligned}$$

the angles of rotation γ_1 and γ_2 of the normal $\mathbf{n}$ are equal to

(2.6.5)
$$\gamma_1 = -\frac{1}{A_1}\frac{\partial w}{\partial \alpha_1} - \frac{u_1}{R_1}, \qquad (1 \leftrightarrow 2).$$

The bending surface deformations, κ_1, κ_2 and τ, are

(2.6.6)
$$\begin{aligned} \kappa_1 &= -\frac{1}{A_1}\frac{\partial \gamma_1}{\partial \alpha_1} - \frac{1}{A_1 A_2}\frac{\partial A_1}{\partial \alpha_2}\gamma_2, \qquad (1 \leftrightarrow 2), \\ \tau &= -\frac{1}{A_2}\frac{\partial \gamma_1}{\partial \alpha_2} + \frac{1}{A_1 A_2}\frac{\partial A_2}{\partial \alpha_1}\gamma_2 + \frac{\omega_1}{R_2}. \end{aligned}$$

2.6.2. Equilibrium equations and elasticity relations. The equations of equilibrium of an element of the neutral surface for small deformations are

(2.6.7)
$$\begin{gathered} \frac{\partial(A_2 T_1)}{\partial \alpha_1} - \frac{\partial A_2}{\partial \alpha_1} T_2 + \frac{\partial(A_1 S_2)}{\partial \alpha_2} + \frac{\partial A_1}{\partial \alpha_2} S_1 - \frac{A_1 A_2}{R_1} Q_1 + A_1 A_2 F_1 = 0, \qquad (1 \leftrightarrow 2), \\ \frac{\partial(A_2 Q_1)}{\partial \alpha_1} + \frac{\partial(A_1 Q_2)}{\partial \alpha_2} + A_1 A_2 \left(\frac{T_1}{R_1} + \frac{T_2}{R_2} + F_n \right) = 0, \\ A_1 A_2 Q_2 + \frac{\partial(A_2 H_1)}{\partial \alpha_1} + \frac{\partial(A_1 M_2)}{\partial \alpha_2} + \frac{\partial A_2}{\partial \alpha_1} H_2 = 0, \\ S_1 - S_2 + \frac{H_1}{R_1} - \frac{H_2}{R_2} = 0, \end{gathered}$$

where T_i, S_i and Q_i are the projections of the stress-resultant of the internal forces acting in the cross-section $\alpha_i = \text{const}$, on the unit vectors $\mathbf{e}_1$, $\mathbf{e}_2$ and $\mathbf{n}$, H_i and M_i are the projections of the stress-couples of the internal forces, and F_1, F_2 and F_n are the projections of the distributed external load.

Relations (2.6.1)–(2.6.7) are the same for all linear two-dimensional shell theories. The differences are in the formulae which connect forces to deformations. The Kirchhoff–Love's hypotheses, which generally assume that a shell element, normal to the neutral surface before deformation, does not change length and remains normal to the neutral surface after deformation, lead to the elasticity relations introduced by V. V. Novozhilov [**143**]

(2.6.8)
$$\begin{gathered} T_1 = K(\varepsilon_1 + \nu\varepsilon_2), \qquad S_1 = \frac{K(1-\nu)}{2}\left(\omega + \frac{h^2\tau}{6R_2}\right), \qquad (1 \leftrightarrow 2), \\ M_1 = D(\kappa_1 + \nu\kappa_2), \qquad H_1 = H = D(1-\nu)\tau, \\ K = \frac{Eh}{1-\nu^2}, \qquad D = \frac{Eh^3}{12(1-\nu^2)}, \end{gathered}$$

where E is Young's modulus, ν is Poisson's ratio (here it is supposed that the shell material is homogeneous and isotropic).

The small parameter μ is introduced into the system of equations precisely through the elasticity relations (2.6.8). Indeed, K is proportional to the shell thickness h and D is proportional to h^3. Assuming that the shell thickness h is small compared to the characteristic shell length R, after transition to non-dimensional variables, we get the small parameter μ connected with the ratio h/R by formula (2.3.18). This parameter means that the elongation stiffness is much larger than the bending stiffness.

In most problems, the same accuracy is provided by simpler elasticity relations proposed by A. E. H. Love [**119**], where

$$S_1 = S_2 = S = \frac{1}{2}K(1-\nu)\omega. \tag{2.6.9}$$

But these relations have the disadvantage that the last of the equilibrium equations (2.6.7) is not accurately fulfilled.

Elasticity relations more complex than (2.6.8) are given in [**67, 121**].

Since the system of equations (2.6.7) is of the eighth order, we need to supply four boundary conditions at each edge of the shell. The simplest variants of the boundary conditions at the edge $\alpha_1 = \alpha_1^0$ are to set the generalized displacements $(u_1,\, u_2,\, w,\, \gamma_1)$ or the generalized forces corresponding to them, as follows:

$$\begin{aligned}
&u_1 = u_1^0 &&\text{or} && T_1 = T_1^0,\\
&u_2 = u_2^0 &&\text{or} && S_1 + \frac{H}{R_2} = S_1^0 + \frac{H^0}{R_2},\\
&w = w^0 &&\text{or} && Q_1 - \frac{1}{A_2}\frac{\partial H}{\partial \alpha_2} = Q_1^0 - \frac{1}{A_2}\frac{\partial H^0}{\partial \alpha_2},\\
&\gamma_1 = \gamma_1^0 &&\text{or} && M_1 = M_1^0.
\end{aligned} \tag{2.6.10}$$

2.6.3. Shallow shell equations. As one can see, the full system of equations of the general shell theory is rather complex. But, in some cases, these equations may be simplified. For example, if buckling is accompanied by small waves (the sizes of which, at least in one direction, being small compared with the characteristic sizes of the shell), then the shallow shell equations can be applied. In this case, the equilibrium equations and the compatibility equations of the deformations reduce to the following pair of equations for the deflection function, w, and the force function, Φ, (the derivation may be found in Donnell's monograph [**41**]):

$$\begin{aligned}
&D\Delta^2 w - \Delta_k \Phi + F_n = 0,\\
&\frac{1}{Eh}\Delta^2\Phi + \Delta_k w = 0,
\end{aligned} \tag{2.6.11}$$

where

$$\begin{aligned}
\Delta &= \frac{1}{A_1A_2}\left[\frac{\partial}{\partial\alpha_1}\left(\frac{A_2}{A_1}\frac{\partial}{\partial\alpha_1}\right) + \frac{\partial}{\partial\alpha_2}\left(\frac{A_1}{A_2}\frac{\partial}{\partial\alpha_2}\right)\right],\\
\Delta_k &= \frac{1}{A_1A_2}\left[\frac{\partial}{\partial\alpha_1}\left(\frac{1}{R_2}\frac{A_2}{A_1}\frac{\partial}{\partial\alpha_1}\right) + \frac{\partial}{\partial\alpha_2}\left(\frac{1}{R_1}\frac{A_1}{A_2}\frac{\partial}{\partial\alpha_2}\right)\right].
\end{aligned} \tag{2.6.12}$$

The stress-resultants are connected with the force function Φ by the following relations:

$$\begin{aligned}
T_1 &= \frac{1}{A_2}\frac{\partial}{\partial\alpha_2}\left(\frac{1}{A_2}\frac{\partial\Phi}{\partial\alpha_2}\right) + \frac{1}{A_1^2A_2}\frac{\partial A_2}{\partial\alpha_1}\frac{\partial\Phi}{\partial\alpha_1}, \qquad (1\leftrightarrow 2),\\
S &= -\frac{1}{2}\left[\frac{A_2}{A_1}\frac{\partial}{\partial\alpha_1}\left(\frac{1}{A_2^2}\frac{\partial\Phi}{\partial\alpha_2}\right) + \frac{A_1}{A_2}\frac{\partial}{\partial\alpha_2}\left(\frac{1}{A_1^2}\frac{\partial\Phi}{\partial\alpha_1}\right)\right].
\end{aligned}$$

There are no such relations between the equations of (2.6.8) from which the transverse forces Q_1 and Q_2 may be determined. However, these may be found from equations (2.6.7).

If we do not accept the Kirchhoff–Love hypotheses and introduce the shift angles δ_1 and δ_2, then there are five main unknowns, i.e. $(u_1, u_2, w, \delta_1, \delta_2)$. In this case, we add two more relations to the elasticity relations of type (2.6.8), namely

$$Q_1 = \frac{Eh}{2(1+\nu)}\delta_1, \qquad Q_2 = \frac{Eh}{2(1+\nu)}h\delta_2. \tag{2.6.13}$$

Thus, the order of the system increases from 8 to 10. This theory is usually associated with the name of S. P. Timoshenko [**210**]. Here it is assumed that the shell element normal to the neutral surface before deformation does not change length. There exist theories, described by systems of the 12-th order, which allow element elongation (see, for example, [**176**]). The two-dimensional shell theories of higher order based on series expansions in powers of the thickness [**93, 256**] are not widely used. In [**269**] a two-dimensional shell theory is proposed in which a shell is represented by the directed surface, each point of which has six degrees of freedom (three translational and three rotational).

In Section 4.5, we shall consider the error analysis of the two-dimensional shell equations, as compared with the three-dimensional elasticity theory.

2.6.4. Shell dynamics and stability equations. If we take into account the relations (2.6.4)–(2.6.8), the system of equations of shell theory may be reduced to a form in which only the displacements u_1, u_2 and w are unknown (see A. L. Gol′denveizer [**67**]). Thus we have

$$\sum_{j=1}^{3} \left(\mu^4 N_{ij} + L_{ij}\right) u_j + F_i = 0, \qquad i = 1, 2, 3, \tag{2.6.14}$$

where the functions $u_j(\alpha_1, \alpha_2)$ denote the displacement projections ($u_3 = w$), $N_{ij}(\alpha_1, \alpha_2)$ and $L_{ij}(\alpha_1, \alpha_2)$ are linear differential operators (generally with variable coefficients in α_1 and α_2), the independent variables $(\alpha_1, \alpha_2) \in S$ are orthogonal curvilinear coordinates on the neutral surface coinciding with the lines of curvature, $\mu > 0$ is a small geometrical parameter (see (2.3.18)) and F_i are the load projections.

We consider the problems in which F_i are proportional to the eigenvalue Λ. In the case of vibrations,

$$F_i = -\Lambda u_i, \qquad \omega^2 = \frac{E\Lambda}{\rho R^2}, \tag{2.6.15}$$

where ρ is the density, E is Young's modulus, R is the characteristic linear size and ω is the unknown natural frequency.

For the stability problem of the momentless stress state, described by the stress-resultants T_1^0, T_2^0 and S^0, we have

$$\begin{gathered} F_1 = F_2 = 0, \qquad F_3 = \Lambda \Delta_t w, \\ \Delta_t w = \frac{1}{A_1 A_2} \sum_{i=1,2} \frac{\partial}{\partial \alpha_i}\left(\frac{A_j t_i}{A_i}\frac{\partial w}{\partial \alpha_i} + t_3 \frac{\partial w}{\partial \alpha_j}\right), \qquad i \neq j, \\ \left(T_1^0, T_2^0, S^0\right) = -\Lambda E h \mu^2 (t_1, t_2, t_3), \end{gathered} \tag{2.6.16}$$

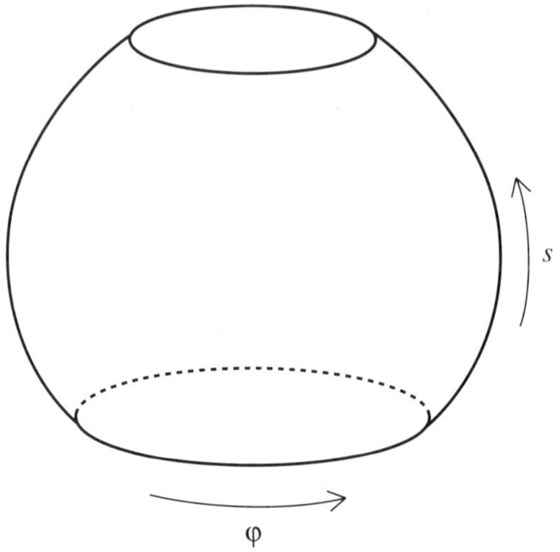

FIGURE 10. A shell of revolution.

where A_i are Lamé's coefficients, $w = u_3$ is the deflection (normal displacement) under buckling and $\Lambda > 0$ is the load parameter.

The best way to analyze the above equations is to use asymptotic integration, which is central in the theory of thin shells (see [**8, 32, 66, 67, 69, 73, 144, 146**]).

2.7 One-dimensional problems of shell vibrations and buckling

2.7.1. Momentless systems. If $\mu \neq 0$, (2.6.14) is an elliptic system of order eight. Four homogeneous boundary conditions are specified on the boundary, Γ, of S. If, under these conditions, the system is self-adjoint, then the spectrum of problem (2.6.14) is discrete. We note that, if the conditions (2.6.10) are homogeneous, then the system is self-adjoint.

If $\mu = 0$, system (2.6.14) (called, in this case, *momentless system*) is of the fourth order and the number of boundary conditions reduces to two. Depending on the sign of the Gaussian curvature, K, of the neutral surface, the system (2.6.14) may be elliptic, hyperbolic or parabolic, as $K > 0$, $K < 0$ or $K = 0$, respectively. The spectrum of the boundary value problem for a momentless system may be continuous on intervals and this fact may lead to difficulties in the limit as $\mu \to 0$ (see [**73, 111**]).

2.7.2. Shells of revolution. To describe the vibrations of shells of revolution, it is convenient to introduce as curvilinear coordinates, α_1 and α_2, the arc length, s, of the generator and the angle, φ, in the circumferential direction (see Fig. 10).

In this case,

$$A_1 = 1, \qquad A_2 = B, \qquad R_1 = \frac{d\theta}{ds}, \qquad R_2 = \frac{B}{\sin\theta},$$

where B is the distance between a point on the neutral surface and the axis of symmetry, θ is the angle between the normal to the surface and the axis of symmetry. Since the functions B, R_1 and R_2 depend only on s, the coefficients of system

(2.6.14) do not depend upon φ. If the shell is bounded by two parallels or is a cupola, we can separate the variables,

$$w(s,\varphi) = w(s)\, e^{im\varphi}, \qquad m = 0, 1, \ldots \tag{2.7.1}$$

The remaining unknowns also contain the factor $e^{im\varphi\neg}$, which cancels by substitution into (2.6.14). As a result, we get a system of ordinary differential equations with parameters μ, m, Λ. The structure of the asymptotics expansions of the solutions of system (2.6.14) depends upon the relations between the orders of these parameters.

For a further analysis of specific cases, we shall use system (2.6.14) (without writing its coefficients explicitely because of their complexity) and also the less accurate system

$$\begin{gathered} \mu^4 \Delta\Delta w - \Delta_k \Phi - \Lambda w = 0, \qquad \Delta\Delta\Phi + \Delta_k w = 0, \\ \Delta w = \frac{1}{B}\frac{d}{ds}\left(B\frac{dw}{ds}\right) - \frac{m^2 w}{B^2}, \qquad \Delta_k w = \frac{1}{B}\frac{d}{ds}\left(\frac{B}{R_2}\frac{dw}{ds}\right) - \frac{m^2 w}{B^2 R_1}, \end{gathered} \tag{2.7.2}$$

obtained from (2.6.11) after separation of variables in (2.7.1). System (2.7.2) allows us to determine precisely the main terms of the asymptotic expansions of the solutions in all cases considered below, except the momentless integrals (see Section 3.3).

The cases $m = 0$, $m \sim 1$ and $\mu m \ll 1$ will be considered in Section 3.3.

2.7.3. Vibrations with large wave number. Now, if we assume that

$$m = \mu^{-1}\rho, \qquad \rho \sim 1, \qquad \Lambda \sim 1, \tag{2.7.3}$$

then, all solutions of system (2.6.14) have a large index of variation, $t = 1$, (see Section 1.7). System (2.6.14) may be transformed into the form (2.1.1) and its solutions have the form (2.2.1), i.e.

$$w(s,\mu) \simeq \sum_{k=0}^{\infty} w_k(s)\mu^k \exp\left\{\frac{1}{\mu}\int_{s_0}^{s} \lambda(s)ds\right\}, \tag{2.7.4}$$

where $\lambda(s)$ satisfies the characteristic equation, which has eight roots and may be written in an already resolved form for Λ,

$$\Lambda = f(p,s) = \frac{(p^2+q^2)^4 + (k_2 p^2 + k_1 q^2)^2}{(p^2+q^2)^2}, \tag{2.7.5}$$

where

$$\lambda = ip, \qquad q = \frac{\rho}{B}.$$

The behaviour of the roots of (2.7.5) depends upon the form of the neutral surface. Different cases of this behaviour are studied in [**73**] (see also [**215, 216, 225**]).

2.7.4. Existence of oscillating solutions. Oscillating solutions of system (2.6.14) exist only if equation (2.7.5) has real roots, p. We find Λ_0 by formula (2.5.11),

$$\Lambda_0 = \min_{\substack{p,x \\ f>0}}\{f\} = f(p_0, s_0), \tag{2.7.6}$$

and assume that

$$\Lambda^+ = \max_s f(0, s). \tag{2.7.7}$$

Then for $\Lambda < \Lambda_0$ there are no spectral points. If $\Lambda_0 < \Lambda < \Lambda^+$, there are turning points inside the interval of integration. If, lastly, $\Lambda > \Lambda^+$, then system (2.6.14) has one pair of oscillating solutions and the frequency equations take the form (2.4.13) (see [**73, 215**]).

2.7.5. Cases with turning points. We make some remarks about the cases with turning points. For conic shells and shells in the form of paraboloids of revolution, in (2.7.6) we have $p_0 = 0$ and the initial parallel, s_0, coincides with the final parallel, $s_0 = s_2$. For these shells, if $\Lambda \in (\Lambda_0, \Lambda^+)$, then there exists one simple turning point, s_* ($\Lambda = f(0, s_*)$), and the frequency equation reduces to (2.4.15) (see [**73, 216**]). For shells in the form of stretched ellipsoids of revolution, $p_0 = 0$ as before, but s_0 coincides with the equator. The frequencies of the free vibrations close to Λ_0 can be found by formulae (2.5.9) and (2.5.13).

2.7.6. Frequency distribution density. As noted above, the distance between the roots of equations (2.4.13) and (2.4.15) is of the order μ. For a given m, the average distance between the vibration frequencies is called the *frequency distribution density*. Summing the frequency densities over all m, we obtain the entire frequency distribution density. For this purpose, it is enough to consider the case $m \sim \mu^{-1}$ and apply formulae (2.4.13) and (2.4.15) (see [**223**]). The frequency distribution density is considered in [**27**] by V. V. Bolotin for shallow shells and in [**10**] by A. G. Aslanyan and V. B. Lidsky for shells of arbitrary geometry. The frequency density for different elastic systems is analyzed by E. Skudzyk in [**197**].

2.7.7. Buckling of shells of revolution. The general buckling problem for a shell is discussed in Section 4.4. Here we consider the buckling of shells of revolution. For these shells, the study of the buckling of an initial axisymmetric state under bifurcation into a nonaxisymmetric form with m waves in the circumferential direction is remarkably similar to the study of the shell nonaxisymmetric vibrations. The differences from the vibrations problem are the following:

(1) equation (2.7.5) changes to the equation

$$\Lambda = f(p, s, \rho) = \frac{(p^2 + q^2)^4 + (k_2 p^2 + k_1 q^2)^2}{(p^2 + q^2)^2 (t_1 p^2 + 2t_3 pq + t_2 q^2)}, \tag{2.7.8}$$

where the notation is as in (2.7.5);

(2) when analyzing buckling problems, we are only interested in the lowest eigenvalue of the load parameter, Λ, (and perhaps in neighbouring eigenvalues).

Thus, a necessary step in the analysis of the buckling problem is the minimization of Λ over the wave number, m, in the circumferential direction (or over the parameter ρ).

As in (2.7.6), we find

$$\Lambda_0 = \min_{\substack{p,s,\rho \\ f>0}} \{f\} = f(p_0, s_0, \rho_0), \tag{2.7.9}$$

where p is real and the minimum is evaluated, as in (2.5.11), only for those p, s and ρ for which $f > 0$.

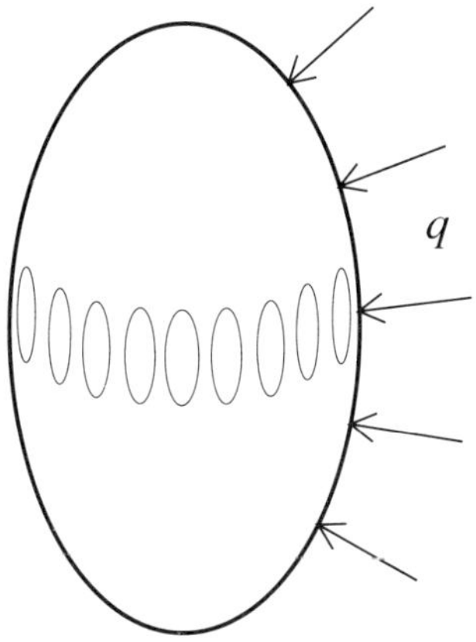

FIGURE 11. Buckling of an elongated ellipsoid.

For the existence of a positive minimum, $\Lambda_0 > 0$, we need to impose the following extra conditions (see [**70**]):

(a) There must be directions along which the initial stresses are compressing, i.e., in mathematical terms, the quadratic form

$$t_1 p^2 + 2t_3 pq + t_2 q^2$$

is not negative definite.

(b) We must exclude shells of negative Gaussian curvature ($k_1 k_2 < 0$) and some shells of zero curvature ($k_1 = 0$), because, in these cases, $\Lambda_0 = 0$ for $p_0 = q_0 = 0$, and this violates our initial assumptions, in particular, $m \sim \mu^{-1}$. These cases will be considered in Section 3.3 (see also (2.7.15) and (2.7.16)).

We shall assume that these extra conditions are satisfied.

2.7.8. Weak supports. For the particular boundary conditions, called *weak supports* of the edge $s = s_2$, buckling occurs in a neighbourhood of this edge for $\Lambda < \Lambda_0$ (see [**237, 238**]). In this case, (2.6.14) or (2.7.2) has no oscillating solutions and the buckling mode is a linear combination of four integrals of the edge effect near the edge $s = s_2$. To determine Λ we use equality (2.4.12). In particular, the buckling mode of a circular cylindrical shell with weak edge supports under axial compression has this form (see, for example, [**77, 138**]).

We note that not only the boundary conditions but also the form of the shell and the character of the initial loading in a neighbourhood of the edge $s = s_2$ determine whether the support is weak or not. A free edge is always a weak support.

2.7.9. Weakest parallel. If the support is not weak and if Λ_0 has an isolated minimum at s_0 in (2.7.9), then the parallel $s = s_0$ is called the *weakest parallel* because the buckling mode is localized in a neighbourhood of this parallel (see [**220, 238**]). If the parallel s_0 is located far enough from the shell edges, the buckling mode and the critical load do not depend upon the boundary conditions and are determined by formulae (2.5.9). This is the case, in particular, for the buckling of elongated ellipsoids of revolution under external pressure (see [**38, 160, 220, 238, 261**] and Fig. 11 in which the buckling mode is plotted).

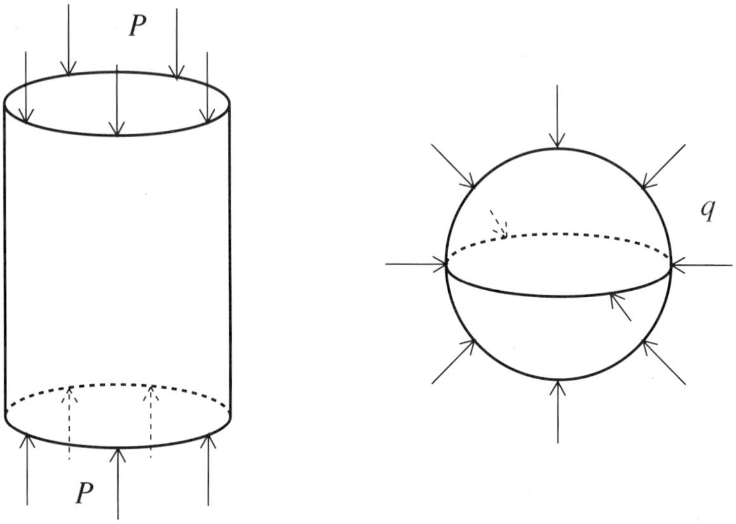

FIGURE 12. Two classical problems of shell buckling.

If the parallel s_0 coincides with the edge ($s_0 = s_2$), then there exists a turning point $s = s_*$ in a neighbourhood of this edge and we use equation (2.4.17) to determine the critical load. In particular, for the buckling of a convex shell of revolution under torsion moments applied to its edges, the weakest parallel is always the last parallel (see [**220, 238**]). In this case, the critical value of the load parameter, Λ, is represented by a power series in $\mu^{1/3}$,

$$\Lambda = \Lambda_0 + \mu^{2/3}\Lambda_2 + \mu\Lambda_3 + O\left(\mu^{4/3}\right), \tag{2.7.10}$$

where Λ_0 is determined by formula (2.7.9), Λ_2 depends only upon the function (2.7.8) and Λ_3 depends upon both (2.7.8) and the boundary conditions for $s = s_2$.

2.7.10. Shells without weakest parallel. It is interesting to consider two buckling problems, for which the function (2.7.8) does not depend upon s, i.e. there is no weakest parallel, and all the shell points are equally predisposed to buckling. Here, we consider (*a*) circular cylindrical shells under uniform axial compression and (*b*) spherical shells under uniform external pressure (see Fig. 12).

These problems played a decisive role in the development of the theory of shell buckling, firstly, because of their wide applications to the analysis of engineering structures, and, secondly, because, despite their simple formulation, their strict solution (in the nonlinear formulation) is faced with all the main difficulties of shell buckling problems (see, for example, [**30, 41, 62, 77, 160, 173, 261, 267**]).

By a proper choice of the non-dimensional variables for cylindrical shells under axial compression and for spherical shells under external pressure, the function (2.7.8) becomes, respectively,

$$f_C = \left(\frac{p^2+q^2}{p}\right)^2 + \left(\frac{p}{p^2+q^2}\right)^2, \qquad f_S = p^2 + q^2 + \frac{1}{p^2+q^2}. \tag{2.7.11}$$

Then Λ_0 attains its minimum, $\Lambda_0 = 2$, at a number of pairs (p, q), each of which corresponding to some buckling mode. Therefore, in this case the buckling mode is nonunique. For example, for a cylindrical shell with freely supported edges the buckling mode has the form

$$w_{mn} = w_0 \sin \frac{p_n s}{\mu} \sin \frac{q_m \varphi}{\mu}, \qquad q_m = \mu m, \qquad p_n = \frac{\mu n \pi R}{L}, \tag{2.7.12}$$

where L is the shell length and the integers m and n satisfy the following approximate equality,

$$p_n^2 + q_m^2 \simeq p_n.$$

We also write the critical loads for these problems in dimensional variables

$$T_1^0 = -\frac{Eh^2}{\sqrt{3(1-\nu^2)}\,R}, \qquad q^0 = \frac{2Eh^2}{\sqrt{3(1-\nu^2)}\,R^2}, \tag{2.7.13}$$

where T_1^0 is the buckling critical axial stress-resultant (see [**118, 210**]) for a cylindrical shell and q^0 is the critical pressure for a sphere (see [**173**]). The numbers T_1^0 and q^0 in (2.7.13) are usually called the *classical values*. The later development of the theory of shell buckling is connected, to a marked degree, with the analysis of different perturbation factors (initial imperfections, boundary conditions, initial stress moments, nonlinearity), which affect in an essential way the classical values (2.7.13), and in general reduce them (see Section 4.4).

2.7.11. Freely supported shells of constant length. We consider one more class of one-dimensional problems of the shell theory, which reduce to equation (2.1.1). This class consists of cylindrical shells of constant length, L, with freely supported curvilinear edges. In this case, the coefficients of systems (2.6.14) or (2.7.2) do not depend upon the coordinate, s, along the generator, and after separating variables:

$$w(s,x) = w(x) \sin \beta_n s, \qquad \beta_n = \frac{n\pi R}{L}, \qquad n = 1, 2, \dots, \tag{2.7.14}$$

we have a system of ordinary differential equations in the independent variable x $(0 \le x \le x_0)$ in the circumferencial direction. For a shell which is closed in the circumferential direction, the boundary conditions are replaced by periodic conditions.

To study the lower part of the spectrum of the natural frequencies, ω, we rewrite system (2.7.2) in the form (see [**24, 73**])

$$\begin{aligned} \mu_1^4 \Delta\Delta w + \beta_n^2 k_2(x) \Phi + F_3 &= 0, \\ \mu_1^4 \Delta\Delta \Phi - \beta_n^2 k_2(x) w &= 0, \\ \Delta w = \frac{d^2 w}{dx^2} - \beta_n^2 w&, \end{aligned} \tag{2.7.15}$$

where

$$\mu_1^2 = \mu, \qquad F_3 = \Lambda w, \qquad \omega^2 = \frac{\mu_1^4 \Lambda E}{\rho R^2}.$$

Here $\mu_1 > 0$ is a small parameter, $k_2(x)$ is the curvature of the generator (which is generally variable) and Λ is the unknown eigenvalue. Because of the differences between the expressions for ω^2 in (2.6.15) and (2.7.15), respectively, the order of the frequency is reduced.

2.7.12. The buckling problem. The buckling problem for the shell considered in the above Subsection 2.7.11, under the initial stresses T_1^0 and T_2^0, is described by the same system of equation (2.7.15), in which

$$F_3 = \Lambda \left\{ \frac{d}{dx} \left[t_2(x) \frac{dw}{dx} \right] - \beta_n^2 t_1(x) w \right\}, \quad (T_1^0, T_2^0) = -\Lambda E h \mu_1^6 (t_1, t_2). \tag{2.7.16}$$

In particular, in the case of uniform external normal pressure, $t_1 = 0$ and $t_2 = k_2(x)$.

The problem of cylindrical shell buckling under nonuniform axial compression is described by system (2.7.2) with

$$\Delta w = \frac{d^2 w}{dx^2} - \beta_n^2 w, \qquad F_3 = -\Lambda t_1(x) \beta_n^2 w, \qquad T_1^0 = -\Lambda E h t_1 \mu^2. \tag{2.7.17}$$

In all the above mentioned cases, system (2.7.15) or (2.7.2) may be represented in the form (2.1.1). The behaviour of its solutions and the existence of localized solutions are determined by the existence of pure imaginary roots, λ, of the characteristic equation (2.1.6). If we assume that $\lambda = ip$ and solve this equation for Λ, then, in the study of shell vibrations we have

$$\Lambda = f_1(p, x) = p^4 + \left[\frac{\beta_n^4 k_2^2(x)}{p^4} \right], \qquad \beta_n \sim 1, \tag{2.7.18}$$

and in the study of shell bucklings we have

$$\Lambda = f_2(p, x, \beta_n) = \left[p^2 + \frac{\beta_n^4 k_2^2(x)}{p^6} \right] \frac{1}{t_2(x)}, \tag{2.7.19}$$

under normal pressure, and

$$\Lambda = f_3(p, x, q_n) = \frac{(p^2 + q_n^2)^4 + k_2^2(x) q_n^4}{t_1(x) q_n^2 (p^2 + q_n^2)^2}, \tag{2.7.20}$$

under axial compression, where $q_n = \mu \beta_n$.

In all these cases, the minimum of the functions f_i $(i = 1, 2, 3)$ over (p, x) (see (2.5.11)) is attained for $p = p_0 > 0$. If the minimum is reached for an isolated value x_0 of x (called the *weakest generator*), and inequality (2.5.12) is satisfied, then the eigenvalues $\Lambda^{(m)}$ are asymptotically double (see (2.5.17)), and the eigenfunctions have the form (2.5.16). We present some results in the following and last subsection of this chapter.

2.7.13. Noncircular cylindrical shells. The lower part of the spectrum of asymptotically double natural frequencies of noncircular cylindrical shells is determined by the formula [**24, 73**]

$$\omega_{mn}^2 = \frac{E \mu_1^4}{\rho R^2} 2 \beta_n^2 k_{20} \left[1 + 4 \left(m + \frac{1}{2} \right) \frac{\mu_1 \sqrt{k_{20}''}}{\beta_n k_{20}^{3/4}} + O\left(\mu_1^2\right) \right], \quad m = 0, 1, 2, \ldots, \qquad n = 1, 2, \ldots, \tag{2.7.21}$$

where k_{20} and k_{20}'' are the values of the curvature $k_2(x)$ and its second derivative, respectively, evaluated at the weakest generator, $x = x_0$, where the curvature, $k_2(x)$, is minimal. The lowest frequency is obtained for $m = 0$, $n = 1$.

The critical value of the external pressure, p_0, for the buckling of a noncircular cylindrical shell of middle length [**228**] is equal to

$$p_0 = \frac{4\pi\mu_1^6 Eh}{3^{3/4}L}\left[1 + 3^{7/8}\mu_1\sqrt{\frac{Lk_{20}''}{2\pi R}} + O\left(\mu_1^2\right)\right]. \tag{2.7.22}$$

In this case, one deformation semi-wave develops along the longitudinal direction ($\beta_n = \beta_1$ in (2.7.19)).

The critical value of the axial compression, T_1, for the buckling of a noncircular cylindrical shell under nonuniform axial compression [**208, 209, 238**] is equal to

$$T_1 = T_1^0\left[1 + \mu\sqrt{\frac{8(1-q_n)}{q_n}\gamma_0''} + O\left(\mu^2\right)\right], \qquad \gamma = \frac{k_2}{t_1}, \tag{2.7.23}$$

where T_1^0 is the classical value of the critical load for a circular shell under uniform compression (see (2.7.13)). The function $\gamma(x)$ attains its minimum along the weakest generator, $x = x_0$. Formula (2.7.23) is valid only for $q_n < 1$. The cases $q_n \approx 1$ and $q_n > 1$ require special considerations (see [**238**]). The problem of shell bending by moments is considered in [**208**], where the critical loads for the odd and even buckling modes are obtained separately by integrating the system of equations numerically. These critical loads happen to be very close to each other. This illustrates the correctness of estimate (2.5.17) for asymptotically double eigenvalues (see also [**238**]).

To go from noncircular to circular cylindrical shells, we need only replace the brackets in formulae (2.7.21), (2.7.22) and (2.7.23) by the number 1. In particular, formula (2.7.22) gives the classical value obtained by R. Southwell [**201**] of the critical pressure for the buckling of a circular cylindrical shell of middle length.

In Chapter 4, we shall resume the discussion of these problems in the cases where the variables α_1 and α_2 in (2.6.14) or (2.7.2) do not separate.

CHAPTER 3

Degenerate Boundary Value Problems

In this chapter, we consider asymptotic methods for the solution of linear one-dimensional singular problems which degenerate to lower-order problems. In Section 3.1, we focus our attention on the regular Vishik–Lyusternik degeneracy. Nonregular degeneracy is discussed in Section 3.2. Cases where the system of equations has integrals with different indices of variation are considered in Section 3.3. Finally, applications to problems of shell vibrations and buckling and other problems of solid mechanics are analyzed in Sections 3.3 and 3.4.

3.1. Regular degeneracy of boundary value problems

3.1.1. Regular degeneracy of A_μ to A_0. We start with a short description of a one-dimensional variant of Vishik–Lyusternik's asymptotic method [**258**] (see also [**242**]). Let us consider the boundary value problem A_μ, which consists of the equation

$$\sum_{k=1}^{n} \mu^k a_{k+m} \frac{d^{k+m} y}{dx^{k+m}} + \sum_{k=0}^{m} a_k \frac{d^k y}{dx^k} = f(x), \qquad x \in [x_1, x_2], \tag{3.1.1}$$

with the boundary conditions
(3.1.2)

$$\frac{d^k y}{dx^k} = q_k^{(j)}, \qquad \text{for } x = x_j,$$
$$k = 0, 1, \ldots, m_j + n_j - 1, \qquad j = 1, 2, \qquad m_1 + m_2 = m, \qquad n_1 + n_2 = n.$$

We shall impose various conditions on the coefficients a_k, but in all cases $a_m \neq 0$ and $a_{m+n} \neq 0$. We consider also the degenerate problem A_0:

$$\sum_{k=0}^{m} a_k \frac{d^k y_0}{dx^k} = f(x), \qquad x \in [x_1, x_2], \tag{3.1.3}$$

with the boundary conditions

$$\frac{d^k y_0}{dx_k} = q_k^{(j)}, \qquad \text{for } x = x_j, \qquad k = 0, 1, \ldots, m_j - 1, \qquad j = 1, 2. \tag{3.1.4}$$

At first, we assume that the coefficients a_k are constant and consider the *additional* characteristic equation (compare with (2.2.10))

$$\sum_{k=0}^{n} a_{k+m} \lambda^k = 0. \tag{3.1.5}$$

DEFINITION 3.1. The degeneracy of problem A_μ into problem A_0 as $\mu \to 0$ ($\mu > 0$) is called *regular*, if equation (3.1.5) has n_1 roots with negative real parts and n_2 roots with positive real parts ($n_1 + n_2 = n$).

For a regular degeneracy, the number of boundary conditions which are dropped at the left end (right end) in the transition from problem A_μ to problem A_0 is equal to the number of roots of equation (3.1.5) with negative (positive) real part. For example, if equation (3.1.5) has one root with negative real part and two roots with positive real part, then one boundary condition is dropped at the left edge and two at the right edge.

If equation (3.1.5) has pure imaginary roots, the degeneracy is *nonregular*.

3.1.2. The Vishik–Lyusternik algorithm. We describe the structure of the asymptotic solution of (3.1.1). If problem A_0 has a unique solution for any $f(x)$ and $q_k^{(j)}$, then the solution of problem A_μ is represented as the sum of three asymptotic series,

$$y(x,\mu) \simeq y^{(0)}(x,\mu) + z^{(1)}(\eta_1,\mu) + z^{(2)}(\eta_2,\mu), \tag{3.1.6}$$

where

$$y^{(0)}(x,\mu) \simeq \sum_{k=0}^{\infty} \mu^k y_k(x)$$

and

$$z^{(j)}(\eta_j,\mu) \simeq \mu^{m_j} \sum_{k=0}^{\infty} \mu^k z_k^{(j)}(\eta_j), \qquad \eta_j = \mu^{-1}(x - x_j), \qquad j = 1, 2.$$

The functions $z^{(j)}$ are the integrals of the boundary layer (or of the edge effect) in neighbourhoods of the edges $x = x_j$, i.e.

$$\begin{aligned} &z_k^{(1)}(\eta_1) \to 0, \qquad \text{as } \eta_1 \to \infty, \\ &z_k^{(2)}(\eta_2) \to 0, \qquad \text{as } \eta_2 \to -\infty, \end{aligned} \qquad k = 0, 1, 2, \dots \tag{3.1.7}$$

The functions y_k and $z_k^{(j)}$ are uniquely defined by the substitution of (3.1.6) into (3.1.1) and (3.1.2).

The function y_0 is the solution of problem A_0. The functions y_k ($k > 0$) are determined from nonhomogeneous boundary value problems of type A_0, with right-hand sides depending on the previously determined functions. The construction of the functions y_k is sometimes called the *first* or the *main* iteration process.

The functions $z_0^{(j)}$ are the solutions of the boundary value problems

$$\begin{aligned} &Lz_0^{(j)} := \sum_{k=0}^{n} a_{k+m} \frac{d^{k+m} z_0^{(j)}}{d\eta_j^{k+m}} = 0, \qquad j = 1, 2, \\ &\frac{d^k z_0^{(j)}}{d\eta_j^k} = \begin{cases} q_k^{(j)} - \dfrac{d^k y_0}{dx_j^k}, & k = m_j, \\ 0, & k = m_j + 1, \dots, m_j + n_j - 1, \end{cases} \qquad \text{for } \eta_j = 0, \end{aligned} \tag{3.1.8}$$

with the additional conditions (3.1.7). The next approximations, $z_k^{(j)}$, are uniquely determined by nonhomogeneous problems similar to (3.1.8), with right-hand sides depending on the previous approximations. The construction of the functions $z_k^{(j)}$

is called the *second* or the *additional* iteration process. The interaction between the first and second iteration processes occurs only through the boundary conditions.

The functions $z_k^{(j)}$ have the form

$$z_k^{(j)}(\eta_j) = \sum_l P_{kl}^{(j)}(\eta_j)\, e^{\lambda_l \eta_j}, \tag{3.1.9}$$

where the degree of each polynomial $P_{kl}^{(j)}$ increases with k and the summation is over all the roots λ_l of equation (3.1.5) with the corresponding signs of the real parts. For this, the roots λ_l need not be simple.

3.1.3. Equations with variable coefficients. This algorithm may be used, practically without any change, if the coefficients a_k in (3.1.1) are not constant and depend on x and μ,

$$a_k(x, \mu) = \sum_{l=0}^{\infty} \mu^l a_{kl}(x), \tag{3.1.10}$$

where $a_{kl}(x)$ are analytic. In this case, we assume that $a_k = a_{k0}(x)$ in equation (3.1.5) and require that, for any x, equation (3.1.5) has n_1 (n_2) roots with negative (positive) real parts. This last requirement, however, may be relaxed. In formula (3.1.9), we take $\lambda_l = \lambda_l(x_j)$.

3.1.4. Eigenvalue problems. The Vishik–Lyusternik algorithm may be applied also to the solution of eigenvalue problems. In this case, the coefficients a_k in (3.1.1) depend on the parameter Λ, and equation (3.1.1) and the boundary conditions (3.1.2) are homogeneous ($f = q_k^{(j)} = 0$).

Let $\Lambda = \Lambda_0$ be an eigenvalue of the degenerate problem A_0. The construction of the eigenvalue Λ for problem A_μ, which is close to Λ_0, and the corresponding eigenfunction may be carried out by applying this algorithm, with the difference that, in determining the function $y_k(x)$ in (3.1.6), we have nonhomogeneous boundary value problems (of types (3.1.3) and (3.1.4)) where Λ is an eigenvalue. The compatibility condition of these problems allows us to determine the correction terms in the series expansion of Λ in powers of μ. In particular, if Λ_0 is a simple eigenvalue for problem A_0, then the expansion of Λ is in integral powers of μ,

$$\Lambda = \sum_{k=0}^{\infty} \mu^k \Lambda_k. \tag{3.1.11}$$

3.1.5. Example of regular degeneracy. As an example, we find the natural frequencies of a nonabsolutely elastic string of length L with clamped edges. We have the boundary value problem A_μ, in non-dimensional variables,

$$-\mu^2 \frac{d^4 y}{dx^4} + \frac{d^2 y}{dx^2} + \Lambda y = 0; \qquad y = \frac{dy}{dx} = 0 \qquad \text{at } x = 0, \pi, \tag{3.1.12}$$

where

$$\mu^2 = \frac{EJ\pi^2}{TL^2}, \qquad \omega^2 = \frac{\Lambda T \pi^2}{\rho L^2}, \tag{3.1.13}$$

EJ is the bending stiffness, T is the stress, ρ is the linear string density and ω is the frequency.

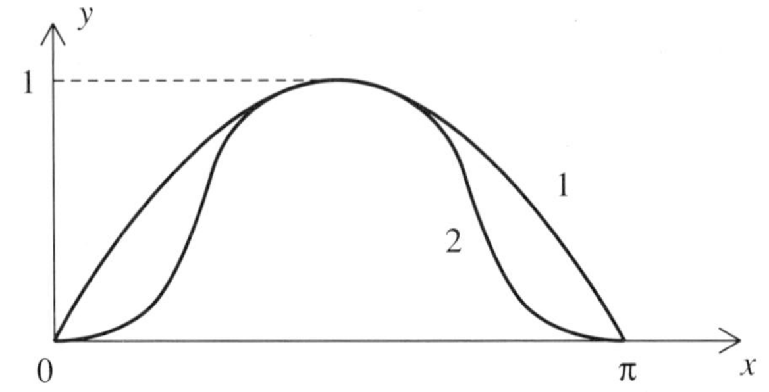

FIGURE 13. Vibration modes of a string.

As $\mu \to 0$, problem A_μ degenerates into problem A_0,

$$\frac{d^2 y_0}{dx^2} + \Lambda_0 y_0 = 0; \qquad y_0 = 0 \qquad \text{for } x = 0, \pi, \tag{3.1.14}$$

the solutions of which are $\Lambda_0 = n^2$ and $y_0 = \sin nx$, $n = 1, 2, \ldots$

The degeneracy of problem A_μ into problem A_0 is regular since $m_1 = n_1 = m_2 = n_2 = 1$ and equation (3.1.5), $-\lambda^2 + 1 = 0$, has one positive and one negative root.

We seek Λ_1 in (3.1.11). In the first approximation, the functions of the boundary layer are

$$z_0^{(1)} = ne^{-\eta_1}, \qquad z_0^{(2)} = (-1)^{n+1} ne^{\eta_2}. \tag{3.1.15}$$

For y_1, we get the boundary value problem

$$\frac{d^2 y_1}{dx^2} + n^2 y_1 = -\Lambda_1 y_0, \qquad y_1(0) = -n, \qquad y_1(\pi) = (-1)^n n, \tag{3.1.16}$$

the compatibility condition of which gives $\Lambda_1 = 4n^2\pi^{-1}$.

The first mode for an absolutely elastic (curve 1) and a nonabsolutely elastic string (curve 2) with clamped edges, respectively, is shown in Fig. 13.

3.1.6. Splitting of boundary conditions. We consider a system of first-order equations with a small parameter in some of the derivatives,

$$\frac{dy}{dx} = A_{11} y + A_{12} z, \qquad \mu \frac{dz}{dx} = A_{21} y + A_{22} z, \tag{3.1.17}$$

where y and z are m- and n-vectors, respectively, and the matrices A_{ij}, of compatible dimensions, depend regularly on μ,

$$A_{ij} = A_{ij}(x, \mu) = \sum_{k=0}^{\infty} \mu^k A_{ij}^{(k)}(x). \tag{3.1.18}$$

If $\det(A_{22}^0) \neq 0$ for all x, system (3.1.17) for $\mu = 0$ degenerates into a system of order m,

$$\frac{dy_0}{dx} = \left(A_{11}^{(0)} - A_{12}^{(0)} A_{22}^{(0)-1} A_{21}^{(0)} \right) y_0, \qquad z^{(0)} = -A_{22}^{(0)-1} A_{21}^{(0)} y_0. \tag{3.1.19}$$

If the characteristic equation,

$$\det\left(A_{22}^{(0)}(x)-\lambda I\right)=0, \tag{3.1.20}$$

has n_1 (n_2) roots with negative (positive) real parts ($n_1+n_2=n$), then the degeneracy of the boundary value problems considered below is regular.

Let problem A_μ consist of system (3.1.17) with the boundary conditions

$$l_k^{(j)}(y,z)=0, \qquad \text{for } x=x_j, \qquad k=1,\dots,m_j+n_j, \qquad j=1,2, \tag{3.1.21}$$

where $l_k^{(j)}$ are linear functions of y and z. Unlike in problem (3.1.1)–(3.1.2), it is not a trivial question to determine which conditions in (3.1.21) have to be kept during the transition to problem A_0, especially if the coefficients in $l_k^{(j)}$ depend on μ. In the general case, problem A_0 consists of system (3.1.19) and m_j linear combinations of the boundary conditions (3.1.21), the construction of which is called the *splitting* of the boundary conditions. The splitting of the boundary conditions for thin shells has been often discussed in the literature (see [**8, 32, 49, 66, 73, 99, 144, 206, 233**]).

3.2. Nonregular degeneracy

3.2.1. Non regular degeneracy without turning points. Here we consider the same boundary value problems A_μ and A_0, as in the previous Section 3.1, but now the degeneracy of problem A_μ into problem A_0 may be nonregular.

If the coefficients a_k in equation (3.1.1) are real and constant, regularity may be lost when equation (3.1.5) has pure imaginary roots.

If the a_k depend on x, equation (3.1.5) may have pure imaginary roots on the entire interval of integration or on part of it. In the last case there will be turning points which seriously complicate the asymptotic integration (see Section 2.3). Below it is assumed that there are no turning points.

We suppose that equation (3.1.1) and the boundary conditions (3.1.2) are homogeneous ($f=q_k^{(j)}=0$), and the coefficients a_k depend on the parameter Λ. We seek the eigenvalues Λ and the corresponding eigenfunctions.

3.2.2. Boundary value problems and eigenvalue problems. Let equation (3.1.5) have l_1 (l_2) roots with negative (positive) real parts and $2r$ pure imaginary roots ($l_1+l_2+2r=n$, $l_1\le n_1$, $l_2\le n_2$). The general solution of (3.1.1) can be written in the form

$$y=\sum_{k=1}^{m+n} C_k y^{(k)}(x,\mu), \tag{3.2.1}$$

where C_k are arbitrary constants. The functions $y^{(k)}(x,\mu)$, for $1\le k\le m$, are of the form (2.2.11), and the remaining functions $y^{(k)}(x,\mu)$, for $m+1\le k\le m+n$, are of the form (2.2.1), moreover l_1+l_2 of them are edge effect integrals and $2r$ functions oscillate.

The substitution of solution (3.2.1) into the boundary conditions (3.1.2) yields the equation

$$\Delta_{m+n}(\Lambda,\mu)=0, \tag{3.2.2}$$

where the elements of the p-th order determinant, Δ_p, are the functions $y^k(x,\mu)$ and their derivatives at $x = x_1, x_2$.

We also consider the equation

$$\Delta_m^0(\Lambda_0) = 0, \tag{3.2.3}$$

which corresponds to problem A_0 and depends on the functions $y^{(k)}(x,0)$, for $1 \leq k \leq m$.

For $m_1 = m_2$, equation (3.2.2) may be represented in the form (see also [**211**])

$$\Delta_{m+n} = \Delta_m^0(\Lambda)\Delta_n(\Lambda,\mu) + o(1) = 0, \qquad \text{as } \mu \to 0, \tag{3.2.4}$$

where the determinant Δ_n consists of the integrals (2.2.1).

If, for some $\mu_0 > 0$,

$$\Delta_n(\Lambda,\mu) \geq a > 0, \qquad \text{for} \qquad 0 \leq \mu \leq \mu_0, \tag{3.2.5}$$

then, for the eigenvalues of problems A_μ and A_0, the following limit holds:

$$\lim_{\mu\to 0} \Lambda(\mu) = \Lambda_0. \tag{3.2.6}$$

If Λ_0 is a simple eigenvalue, then we can write the following asymptotic expansions in integral powers of μ,

$$\Lambda(\mu) \simeq \sum_{k=0}^{\infty} \Lambda_k \mu^k, \qquad \text{as } \mu \to 0. \tag{3.2.7}$$

In particular, in the case of regular degeneracy, condition (3.2.5) is satisfied.

3.2.3. Additional eigenvalues. If condition (3.2.5) is not satisfied, problem A_μ can have additional eigenvalues which do not converge to the eigenvalues of problem A_0, as $\mu \to 0$. The structure of the additional spectrum of problem A_μ depend on the character of the equation

$$\Delta_n(\Lambda,\mu) = 0, \tag{3.2.8}$$

as $\mu \to 0$. We note two possibilities, which occur in the applications.

If equation (3.2.8) can be written in the form

$$\Delta_n(\Lambda,\mu) = \Delta_n^0(\Lambda) + o(1) = 0, \qquad \text{as } \mu \to 0, \tag{3.2.9}$$

and the equation $\Delta_n^0(\Lambda) = 0$ has the roots $\Lambda = \Lambda_*$, then problem A_μ has additional eigenvalues similar to (3.2.6) and (3.2.7), and these converge to Λ_* as $\mu \to 0$.

Equation (3.2.8) can also be similar to (2.4.13) (see [**211**]),

$$B_0(\Lambda,\mu) + \sum_{j=1}^{N} B_j(\Lambda,\mu)\cos[\mu^{-1}\psi_j(\Lambda,\mu)] = 0, \tag{3.2.10}$$

where B_j and ψ_j depend regularly on μ. Under some additional assumptions on B_j and ψ_j, the distance between the roots of equation (3.2.10) is of the order μ, i.e. the additional spectrum of the boundary value problem A_μ is dense. Problems with such spectra are considered in Sections 2.4 and 3.3.

3.2.4. An example. We now return to the case where equation (3.2.8) is represented in the form (3.2.9), and consider an example [**212**]. The small free vibrations of a planar spiral spring in its own plane are described by the equation

$$\begin{aligned}\mu^4\frac{d}{dx}\left[\frac{1}{q}\frac{d^4}{dx^4}\left(\frac{1}{q}\frac{dy}{dx}\right)\right]+\mu^2\frac{d}{dx}\left[\frac{1}{q}\frac{d^3}{dx^3}(qy)\right]+\mu^2 q\frac{d^3}{dx^3}\left(\frac{1}{q}\frac{dy}{dx}\right)\\ +q\frac{d^2}{dx^2}(qy)+\Lambda\left[y-\mu^2\frac{d}{dx}\left(\frac{1}{q^2}\frac{dy}{dx}\right)\right]=0, \qquad 0\le x\le 1,\end{aligned} \tag{3.2.11}$$

with the boundary conditions

$$y=\frac{dy}{dx}=\frac{d^2y}{dx^2}=0, \qquad \text{for } x=0,\ x=1.$$

Here x is the non-dimensional length of the elastic line, $q(x)$ is its curvature, $y(x)$ is the projection of the displacement on the normal to the elastic line. Moreover,

$$\mu=\frac{r}{l}, \qquad \Lambda=\frac{\rho S r^2 l^2}{EJ}\omega^2, \tag{3.2.12}$$

where $\mu>0$ is a small parameter, r is the characteristic radius of curvature, l is the length of the elastic line, ρ is the density, S is the cross-section area, EJ is the bending stiffness and ω is the frequency. The boundary conditions for (3.2.11) correspond to clamped edges.

Since the additional characteristic equation (3.1.5), i.e.

$$q^{-2}\lambda^4+2\lambda^2+q^2=0, \tag{3.2.13}$$

has a pair of double pure imaginary roots, $\lambda=\pm iq$, the degeneracy of problem (3.2.11) into the problem

$$q\frac{d^2}{dx^2}(qy_0)+\Lambda_0 y_0=0, \qquad y_0(0)=y_0(1)=0, \tag{3.2.14}$$

is not regular.

Two of the six integrals of equation (3.2.11) have the form (2.2.11) and the remaining four are of the form (2.1.1) and oscillate ($p=3,4,5,6$),

$$y^{(p)}(x,\mu)\simeq\sum_{k=0}^{\infty}y_k^{(p)}(x)\mu^k\exp\left(\frac{1}{\mu}\int_0^x\lambda_p(\tilde{x})\,d\tilde{x},\right), \qquad \lambda_p=\pm iq. \tag{3.2.15}$$

Since the roots of equation (3.2.13) have multiplicity two, the functions $y_k^{(p)}$ are obtained from differential equations of the second order. In particular,

$$2\frac{d}{dx}\left(q^2\frac{dy_0^{(p)}}{dx}\right)+\Lambda_* y_0^{(p)}=0. \tag{3.2.16}$$

The frequency equation (3.2.2) has the form

$$\Delta_0(\Lambda)\Delta_*^2(\Lambda)+o(1)=0, \qquad \text{as } \mu\to 0, \tag{3.2.17}$$

where the equation $\Delta_0(\Lambda)=0$ provides the eigenvalues, Λ_0, of problem (3.2.14) and the equation $\Delta_*(\Lambda)=0$ gives the eigenvalues, Λ_*, of the problem consisting of equation (3.2.16) with the boundary conditions $y(0)=y(1)=0$.

If the roots of the equations $\Delta_0(\Lambda)=0$ and $\Delta_*(\Lambda)=0$ are not equal to each other, then the higher approximations show that, for the eigenvalues of the full boundary value problem (3.2.11), the expansion (3.2.7) holds. In this case, the root Λ_* corresponds to a pair of close eigenvalues of problem (3.2.11), which differ by quantities of the order μ.

3.3 Degenerate thin shell boundary value problems

Since the equations for thin shells contain a small natural parameter related to the relative shell thickness, an asymptotic expansion of the solutions, which is a necessary element of the qualitative analysis, can significantly simplify the construction of approximate numerical solutions.

In this section we consider the shell equation (2.6.14) for such values of the parameters μ, m and Λ for which solutions simultaneously exist with large and zero indices of variation. In such a case, the boundary value problem for system (2.6.14) degenerates into a problem of lower order.

3.3.1. Axisymmetric vibrations. We write the equations of the axisymmetric vibrations of a shell of revolution, $m = 0$, in the form (2.2.12),

$$\mu^4 \sum_{k=1}^{6} a_k(s) \frac{d^k w}{ds^k} + \sum_{k=0}^{2} b_k(s) \frac{d^k w}{ds^k} = 0, \tag{3.3.1}$$

where

$$b_2(s) = \frac{1}{R_2^2(s)} - \Lambda, \qquad a_6 = 1,$$

and the notation of formulae (2.3.18) is used. Three uniform boundary conditions are given at the edges $s = s_1$ and $s = s_2$. The degenerate (or momentless) problem consists in solving the equation

$$\sum_{k=0}^{2} b_k(s) \frac{d^k w_0}{ds^k} = 0, \tag{3.3.2}$$

which satisfies one (tangential) boundary condition at each edge.

DEFINITION 3.2. A condition is called *tangential*, if it restricts the projection of the displacement or stress to the tangential direction to the neutral surface.

For example, in the case of a clamped edge ($u = w = \frac{dw}{ds} = 0$) the condition $u = 0$ is tangential, since it restricts the projection of the displacement u to the direction of the generator.

Since equation (3.1.5) has the form

$$\lambda^4 + b_2(s) = 0, \tag{3.3.3}$$

then, if $b_2(s) > 0$ for all s, the degeneracy of the moment (or full) problem (3.3.1) into the momentless problem (3.3.2) is regular.

In particular, in the static case (i.e. for $\Lambda = 0$), the degeneracy is always regular, provided we exclude toroidal shells, for which $R_2(s) = \infty$ on some parallels. In analyzing the shell static stresses, we separate the stress state into the momentless state and the edge effect (see [**32, 66, 67, 143, 144**]).

The degeneracy is regular if $\Lambda < \Lambda^{(1)}$, where $\Lambda^{(1)}$ is given by formula (2.3.19). For free vibrations, the eigenvalues Λ have expansions (3.2.7), where, in the general case, $\Lambda_1 \neq 0$ despite the fact that the disturbed operator in (3.3.1) contains the multiplier μ^4. The term $\mu\Lambda_1$ in (3.2.7) comes from satisfying the boundary conditions with the edge effect integral.

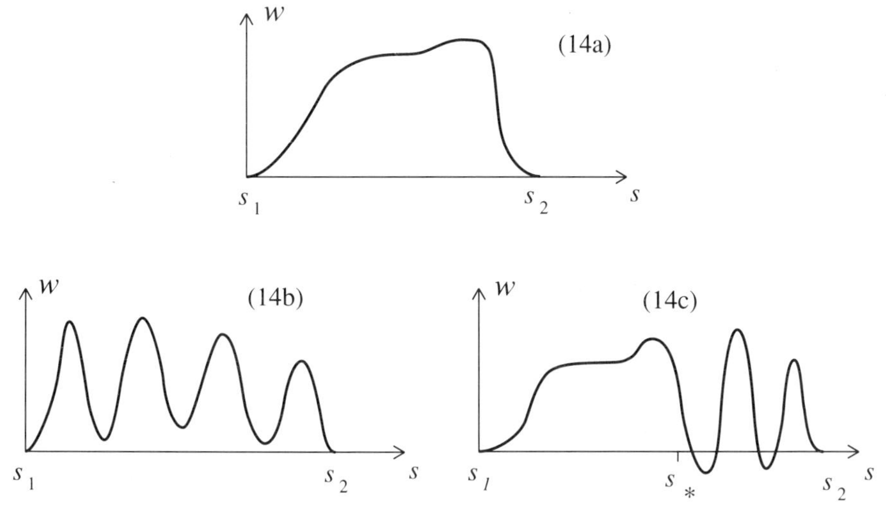

FIGURE 14. Modes of axisymmetric shell vibrations.

3.3.2. Problems with turning points. For $\Lambda^{(1)} \leq \Lambda \leq \Lambda^{(2)}$ (see (2.3.19)), the interval of integration contains a turning point (or points). The set $\{\Lambda : \Lambda \in [\Lambda^{(1)}, \Lambda^{(2)}]\}$ consists of points of the continuous spectrum of the momentless problem (3.3.2) (see [**10, 73**]). For $\Lambda > \Lambda^{(1)}$, in the moment problem (3.3.1), the distance between neighbouring spectral points is of the order μ (see [**73**]).

3.3.3. Nonregular degeneracy. If $\Lambda > \Lambda^{(2)}$, there are no turning points and equation (3.3.3) has one pair of pure imaginary roots since $b_2 < 0$. The degeneracy of (3.3.1) into (3.3.2) is not regular and equation (3.2.2) has the form [**73**]

$$(3.3.4) \qquad \Delta(\Lambda, \mu) = \Delta_0(\Lambda) \cos[\mu^{-1}\delta(\Lambda)] + O(\mu) = 0, \qquad \delta = \int_{s_1}^{s_2} (-b_2)^{1/4}\, ds,$$

where the equation $\Delta_0(\Lambda) = 0$ provides the points Λ_0 of the spectrum of the momentless problem (3.3.2) (this spectrum is discrete for $\Lambda > \Lambda^{(2)}$ and has a limit point for $\Lambda = \infty$). If $\Delta_0(\Lambda)$ is not close to zero (to be precise, $|\mu^{-1}\Delta_0| \gg 1$), equation (3.3.4) which becomes

$$(3.3.5) \qquad \cos\left(\frac{\delta}{\mu}\right) = o(1), \qquad \text{as } \mu \to 0,$$

gives the spectrum and the distance between neighbouring points is of the order μ (see [**73, 165, 182**]).

If $\Lambda - \Lambda_0 \sim \mu$, where $\Delta_0(\Lambda_0) = 0$, it may be shown that, near $\Lambda = \Lambda_0$, the roots (3.3.5) move apart and one extra root appears. It is not always possible to say that the root Λ_0 corresponds to a specific root of equation (3.3.4). The evolution of the shell natural modes, as Λ goes through the point Λ_0, is discussed in [**73**].

In Fig. 14, the modes of axisymmetric shell vibrations are schematically shown for the cases $\Lambda < \Lambda^{(1)}$ (Fig. 14a), $\Lambda > \Lambda^{(2)}$ (Fig. 14b) and $\Lambda^{(1)} < \Lambda < \Lambda^{(2)}$ (Fig. 14c). In the last figure, it is supposed that $R_2' > 0$ and the interval $s_1 \leq s \leq s_2$ contains the turning point $s = s_*$ to the right of which the function $w(s)$ contains the rapidly oscillating term.

3.3.4. Nonaxisymmetric vibrations. For nonaxisymmetric vibrations with a small wave number m ($m \sim 1$, $\mu \to 0$) in the circumferential direction, everything that was said above about the axisymmetric vibrations is still valid (see [**10, 73, 222**]), except that the orders of equations (3.3.1) and (3.3.2) increase by two and the number of tangential boundary conditions increases by one (the number of non-tangential boundary conditions, which are omitted during the transition from problem (3.3.1) to problem (3.3.2) is still two).

3.3.5. Bending and lowering of natural frequencies. We now discuss a series of problems of shell free vibrations, in which $\Lambda(\mu) \to 0$ as $\mu \to 0$, i.e. the order of some of the natural frequencies decreases. This reduction is connected with the weak support of the shell edges, $s = s_1$ and $s = s_2$, and/or with the geometry of the shell neutral surface.

DEFINITION 3.3. Let u, v and w be the projections of the displacements onto the axes of the local coordinate system $\{\mathbf{e}_1, \mathbf{e}_2, \mathbf{n}\}$, defined in (2.6.2). We say that a support is *tangential* if $u = 0$ or $v = 0$ and *non-tangential* if $w = 0$ or $\frac{dw}{ds} = 0$.

Without any supports, the corresponding generalized stress is equal to zero ($T_1 = 0$, $S = 0$, $Q_1 = 0$, $M_1 = 0$).

The reduction effect of a weak edge support on the natural frequencies is studied in [**29, 68, 73, 155, 225, 226**].

DEFINITION 3.4. Shell displacements for which the neutral surface has none of the tangential deformations ε_1, ε_2, ω, are called *bendings*.

Small bendings of shells of revolution with m waves in the circumferential direction satisfy the equations

$$\varepsilon_1 = \varepsilon_2 = \omega = 0, \tag{3.3.6}$$

where

$$\varepsilon_1 = \frac{du}{ds} - \frac{w}{R_1}, \qquad \varepsilon_2 = \frac{B'}{B}u + \frac{mv}{B} - \frac{w}{R_2}, \qquad \omega = B\frac{d}{ds}\left(\frac{v}{B}\right) - \frac{mu}{B}. \tag{3.3.7}$$

If there exists a non-zero solution to system (3.3.6), satisfying all the supports, we say that the shell possesses a *true bending*. The maximal reduction in the vibration frequency occurs for the bending [**73, 119, 172**]

$$\Lambda \sim \mu^4, \tag{3.3.8}$$

and the vibrations with modes close to the bendings are said to be vibrations of the *Rayleigh type*.

The natural frequencies may also be found from the minimal condition for the Rayleigh functional

$$\Lambda = \min_{u,v,w} \frac{\Pi_\varepsilon + \mu^4 \Pi_\kappa}{T}, \tag{3.3.9}$$

where the minimum is taken over the displacements, u, v, w, which satisfy all supports and the values,

$$\Pi_\varepsilon = \frac{1}{1-\nu^2}\int\int(\varepsilon_1^2 + 2\nu\varepsilon_1\varepsilon_2 + \varepsilon_2^2 + \frac{1-\nu}{2}\omega^2)dG,$$

$$\Pi_\kappa = \int\int(\kappa_1^2 + 2\nu\kappa_1\kappa_2 + \kappa_2^2 + 2(1-\nu)\tau^2)dG,$$

$$T = \int\int(u^2 + v^2 + w^2)dG,$$

are proportional to the potential elongation and bending shell energies and its kinetic energy, respectively. Integrations are taken over the entire neutral surface.

We see, from (3.3.9), why the existence of bendings or nearby deformations essentially reduce the natural frequencies.

Rayleigh's energy method has been used in [**73, 172, 226**] to determine the lowest frequencies.

3.3.6. Tangentially possible bending. We define a term proposed by A. L. Gol′denveiser [**67**]).

DEFINITION 3.5. Let system (3.3.6) have a non-zero solution which satisfies all the tangential supports but does not satisfy the non-tangential supports. Such a solution is called a *tangentially possible bending.*

In the case of tangentially possible bending, the most effective way to determine the vibration frequencies and modes is to use Vishik–Lyusternik's method. The solution of system (3.3.6) also satisfies the momentless equations (2.7.2) for $\mu = 0$ and $\Lambda_0 = 0$.

To find the non-zero terms in the expansion of the parameter Λ given by (3.2.7), we construct the edge effect integrals which eliminate the discrepancies in non-tangential boundary conditions. In this case, $\Lambda \sim \mu$ if the tangentially possible bending does not satisfy the condition $w = 0$. On the other hand, $\Lambda \sim \mu^3$ if the tangentially possible bending only fails to satisfy the conditions $w' = 0$.

The above estimates of the order of Λ (i.e. $\Lambda \sim \mu^\alpha$) are nonhomogeneous in m, since $\Lambda \sim 1$ for $m \sim \mu^{-1}$ (see Section 2.7).

3.3.7. Shells of zero Gaussian curvature. The reduction of the orders of the vibration frequencies is relevant to shells of zero and negative Gaussian curvature. To construct the lower part of the spectrum of a "well" supported shell of zero Gaussian curvature ($R_1^{-1} = 0$), we must assume (see [**73, 218**]) that

$$m \sim \mu^{-1/2}, \qquad \Lambda \sim \mu^2. \tag{3.3.10}$$

Under these assumptions, system (2.6.14) or (2.7.2) has four integrals of the edge effect type (2.2.1) and four integrals with the expansions (2.2.11), i.e.

$$w(s,\mu) = \sum_{k=0}^{\infty} \mu^k w_k(s), \tag{3.3.11}$$

where $w_0(s)$ satisfies the following fourth-order differential equation:

$$\frac{1}{R_2}\frac{d^2}{ds^2}\left(\frac{B^4}{R_2}\frac{d^2 w_0}{ds^2}\right) + \left(\frac{\mu^4 m^8}{B^4} - \Lambda m^4\right) w_0 = 0. \tag{3.3.12}$$

By virtue of (3.3.10), $\mu^4 m^8 \sim 1$ and $\Lambda m^4 \sim 1$; thus every coefficient of equation (3.3.12) is of the first order.

For a conical shell, B and R_2 are linear functions of s ($B = R_2 \cos\alpha$, where α is the angle at the top of the cone), and for a cylindrical shell we may take $B = R_2 = 1$ (by properly choosing the non-dimensional variables).

The degeneracy, as $\mu \to 0$, of the boundary value problem (2.7.2) into the problem which consists of equation (3.3.12) and two (usually, tangential) boundary conditions at each of the edges $s = s_1$ and $s = s_2$, is regular. Equation (3.3.12) is usually called *semi-momentless*, since, in deriving it, we keep the term coming from the bending shell stiffness in the circumferential direction (the term which is multiplied by m^8).

3.3.8. Shells of negative Gaussian curvature. The papers [**112–114**] (see also [**73, 228**]) are devoted to the study of the lower part of the spectrum of shells of revolution of negative Gaussian curvature ($R_1 R_2 < 0$). We consider the case where the shell edge support resists bending. In this case, in a neighbourhood of the lower frequencies,

$$m \sim \mu^{2/3}, \qquad \Lambda \sim \mu^{4/3}, \tag{3.3.13}$$

and system (2.6.14) has four edge effect integrals of the form [**73**]

$$\begin{gathered} w^{(p)}(s,\mu) \simeq \sum_{k=0}^{\infty} \mu^{k/3} w_k^{(p)}(s) \exp\left\{ \frac{1}{\mu} \int_{s_0}^{s} \lambda_p(\tilde{s})\, d\tilde{s} + \mu^{-1/3} \int_{s_0}^{s} \lambda_p'(\tilde{s})\, d\tilde{s} \right\}, \\ \lambda_p^4 + \frac{1}{R_2^2(s)} = 0, \qquad w_0^{(p)} = B^{-1/2} \lambda_p^{-3/2}, \qquad p = 1,2,3,4. \end{gathered} \tag{3.3.14}$$

Four other integrals of system (2.6.14) have the index of variation $t = 2/3$ and they oscillate. These may be represented by the asymptotic series

$$w^{(p)}(s,\mu) \simeq \sum_{k=0}^{\infty} \varepsilon^k w_k^{(p)}(s) \exp\left(\frac{1}{\varepsilon} \int_{s_0}^{s} \lambda_p(\tilde{s})\, d\tilde{s} \right), \qquad \varepsilon = \mu^{2/3} > 0, \tag{3.3.15}$$

where

$$\left(\lambda_p^2 - \frac{\rho R_2}{B R_1} \right)^2 = 0, \qquad \rho = m\varepsilon, \qquad p = 5,6,7,8. \tag{3.3.16}$$

Moreover, since equation (3.3.16) has multiple pure imaginary roots, the functions $w_k^{(p)}$ satisfy the following second-order differential equations:

$$\begin{gathered} \frac{d}{ds}\left(g(s) \frac{dy}{ds} \right) + \left(\rho^2 \Lambda_0' f(s) - \frac{\rho^6}{g(s)} \right) y = 0, \\ \Lambda = \varepsilon^2 \Lambda', \qquad w_0^{(p)} = \left| \frac{R_1 R_2}{B} \right|^{1/4} y, \end{gathered} \tag{3.3.17}$$

where

$$f(s) = \frac{\rho R_2}{2|\lambda_p|}, \qquad g(s) = \frac{2R_1^2 B^4 |\lambda_p|}{\rho R_2 (R_2 - R_1)^2}.$$

We obtain an approximate solution by substituting the series (3.3.15) into the tangential boundary conditions. Therefore, the additional functions $y(s)$ are solutions to Sturm–Liouville's problems (see [**73, 112, 113**]) which depend upon the type of tangential boundary conditions considered. By constructing the next approximation and by having the integrals (3.3.14) satisfy the non-tangential boundary conditions, we obtain Λ by the following expansion:

$$\Lambda \simeq \mu^{4/3}(\Lambda_0' + \mu^{2/3}\Lambda_2' + \mu\Lambda_3' + \dots), \tag{3.3.18}$$

where Λ_0' and Λ_2' depend only on the tangential boundary conditions and Λ_3' depends on the non-tangential boundary conditions and on integrals (3.3.14).

By letting the shell parameters approach values for which, for some small m, (true or tangentially possible) bending exists, Λ continues to diminish until it reaches the order (3.3.8) (see [**73, 112, 113, 155, 157**]).

3.3.9. Shells partly of positive and partly of negative Gaussian curvature. It is interesting, from the asymptotic integration point of view, to analyze a shell of revolution which is partly of negative and partly of positive Gaussian curvature.

For $s_1 \le s < s_*$, let $R_1 > 0$ and, for $s_* < s \le s_2$, let $R_1 < 0$. Moreover, suppose that $k_1(s_*) = 0$ and $k_1'(s_*) < 0$, where $k_1(s) = [R_1(s)]^{-1}$. The point s_* where the curvature, k_1, changes sign is a turning point. In a neighbourhood of this point the integrals (3.3.15) are not valid since equation (3.3.17) has a singular point at $s = s_*$ $(g(s_*) = 0)$.

To construct the asymptotic expansions of the four integrals of system (2.6.14) we use the standard Airy function, similar to (2.3.12) (see [**156, 232, 237**]). In these papers, boundary value problems for the free vibrations and buckling of shells of this form are also considered.

3.3.10. Buckling of shells of revolution. Since the study of the buckling of shells of revolution under an axisymmetric loading is quite similar to that of free vibrations, we consider only some of the differences between these two problems and indicate references to the relevant literature.

In the first equation of (2.7.2) the third term, $-\Lambda w$, is replaced with $\mu^2\Lambda\Delta_t w$, where

$$\Delta_t w = -\frac{m^2}{B^2}t_2 + \frac{im}{B}\left(2t_3\frac{dw}{ds} + w\frac{dt_3}{ds}\right) + \frac{1}{B}\frac{d}{ds}\left(Bt_1\frac{dw}{ds}\right), \tag{3.3.19}$$

and the t_i are the momentless initial stresses (see (2.6.16)).

The qualitative study of shell buckling in the linear approximation is carried out, for example, in [**30, 68, 70, 77, 160, 235, 238, 261**]. We discuss some of these results.

3.3.11. Shells with arbitrary neutral surfaces. Suppose a shell, with an arbitrary neutral surface, has a true bending with a small wave number, m, in the circumferential direction. If work is done on the bending by the loading term (3.3.19), then (see [**68, 70, 235**])

$$\Lambda \sim \mu^2. \tag{3.3.20}$$

Comparing the estimates (3.3.20) and (3.3.8) for the vibrations, we find that, in the buckling problem, bending reduces the value of the parameter Λ less than vibrations do. The same fact is also observed for weakly supported shells. In particular, the existence of tangentially possible bendings (in the general case) does not reduce the order of the critical load (see [**235**]). However, if the bending only fails to satisfy the condition $w' = 0$, the order of Λ is reduced.

3.3.12. Shells of revolution of zero curvature. For a shell of revolution of zero curvature, equation (3.3.12) is replaced with equation

$$\frac{1}{R_2}\frac{d^2}{ds^2}\left(\frac{B^4}{R_2}\frac{d^2w_0}{ds^2}\right)+\frac{\mu^4m^8}{B^4}w_0+\mu^2\Lambda m^4\Delta_t w_0=0, \tag{3.3.21}$$

where $\Delta_t w$ is determined by (3.3.19).

Equation (3.3.21) allows us to get the approximate solution of some problems of shell buckling which, as in the case of vibrations, may be defined more precisely by introducing the edge effect integrals.

For a well supported shell, we distinguish three cases. Firstly, in the case of external normal pressure, $t_2 > 0$ and

$$m \sim \mu^{-1/2}, \qquad \Lambda \sim \mu. \tag{3.3.22}$$

Secondly, in the case of torsion by moments applied to the edges, $t_2 = 0$, $t_3 \neq 0$ and

$$m \sim \mu^{-1/2}, \qquad \Lambda \sim \mu^{1/2}. \tag{3.3.23}$$

Thirdly, in the case of axial compressing loading, $t_2 = t_3 = 0$, $t_1 > 0$ and $\Lambda \sim 1$. In this last case, equation (3.3.21) does not provide the approximate solution and we have to proceed from system (2.7.2).

The buckling of a cylindrical shell under external pressure and/or torsion has been studied extensively (see the reviews [**30, 62, 77, 267**]). In [**133**], the buckling of a conical shell under external pressure is considered. In [**233**], one studies the question of the separation of the boundary conditions, i.e. of the choice of two conditions (at each edge) which have to be satisfied by the solution of equation (3.3.21).

We also note that in [**136, 137, 236**] one studies free vibrations and buckling of shells close to shells of zero Gaussian curvature, since the curvature, $k_1 = 1/R_1$, of the generator is small ($k_1 \sim \mu$). The construction of zeroth approximation for the buckling problems is reduced to the following fourth-order equation:

$$\Delta_k(B^4\Delta_k w_0)+\frac{\mu^4m^8}{B^4}w_0+\mu^2m^4\Lambda\Delta_t w_0=0, \tag{3.3.24}$$

where the differential operator Δ_k is determined by (2.7.2). Equation (3.3.24) for $1/R_1 = 0$ transforms into (3.3.21).

3.3.13. Shells of revolution of negative Gaussian curvature. Buckling of shells of revolution of negative Gaussian curvature under axisymmetric loading has been studied in [**5, 113, 115, 155, 156**]. The buckling analysis is similar to that of low frequency vibrations. The construction of the asymptotic solutions for the buckling mode and the critical load in the zeroth approximation reduces to a Sturm–Liouville problem for an equation of type (3.3.17), and

$$m \sim \mu^{-2/3}, \qquad \Lambda \sim \mu^{2/3}. \tag{3.3.25}$$

If, for small m, bendings which satisfy all the boundary conditions (except may be $w' = 0$) are possible, a further reduction of the critical load occurs until the order of (3.3.20) is reached (see [**155, 156**]).

3.4. Vibrations of rotating shells of revolution

In this section, we consider the characteristics of the free vibration spectrum of a rotating thin elastic shell of revolution.

The main characteristics of this spectrum is the bifurcation of the eigenvalues due to rotation. In the nonrotating case, these eigenvalues are double for $m \geq 1$ as it was shown, for example, by Lidsky and Tovstik [**111**]. Moreover, rotation leads to a shift of the natural frequencies.

3.4.1. The equations of a rotating shell. Let us consider the vibrations of a shell of revolution rotating with constant angular velocity, Ω, around its axis of symmetry. Here and later, we assume that the shell is bounded by two parallels or is in the form of a cupola.

The present case differs from that of a nonrotating shell (see Section 2.7). Here, after separation of variables, the particular solutions are represented by a superposition of running waves [**190,199**],

$$
\begin{aligned}
u(s,\varphi,t) &= u(s)\cos(m\varphi - \omega t + \alpha),\\
v(s,\varphi,t) &= v(s)\sin(m\varphi - \omega t + \alpha),\\
w(s,\varphi,t) &= w(s)\cos(m\varphi - \omega t + \alpha),
\end{aligned}
\tag{3.4.1}
$$

where $\mathbf{U} = (u, v, w)^T$ is the displacement vector.

The equations describing such vibrations, after separation of variables, are of the form (2.6.14)–(2.6.15), with additional operators generated by the rotation,

$$
\left(\mu^4 N + L + \omega_*^2 + 2\omega_*\Omega_* L_C + \Omega_*^2 L_{\Omega_*}\right)\mathbf{U} = 0,
\tag{3.4.2}
$$

where ω_* and Ω_* are the non-dimensional vibration frequency and angular velocity connected, respectively, with the corresponding dimensional values by the formulae

$$
\omega_* = \sqrt{\frac{\rho R^2}{E}}\omega, \qquad \Omega_* = \sqrt{\frac{\rho R^2}{E}}\Omega.
$$

In the sequel, we shall omit the subscript $*$. The operators N and L in (3.4.2) are as in (2.6.14). The operator L_C, which takes into account the Coriolis forces of inertia, has the form

$$
\begin{bmatrix}
0 & -B' & 0\\
-B' & 0 & \frac{B}{R_2}\\
0 & \frac{B}{R_2} & 0
\end{bmatrix}.
\tag{3.4.3}
$$

The last term in equation (3.4.2) takes into account the forces of inertia of the transient motion and the initial axisymmetric stresses connected with them. The concrete form of this term depends upon the small terms which are kept. The main part of this operator is obtained by A. L. Smirnov and P. E. Tovstik in [**200**].

Since we consider only small vibrations, the initial deformations lead to $\Omega^2 \ll 1$.

3.4.2. The perturbation method. The eigenvalue problem for nonrotating shells was considered in Sections 2.7 and 3.3. Let ω_0 be a simple eigenvalue and $\mathbf{U}_0$ a corrresponding eigenfunction for that problem.

In the case $0 < \Omega^2 \ll 1$, let ω_1, ω_2 and $\mathbf{U}^{(1)}$, $\mathbf{U}^{(2)}$ denote the eigenvalues and corresponding eigenfunctions which bifurcate from ω_0^2 and $\mathbf{U}_0$ for a nonrotating shell, when $\Omega = 0$. We represent $\omega_{1,2}$ in the following form:

$$\omega_{1,2} = \pm\alpha(\Omega) + \beta(\Omega), \tag{3.4.4}$$

where

$$\alpha(\Omega) = \omega_0 + \sum_{i=1}^{\infty} \Omega^{2i}\alpha_{2i}, \qquad \beta(\Omega) = \sum_{i=1}^{\infty} \Omega^{2i-1}\beta_{2i-1},$$

and

$$\mathbf{U}^{(1)}(s,\Omega) = \mathbf{U}_0(s) + \sum_{i=1}^{\infty} \Omega^i \mathbf{U}_i(s),$$

$$\mathbf{U}^{(2)}(s,\Omega) = \mathbf{U}_0(s) + \sum_{i=1}^{\infty} (-1)^i \Omega^i \mathbf{U}_i(s).$$

Then the perturbation method gives the following formulae for the first coefficients in the above expansions

$$\beta = \beta_1\Omega + O\left(\Omega^3\right), \tag{3.4.5}$$

where

$$\beta_1 = -\frac{(L_C\mathbf{U}_0, \mathbf{U}_0)}{(\mathbf{U}_0, \mathbf{U}_0)},$$

and

$$\alpha_2 = \frac{\beta_1^2}{2\omega_0} - \frac{(L_C\mathbf{U}_1, \mathbf{U}_0) + \beta_1(\mathbf{U}_1, \mathbf{U}_0)}{(\mathbf{U}_0, \mathbf{U}_0)} - \frac{L_\Omega(\mathbf{U}_0, \mathbf{U}_0)}{2\omega_0(\mathbf{U}_0, \mathbf{U}_0)}. \tag{3.4.6}$$

Here (f, g) denotes the L_2 scalar product of f with g,

$$(f, g) = \int_{s_1}^{s_2} \sum_i f_i y_i B\, ds.$$

In the case of axisymmetric vibrations ($m = 0$), $\beta = 0$ and rotation leads only to a frequency shift without bifurcation. The frequency bifurcation in a field of gyroscopic forces and the mode precession phenomenon are similar, for example, to Larmore's precession and vibrations of Foucault's pendulum. This effect is discussed in detail in [**148–150, 272**].

We study the effect of rotation on the different types of vibrations according to the classification given in the monograph [**67**] (see Section 4.3). Thus, we compare the orders of all the variables with the non-dimensional thickness $h_* = h/R$.

For convenience, we define

$$\Omega = h^q, \qquad q > 0, \qquad \Delta_{1,2} = \frac{|\omega_{1,2}| - \omega_0}{\omega_0}.$$

Let p be the index of variation of the shell main stress-strain state and $k = 2$ if $c > 0$ and $k = 1$ if $c < 0$. Then, we have three cases. Firstly, for quasi-transverse vibrations with small index of variation,

$$\Delta \sim h^{p+q} + h^{kc}, \qquad c = q - p, \qquad 0 \leq p < 0.5.$$

Secondly, for quasi-transverse vibrations with large index of variation,

$$\Delta \sim h^{3p+q-1} + h^{kc}, \qquad c = q + p - 1, \qquad 0.5 \le p < 1.$$

Thirdly, for quasi-tangential vibrations,

$$\Delta \sim h^{p+q} + h^{2q}, \qquad 0 \le p < 0.5.$$

In the above estimates, the first terms provide the order of $\beta\omega_0^{-1}$ and the second terms give the value of $(\alpha - \omega_0)\omega_0^{-1}$, respectively. It thus follows that, for the types of vibrations considered here, the difference 2β, between the frequencies ω_1 and $-\omega_2$, is relatively small. Concerning the range of values of β, Vilke [**257**] has shown that $-1 \le \beta \le 0$, and Egarmin [**43**] proved that $\beta = -1$ only if $m = 1$ and only for a ring or a cylinder without any axial displacement ($u = 0$).

For quasi-tangential vibrations, Δ is also small.

For quasi-transverse vibrations, the rotation effect on Δ is determined by the parameter c. For $c > 0$ the frequency shift is small, but for $c \le 0$ the frequency shift is essential. The maximum rotation effect occurs for frequencies of the quasi-transverse vibrations with middle variation index ($p = 0.5$), since, in this case, the value of the parameter c is minimum. This effect increases as the angular velocity, Ω, increases and as the shell thickness, h, decreases.

For $c \le 0$ the perturbation method described above cannot be applied, but for $c > 0$ it provides an estimate for the rotation effect. Let $c > 0$ and suppose that ω_0 and $\mathbf{U}_0(s)$ are known for the nonrotating shell. Then the bifurcation coefficient β is determined from (3.4.5), and to evaluate α_2 by that formula one has to know the solution $\mathbf{U}_1$.

For the quasi-transverse vibrations with $p > 0$, the orders (in h) of the different terms in the expression for α_2 are different and, with an error of the order h^{2p}, we have

$$\alpha_2 = -\frac{(L_\Omega \mathbf{U}_0, \mathbf{U}_0)}{2\omega_0(\mathbf{U}_0, \mathbf{U}_0)}.$$

For the quasi-transverse vibrations with $p = 0$ and any quasi-tangential vibrations, the above simplifications are not valid, but one does not need them, since in those cases $\alpha_2 \sim h^0$ and we may assume that $\alpha = \omega_0$ to an error of the order Ω^2.

3.4.3. Vibrations with superlow frequencies. The low frequency vibrations are of special interest. Among these vibrations we are especially interested in the *superlow* frequencies of the order h^γ, $\gamma > 0$.

The appearance of superlow frequencies may be caused by different factors, one being the existence of *true bendings* (see Subsection 3.3.5). For shells of simple geometry (cones, spheres), the first terms in the expansion of the solution in h have simple analytical representations (see [**73**]).

We write here only the formulae describing the rotation effect on the vibrations of a conical shell of middle length for large m [**199**], a long cylindrical shell [**43, 199**] and a spherical shell [**43, 199, 271**].

In the first case,

$$\omega = \beta_1\Omega \pm \sqrt{\omega_0^2 + m^2\Omega^2}, \qquad \beta_1 = -\frac{2\sin^2\theta}{m}, \qquad \sin\theta = \frac{B}{R_2}, \tag{3.4.7}$$

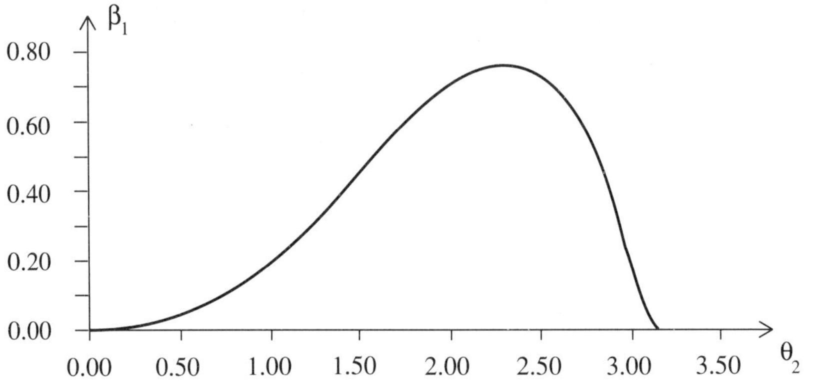

FIGURE 15. The bifurcation coefficient β_1 vs. the spherical angle θ_2 in radian measure.

where, for the cylindrical shell, $\theta = \pi/2$. Note that ω_0 attains its minimum $\omega_0 \sim h^{1/2}$ at $m \sim h^{-1/4}$ (see Section 3.3), and formula (3.4.7) is valid for frequencies in a neighbourhood of the lowest frequency for which $m \sim h^{-p}$, $0 < p < 0.5$.

In the second case ($l \gg 1$),

$$\beta_1 = -\frac{2m}{m^2 + 1 + D^k (ml)^{-2}},$$

where the coefficient D^k depends on the boundary conditions and the wave number along the meridian, $l = L/R$ in the non-dimensional shell length. It is interesting to note that it is only for long cylindrical shells in the case $k = 0$ and $m = 1$ that the vibration frequency may be equal to 0; this means that buckling occurs.

In the third case, V.F. Zhuravlev and A. L. Popov [**271**], who studied the low frequency vibrations of the cupola $B = \sin\theta$ for $0 \le \theta \le \theta_2$, obtained, in the case of pure bending vibrations, the formula

$$\beta_1 = -2m \frac{I_m^1}{I_m^2 + I_m^3 + I_m^4},$$

where I_m^i are functions of θ_2 expressed by quadratures.

In Fig. 15 one can see the dependence of the bifurcation coefficient, β_1, on the spherical angle, θ_2, for $m = 2$.

The effects of the boundary conditions and the edge angle, θ_1, are analyzed in detail in [**43**].

3.4.4. Solutions of the equations of vibration. Here we consider the effect of rotation on the solutions of system (3.4.2). We start with the case of axisymmetric vibrations and vibrations with a small wave number in the circumferential direction (see (3.3.1)). The character of the solutions as $h \to 0$ essentially depends on the following coefficient:

$$b_2(s) = -\omega^2 + \frac{1}{R_2^2} + m^2\Omega^2, \tag{3.4.8}$$

i.e., for a rotating shell, b_2 has one additional term, $m^2\Omega^2$, generated by the initial stresses. If $b_2 = 0$, then we obtain an equation for the turning point s_*. Therefore, in the case of small m, in intervals not containing any turning points and for $B \neq 0$, system (3.4.2) has four regular momentless solutions and four quasi-transverse solutions with a large index of variation, of the form (2.2.1), i.e.

$$x_n(s) = \mu^{\gamma(x)} \sum_{k=1}^{\infty} \mu^k x_n^{(k)}(s) \exp\left(\frac{1}{\mu}\int_{s_1}^{s} \lambda_n \, ds\right),$$

where

$$\lambda_n = b_2^{1/4} e^{(n-1)\pi i/2}, \qquad n = 1, \ldots, 4.$$

Now we examine the nonaxisymmetric vibrations for $m = \mu^{-1}m^*$, where $m^* \sim 1$ and $\omega = \mu^0\omega^*$, where $\omega^* \sim 1$. In this case we look for solutions in the form (2.7.4), i.e.

$$x_{kn}(s, \mu) = \sum_{i=1}^{\infty} \mu^i x_{kn}^{(i)}(s) \exp\left(\frac{1}{\mu}\int_{s_1}^{s} \lambda_n \, ds\right), \qquad k,\, n = 1, \ldots, 8.$$

To determine λ we get an equation of the fourth degree in λ^2 (see (2.1.6)),

$$\det(A^{(0)} - \lambda I) = 0, \tag{3.4.9}$$

where the elements, $a_{kn}^{(0)}$, of the matrix $A^{(0)}$ depend on the parameters s, ω, Ω and m.

Let us analyze the cases where rotation affects the value of λ. If $\Omega = \mu^q\Omega^*$, where $\Omega^* \sim 1$, then we obtain the orders, μ^q and μ^{-2+2q}, of the principal terms of the matrices $2\omega\Omega L_C$ and $\Omega^2 L_\Omega$, respectively. Therefore, for $q > 1$ the rotation does not influence λ. But for $0 < q \leq 1$, the elements of the matrix $\Omega^2 L_\Omega$ have to be taken into account in the construction of the matrix $A^{(0)}$ and the coefficient $a_{85}^{(0)}$ changes.

As a result, for $q = 1$ the roots of equation (3.4.9) are of the order $\lambda \sim 1$ and all solutions have indices of variation 1 (in μ).

In the case $0 < q < 1$ (fast rotation), there are two groups of roots. The roots of the first group are determined by the equation (where the superscript $*$ is omitted)

$$\left(\lambda^2 - \frac{m^2}{B^2}\right)^2 + m^2\Omega^2\mu^{2q-2} = 0.$$

These roots are of the order $\mu^{(q-1)/2}$, and the indices of variation of the solutions are equal to $(3 - q)/2$.

The roots of the second group are obtained from the equation

$$(1 - \nu^2)\left(\frac{m^2}{B^2 R_1}\right)^2 + m^2\Omega^2\mu^{2q-2}\lambda^4 = 0.$$

These roots are of the order $\mu^{(1-q)/2}$, and the indices of variation of the solutions are equal to $(q+1)/2$. This last case is not very interesting since, for such q, there are no pure imaginary roots, and, therefore, no oscillating solutions. For oscillations to exist we need to consider frequencies ω of the order not less than μ^{q-1}. For more detail on the solutions of this equation, see the paper by A. L. Smirnov [**198**].

CHAPTER 4

Asymptotic Solutions of Partial Differential Equations

4.1. Introduction

The choice of asymptotic methods for the integration of partial differential equations depends on the problems considered. We shall limit our discussion to those methods which are applied to problems of solid mechanics, especially to thin shells and plates. We shall not consider any of the methods used in problems of viscous fluids.

We consider the linear partial differential equation, in a domain D of $\mathbb{R}^n$,

$$Lw := \sum_{|k|=0}^{m} a_k(x)\partial_{x_1}^{k_1}\partial_{x_2}^{k_2}\cdots\partial_{x_n}^{k_n} w = f(x), \tag{4.1.1}$$

where

$$x = (x_1, \ldots, x_n), \qquad k = (k_1, \ldots, k_n), \qquad |k| = k_1 + \cdots + k_n,$$

and k_i are nonnegative integers. The main form of the asymptotic expansions of the solutions of equation (4.1.1) is

$$w(x, \mu) \simeq \Phi(x, \mu)e^{\lambda S(x,\mu)}, \tag{4.1.2}$$

where $\mu > 0$ is a small parameter connected with the large parameter $\lambda = \mu^{-t}$, t being the index of variation of $w(x, \mu)$ (see Section 1.8), and the functions Φ and S are expressed by the series

$$\begin{aligned} \Phi(x,\mu) &\simeq \sum_{j=0}^{\infty} \mu^{\alpha_j}\Phi_j(x), \qquad \alpha_{j+1} > \alpha_j, \\ S(x,\mu) &\simeq \sum_{j=0}^{\infty} \mu^{\beta_j} S_j(x), \qquad \beta_0 = 0, \qquad \beta_{j+1} > \beta_j. \end{aligned} \tag{4.1.3}$$

The essential part of the function S consists only in the terms for which $\beta_j < t$. The other terms may be set to zero by modifying the function Φ. The function S may often be taken to be independent of μ.

In some problems, the coefficients a_k in (4.1.1) depend also on the parameter μ; in others, the boundary conditions depend on the parameter μ [**65, 259**]; still in others, the parameter μ is formally introduced in the construction of solutions with large indices of variation.

If $\Re S$ is not constant, the solution $w(x, \mu)$ of (4.1.2) is exponentially increasing or decreasing solutions. On the other hand, if $\Re S = 0$ but $\Im S$ is not constant, then $w(x,\mu)$ is rapidly oscillating. In neighbourhoods of some surfaces or lines, called *caustics*, the representation (4.1.2) is not valid. In the one-dimensional case, caustics correspond to turning points.

To construct the coefficients Φ_j and S_j in (4.1.3), the solution (4.1.2) is substituted into equation (4.1.1) and the coefficients of identical powers of μ are identified. To determine $S_0(x)$, one uses the top order coefficients which provide the following nonlinear partial differential equation in the gradient p of S_0:

$$H(p,x)=0, \qquad p=\left(\frac{\partial S_0}{\partial x_1},\dots,\frac{\partial S_0}{dx_n}\right). \tag{4.1.4}$$

The equations for the determination of S_j $(j>0)$ and Φ_j are obtained from the next approximation, when one equates the coefficients for the next powers of μ. Particular cases of these equations are considered below.

The question whether solutions (4.1.2), which formally satisfy equation (4.1.1), are asymptotic expansions of the exact solution, as $\mu\to 0$, is still open and will not be discussed below any further.

4.2. Helmholtz's equation and its generalization

4.2.1. Eikonal and transport equations. We consider the short-wave asymptotics of the solutions of Helmholtz's equations in $\mathbb{R}^n$,

$$\Delta w+\frac{\lambda^2}{c^2}w=0, \qquad \Delta=\frac{\partial^2}{\partial x_1^2}+\dots+\frac{\partial^2}{\partial x_n}, \tag{4.2.1}$$

as a simple example which reflects the main features of the asymptotic expansions (4.1.2). Here λ is a large parameter and the function $c=c(x)$ is assumed to be infinitely differentiable.

For $n=3$, equation (4.2.1) describes sound propagation with variable velocity $c(x)$ and, for $n=2$, it describes stable vibrations of a membrane with variable density $\rho(x)=[c(x)]^{-2}$. In both cases, λ is the vibration frequency.

The formal asymptotic solution of equation (4.2.1) (see the monographs by V. M. Babich and V. S. Buldyrev [**12**] and B. R. Vainberg [**247**]) is

$$w(x,\mu)=\sum_{k=0}^{\infty}(-i\mu)^k w_k(x)e^{i\lambda\tau(x)}, \qquad \mu=\lambda^{-1}, \tag{4.2.2}$$

where $\tau(x)$ satisfies the *eikonal equation*,

$$\nabla\tau\cdot\nabla\tau=\frac{1}{c^2}, \qquad \nabla=\left(\frac{\partial}{\partial x_1},\dots,\frac{\partial}{\partial x_n}\right), \tag{4.2.3}$$

and the coefficients $w_k(x)$ can be found from the *transport equations*

$$2\nabla\tau\cdot\nabla w_k+w_k\Delta\tau=\Delta w_{k-1}, \qquad k=0,1,2,\dots, \qquad w_{-1}=0. \tag{4.2.4}$$

We introduce *rays* as functions of the orthogonal curvilinear *coordinates* $\alpha=(\alpha_1,\dots,\alpha_{n-1})$ and τ by the transformations

$$x_i=x_i(\alpha,\tau), \qquad i=1,2,\dots,n. \tag{4.2.5}$$

For fixed α_j, formulae (4.2.5) describe lines, called *rays*, which are extremals of Fermi's functional [**2**],

$$\tau = \int_{x_0}^{x} \frac{ds}{c(s)}, \qquad ds^2 = dx_1^2 + \cdots + dx_n^2. \tag{4.2.6}$$

The integration of equation (4.2.4), for $k = 0$, (see [**12**]) gives

$$w_0(x) = \psi(\alpha)c(x)J^{-1/2}, \tag{4.2.7}$$

where the arbitrary function $\psi(\alpha)$ is determined by the boundary conditions and J is the Jacobian of the transformation (4.2.5). In [**12**], the coefficients $w_k(x)$, for $k > 0$, are also obtained by quadratures.

On the envelope of the set of rays (4.2.5), $J = 0$; hence, in this case, the representation (4.2.2) is invalid. Below, we shall consider particular cases, in which, instead of (4.2.2), other expansions are constructed.

Solutions (4.2.2) by means of rays describe a rather wide class of solutions of equation (4.2.1). As usual, in choosing the curvilinear coordinates (4.2.5), one takes into account the form of the domain under consideration. The boundary conditions are satisfied under integration of equation (4.2.4).

4.2.2. Two-dimensional Dirichlet eigenvalue problems. We consider the Dirichlet eigenvalue problem in the interior of the plane domain D ($n = 2$), bounded by a closed smooth convex curve Γ,

$$\frac{\partial^2 w}{\partial x_1^2} + \frac{\partial^2 w}{\partial x_2^2} + \frac{\lambda^2}{c^2}\, w = 0,$$

with Dirichlet boundary conditions,

$$w = 0, \qquad \text{for } (x_1, x_2) \in \Gamma.$$

This problem has no analytical solution even when $c(x) = \text{const}$, except when the variables x_1 and x_2 separate. The behaviour of the n-th eigenfunction, $w^{(n)}(x_1, x_2)$, as $n \to \infty$, is unknown. Below, we consider the asymptotics of a pair of two-parameter sets of eigenfunctions, which are characterized by the fact that they are localized in a neighbourhood of some line Γ_0 and they decay exponentially away from this line.

In the first case, the eigenfunctions $w^{(m,p)}(x_1, x_2)$ are localized in a neighbourhood of the boundary, Γ, of D (i.e. $\Gamma_0 = \Gamma$). Here m is the wave number along the boundary and p is the semi-wave number in the direction orthogonal to the boundary. If $c = \text{const}$ and D is a ring, the variables separate and the problem has an exact solution (see Subsection 2.4.10). Such eigenfunctions are called functions of the *whisper gallery* type and are considered in [**12, 108**]. The term *whisper gallery* comes from the fact that, under some conditions, sound propagates well along the walls of a convex hall and does not propagate through its centre.

We introduce the curvilinear coordinates (s, n) connected with the curve Γ by taking s as the arc length of Γ and $n > 0$ as the distance from Γ along the outward normal, $\mathbf{n}$ to Γ; inside D we take $n < 0$ (see Fig. 16).

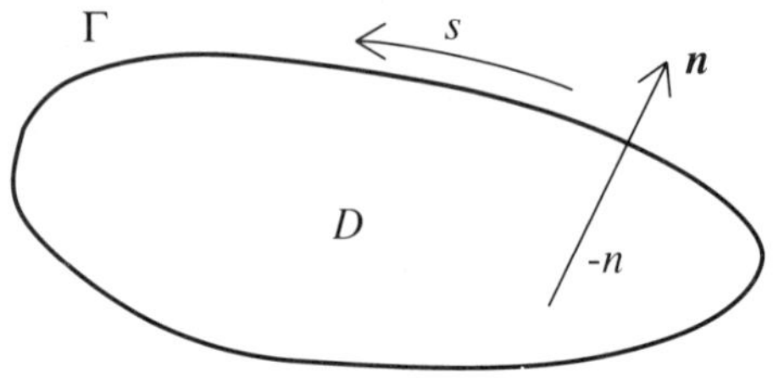

FIGURE 16. The system (s, n) of curvilinear coordinates.

Then the natural frequencies, $\lambda_{m,p}$, and modes, $w^{(m,p)}$, of the set of free vibrations can be represented in the form

$$
\begin{aligned}
&\lambda_{m,p} \simeq \lambda_m \sum_{k=0}^{\infty} a_k \lambda_m^{-k/3}, \qquad \text{as } m \to \infty, \\
&\lambda_m = 2m\pi b, \qquad b = \left[\int_0^L \frac{ds}{c_0(s)}\right]^{-1}, \\
&a_0 = 1, \qquad a_1 = a_3 = 0, \qquad a_2 = \frac{b|\eta_p|}{2^{1/3}} \int_0^L \frac{ds}{c_0^{1/3}(s) P^{2/3}(s)}, \\
&c(s,n) = c_0(s) + n c_1(s) + \dots, \qquad P(s) = \left[\frac{1}{\rho(s)} - \frac{c_1(s)}{c_0(s)}\right]^{-1}, \\
&w^{(m,p)} \simeq e^{i\lambda\alpha} \operatorname{Ai}(\beta), \qquad \text{as } m \to \infty, \\
&\alpha = \alpha(s,n,\lambda) = \sum_{k=0}^{\infty} \alpha_k(s,\nu)\lambda^{-k/3}, \qquad \beta = \sum_{k=0}^{\infty} \beta_k(s,\nu)\lambda^{-k/3}, \\
&\nu = \lambda^{2/3} n, \qquad \lambda = \lambda_{m,p}, \qquad p = 1, 2, \dots .
\end{aligned}
\tag{4.2.8}
$$

Here $m \to \infty$ and p is fixed. As in Section 2.4, the solution $w^{(m,p)}$ is represented by means of the standard Airy functions, $\operatorname{Ai}(\eta)$ and η_p is the p-th root of the equation $\operatorname{Ai}(\eta) = 0$. As a result of the scaling $\nu = \lambda^{2/3} n$ and the expansion into a power series in ν, the variables asymptotically separate in a neighbourhood of Γ. Here α_k and β_k are polynomials in ν, obtained in [**12**]. Finally, in (4.2.8), L is the arc length of Γ and $\rho(s)$ is the radius of curvature. It is assumed that $P > 0$.

The modes $w^{(m,p)}$, to a certain degree, are similar to the shell vibrations and buckling modes localized in a neighbourhood of the edge (see Subsections 2.7.6 and 2.7.10).

4.2.3. A second example: the bouncing ball. Another two-parameter set of frequencies $\lambda_{m,p}$ and modes $w^{(m,p)}$, for which the asymptotic expansion, as $\lambda \to \infty$, is known, is the *bouncing ball* mode [**12, 109**]. In particular, such modes are possible in a neighbourhood of the smallest axis of an elliptical membrane for $c(x_1, x_2) = \text{const}$. A necessary condition for the existence of such modes is the existence of a ray, Γ_0, which cuts the boundary Γ of D in two points at right angle. The term bouncing ball is connected with the fact that the unknown vibration mode may be constructed by the methods of geometrical optics by successive reflections of the rays on the boundary of the domain. The trajectory of a ray follows the trajectory of a perfectly bouncing ball.

We start with the easiest example in which the variables separate and for which we can write the exact solution. Let the domain D be the infinite strip

$$0 \le s \le L, \qquad -\infty < n < \infty, \tag{4.2.9}$$

and consider the boundary value problem (4.2.1) in the form

$$\frac{\partial^2 w}{\partial s^2} + \frac{\partial^2 w}{\partial n^2} + \lambda^2 \left[1 - \left(\frac{n}{n_0}\right)^2\right] w = 0, \tag{4.2.10}$$

with the boundary conditions

$$w(0, n) - w(L, n) = 0.$$

After separating the variables,

$$w(s, n) = w(n) \sin \frac{m\pi s}{L},$$

we obtain Weber's equation,

$$\frac{d^2 w}{dn^2} + \left[\lambda^2 - \left(\frac{m\pi}{L}\right)^2 - \lambda^2 \frac{n^2}{n_0^2}\right] w = 0, \tag{4.2.11}$$

the solution of which can be expressed by means of the parabolic cylinder functions [**1**] (see (2.5.2) and (2.5.3)),

$$\begin{gathered} w^{(m,p)}(n) = D_p\left(\sqrt{\frac{2\lambda}{n_0}} n\right), \qquad p = 0, 1, 2, \dots, \\ \frac{d^2 D_p}{d\zeta^2} + \left(p + \frac{1}{2} - \frac{\zeta^2}{4}\right) D_p = 0, \qquad D_p\left(\zeta\sqrt{2}\right) = e^{-\zeta^2/2} H_p(\zeta), \end{gathered} \tag{4.2.12}$$

and $\lambda = \lambda_{m,p}$ may be obtained from the quadratic equation

$$\lambda_{m,p}^2 = \left(\frac{m\pi}{L}\right)^2 + (2p+1)\frac{\lambda_{m,p}}{n_0}. \tag{4.2.13}$$

In the general case, but under some stability conditions, the eigenfunctions can also be expressed by the parabolic cylinder functions. To achieve this, we introduce the curvilinear coordinates, s and n, connected with the ray Γ_0; then

$$\begin{gathered} w^{(m,p)}(s, n) = e^{i\lambda\alpha} D_p(\beta\sqrt{2}), \qquad \text{as } m \to \infty, \\ \alpha \simeq \sum_{k=0}^{\infty} \lambda^{-k/2} \alpha_k(s, \nu), \qquad \beta \simeq \sum_{k=0}^{\infty} \lambda^{-k/2} \beta_k(s, \nu), \qquad \nu = \lambda^{1/2} n, \\ \lambda = \lambda_{m,p} \simeq \lambda_m \sum_{k=0}^{\infty} a_k \lambda_m^{-k}, \qquad \lambda_m = \frac{m\pi}{L}, \qquad L = \int_{s_0}^{s_1} \frac{ds}{c_0(s)}, \qquad a_0 = 1, \end{gathered} \tag{4.2.14}$$

where, as in (4.2.8), α_k and β_k are polynomials in ν.

We compare the expansions of the functions w, given by formulae (4.2.2), (4.2.8) and (4.2.14). The indices of variation of these functions are $t = 1$. A more detailed consideration shows that the indices of variation of these functions are different in different directions. In the n-direction the functions w are changing more slowly then in the s-direction. We let t_1 denote the partial index of variation in the n-direction orthogonal to the ray (see Section 1.7). Then $t_1 = 0$ for the function (4.2.2), $t_1 = 2/3$ for the function (4.2.8) and $t_1 = 1/2$ for the function (4.2.15).

4.2.4. A third example: a radiation field. We consider a problem connected with the construction of the high-frequency asymptotics of a wave field scattered by a body [**12**]. Since here we consider an infinite domain, D, besides the conditions on the surface of the body, we have to introduce conditions on the reflected field at infinity. In the two-dimensional case, these conditions are

$$\sqrt{r}\left(\frac{\partial w}{\partial r} - i\lambda w\right) \to 0, \qquad \text{as } r \to \infty. \tag{4.2.15}$$

The asymptotics of the reflected wave field, both in the sunlit zone and the shadow zone, are constructed in [**12**]. The construction of the asymptotics on the boundary between light and shadow is the most difficult. Here the standard Airy functions are used.

This problem has numerous applications in hydro-elasticity [**142, 189, 255**]. Unlike the reflection from rigid bodies, the reflection of sound from elastic bodies and, in particular, from thin elastic shells, is much more complicated, since the elastic vibrations generated in the body, in their turn, generate additional sound waves in the liquid. From these waves we may draw conclusions about the parameters of the body.

The paper by Shcheviev [**188**] of this collection is devoted to a simplified problem of sound propagation through elastic shells.

Many publications are devoted to the study of the natural frequency spectrum of thin shells immersed in a liquid or containing a liquid [**11, 74, 111, 163, 164, 168, 252**]. The spectrum of the free vibrations of a shell containing a liquid is real and discrete, but the spectrum of free vibrations of a shell immersed in a liquid is complex. These vibrations are damped since vibrational energy is transferred to infinity by the waves. But there exists a large class of modes for which this damping is very small.

4.2.5. Three-dimensional elasticity equations. The method of rays is also applied to dynamical problems of elastic bodies [**14, 15, 263**]. In rigid bodies, the strain waves and shift waves spread with different velocities. Besides, there may be Rayleigh waves spreading along the surface of the body [**119**]. Asymptotic integration is also applied to seismology and seismic prospecting [**13**].

To study the vibrations of thin bars, plates and shells, traditionally, one uses one-dimensional and two-dimensional equations (see Section 2.6). However for high frequencies, one- or two-dimensional equations gives erroneous results and one has to employ the three-dimensional elasticity theory. Three-dimensional equations have to be used if the wave-length of the deformation is commensurable with, or smaller than, the plate (bar, shell) thickness. In [**22, 72, 87, 88, 239**], asymptotic expansions are used in the study of the high-frequency vibrations of plates and shells by taking into account their small relative thickness.

In [**239**], one constructs a three-parameter family of frequencies and modes which are localized in a neighbourhood of the point of maximum shell thickness for an anisotropic plate of nonconstant thickness. These modes are generalizations of the modes (4.2.14) and the parabolic cylinder functions are used as the standard functions.

The asymptotics distribution of the natural frequencies of a body is an interesting problem. This question was considered in Section 2.7 for the one-dimensional problem. The frequency distribution for two- and three-dimensional elastic bodies is discussed in the review [**251**].

4.2.6. Maslov's canonical operators. Returning to the asymptotic expansions (4.2.2), (4.2.8) and (4.2.14), we note that other methods of construction of asymptotic solutions, like the method of *Maslov's canonical operators*, are described in [**124, 125, 126, 247**]. In these references, the solution is found in the form (4.1.2) and an effective method for constructing the functions Φ and S is developed. This method may also be used for non-linear equations with a parameter. Its main sphere of application is quantum mechanics. In Section 4.4, a simplified variant of this method is applied to problems of local shell buckling.

4.3. Degeneracy of two-dimensional boundary value problems

4.3.1. The main and additional solutions. We consider the boundary value problem A_μ:

$$\mu^n N_{m+p} w + L_m w = f(x), \qquad x = (x_1, x_2) \in D,$$
$$m_i w = m_i^0, \qquad i = 1, 2, \dots, \frac{m+p}{2}, \qquad x \in \Gamma, \tag{4.3.1}$$

in the two-dimensional domain D with boundary Γ. Here $\mu > 0$ is a small parameter, L_m and N_{m+p} are linear differential operators of type (4.1.1) of the orders m and $m+p$, respectively (m and p are even integers), m_i are differential operators of the boundary conditions.

For $\mu = 0$, problem A_μ degenerates into problem A_0:

$$L_m w_0 = f(x); \qquad l_i w_0 = l_i^0, \qquad i = 1, 2, \dots, \frac{m}{2}, \qquad x \in \Gamma, \tag{4.3.2}$$

where the boundary conditions may be obtained either by neglecting part of the boundary conditions (4.3.1) or as a linear combination of conditions (4.3.1). Other variants of the degenerate problems will also be considered below.

The asymptotic methods in the theory of elasticity are discussed by S. A. Nazarov [**141**].

Asymptotic integration of thin shell problems are considered in [**66, 67, 69**]. In this case, $m = p = 4$ and the operator N_{m+p} is elliptic. The type of the operator L_m depends upon the sign of the Gaussian curvature, K, of the neutral surface; L_m is elliptic, parabolic or hyperbolic if $K > 0$, $K = 0$ or $K < 0$, respectively.

As in Section 3.1, we can introduce here the notion of *regular degeneracy* [**258**]. To introduce this concept for equation (4.3.1) in a neighbourhood of the curve Γ, we use the local coordinates s and n, where n is the interior normal to Γ and s is arc length along Γ. Let the highest order derivative with respect to n in the operators $\mu^n N_{m+p}$ and L_m be

$$\mu^p a_{m+p} \frac{\partial^{m+p} w}{\partial n^{m+p}} + a_m \frac{\partial^m w}{\partial n^m}. \tag{4.3.3}$$

As in (3.1.5), we introduce the additional characteristic equation

$$a_{m+p}(s, 0)\lambda^p + a_m(s, 0) = 0, \tag{4.3.4}$$

where the coefficients depend on the parameter s. The degeneracy of problem A_μ into problem A_0 is said to be *regular*, if equation (4.3.4) has $p/2$ roots with negative real parts (in [**258**], a more general case than (4.3.1) is studied).

Let the solution of problem A_0 exists and be unique for any $f(x)$ and l_i^0. Then, as in (3.1.6), we seek the solution to problem A_μ as the sum of two formal series,

$$w(x,\mu) \simeq \sum_{k=o}^{\infty} \mu^k w_k(x) + \mu^\alpha \sum_{k=0}^{\infty} \mu^k z_k(s,\eta), \qquad \eta = \frac{n}{\mu}, \tag{4.3.5}$$

where the functions z_k satisfy the conditions

$$z_k(s,\eta) \to 0, \qquad \text{as } \eta \to \infty, \qquad k = 0,1,2,\dots \tag{4.3.6}$$

Solution (4.3.5) is made of the *main* solution (i.e. the first sum) and the *additional* solution (i.e. the second sum), which has the character of a *boundary layer* or *edge effect.* The main solution is determined by boundary value problems of type (4.3.2). The additional solution, similar to (3.1.8), is determined from the boundary value problems for the following ordinary differential equation with constant coefficients:

$$a_{m+p}(s,0)\frac{d^{m+p}z_k}{d\eta^{m+p}} + a_m(s,0)\frac{d^m z_k}{d\eta^m} = F_k(s,\eta), \qquad k = 0,1,2,\dots, \tag{4.3.7}$$

where s is a parameter, $F_0 \equiv 0$, and the functions F_k $(k > 0)$ depend on the functions z_j $(j < k)$ obtained before. We seek a solution of equation (4.3.7) which satisfies conditions (4.3.6) and, after the addition of the main solution, also those conditions (4.3.1) which are not included in (4.3.2).

4.3.2. Non-absolutely flexible membranes. As an example, we consider two problems for non-absolutely flexible membranes. The equilibrium of a clamped membrane in non-dimensional form is described by the boundary value problem

$$-\mu^2\Delta^2 w + \Delta w = f(x); \qquad w = \frac{\partial w}{\partial n} = 0, \qquad \text{for } x \in \Gamma, \tag{4.3.8}$$

where, as in Subsection 3.1.5, $\mu > 0$ is a small parameter related to the bending stiffness, w is the deflection and $f(x)$ is the density of the surface load. The problem is to be solved in the domain D with smooth boundary Γ (see Fig. 16).

In this problem the degeneracy is regular and the degenerate problem,

$$\Delta w_0 = f(x); \qquad w_0 = 0, \qquad \text{for } x \in \Gamma, \tag{4.3.9}$$

corresponds to an absolutely flexible membrane.

If the solution $w_0(x)$ of problem (4.3.9) is known, then the solution of the complete problem (4.3.8) has the form (4.3.5). The first terms of the series in (4.3.5) are

$$w(x,\mu) = w_0(x) + \mu w_1(x) + \mu\left(\frac{\partial w_0}{\partial n}\right)_\Gamma e^{-n/\mu} + O\left(\mu^2\right), \tag{4.3.10}$$

where $w_1(x)$ is the solution of the Dirichlet problem

$$\Delta w_1 = 0; \qquad w_1 = -\frac{\partial w_0}{\partial n}, \qquad \text{for } x \in \Gamma, \tag{4.3.11}$$

and the notation $(\partial w_0/\partial n)_\Gamma$ means that the partial derivative is evaluated along the curve Γ.

The free vibrations of this membrane may be obtained from the solution of the boundary value problem

$$(4.3.12) \qquad -\mu^2\Delta^2 w + \Delta w + \Lambda w = 0; \qquad w = \frac{\partial w}{\partial n} = 0, \qquad \text{for } x \in \Gamma.$$

As $\mu \to 0$, this problem degenerates into a problem for an absolutely flexible membrane

$$(4.3.13) \qquad \Delta w_0 + \Lambda_0 w_0 = 0; \qquad w_0 = 0, \qquad \text{for } x \in \Gamma.$$

Here Λ is an unknown parameter, proportional to the second power of the natural frequency (see (2.4.25)).

Let Λ_0 be a simple eigenvalue of problem (4.3.13) with eigenfunction w_0. Then the degeneracy of problem (4.3.12) into problem (4.3.13) is regular and

$$(4.3.14) \qquad \Lambda = \Lambda_0 + \mu\Lambda_1 + O\left(\mu^2\right).$$

Moreover,

$$(4.3.15) \qquad \Lambda_1 = \left(\iint_D w_0^2 dS\right)^{-1} \int_\Gamma \left(\frac{\partial w_0}{\partial n}\right)^2 ds$$

is obtained from the compatibility condition for the spectral boundary value problem for the function $w_1(x)$.

4.3.3. Applications to thin shells. The thin shell equations may be schematically represented in the form (4.3.1) with $m = p = 4$. The degeneracy of the complete problem A_μ into the momentless problem A_0 for shells is discussed in the survey [**69**].

The problem for regular or nonregular degeneracy is rather easily solved. Let Γ be the boundary of the domain D containing the neutral surface. We introduce the local coordinates s and n connected with the curve Γ and we let K_2 be the normal curvature of the surface in the direction Γ. Then the degeneracy is regular if

$$(4.3.16) \qquad K_2^2(s) - \Lambda > 0$$

at each points of Γ. Here Λ is connected with the natural frequency ω by formula (2.6.15).

In static problems ($\Lambda = 0$), the regularity of the degeneracy breaks down only when $K_2 = 0$, namely, at the asymptotic lines of shells of zero and negative Gaussian curvature. The construction of the solution in a neighbourhood of the edge of a shell of zero or negative Gaussian curvature, if this edge coincides with the asymptotic line, is carried out in [**177, 260**]. The construction is more complicated [**120**] if the edge is tangent to the asymptotic lines at some points.

As it was mentioned before (see Subsection 4.3.1), the substitution of problem A_0 for problem A_μ in the zeroth approximation requires that, besides a regular degeneracy, problem A_0 be uniquely solvable for any right-hand sides in (4.3.2). The conditions for which problem A_0 is uniquely solvable are not always satisfied in typical problems of the shell theory. Its resolution is connected with the sign of the Gaussian curvature of the neutral surface, the character of the boundary conditions, the presence of bendings or pseudo-bendings (see A. L. Gol′denveizer [**68, 69**]).

Pseudo-bendings are displacements which satisfy the boundary conditions and for which the tangential deformations of the neutral surface are relatively small compared with transverse deformations.

In thin shell dynamics, an analysis by the momentless theory (i.e. by the problem A_0) is even less possible than in statics. Above some frequency, the degeneracy of problem A_μ into problem A_0 ceases to be regular. Special difficulties occur [**69, 73, 111**] if the frequency, Λ, lies in the interval

$$\Lambda^- \le \Lambda \le \Lambda^+, \tag{4.3.17}$$

where

$$\Lambda^- = \min_{x_1, x_2, \theta} \{K\}, \qquad \Lambda^+ = \max_{x_1, x_2, \theta} \{K\}, \qquad K = k_1 \cos^2\theta + k_2 \sin^2\theta,$$

and k_1 and k_2 are the principal curvatures of the neutral surface.

The interval (4.3.17) coincides with the continuous spectrum of the homogeneous problem A_0. Therefore, *transition lines* occur in the domain of integration D. Across these lines, the type of the momentless operator L_m changes from elliptic to hyperbolic. These lines generalize the turning points in the two-dimensional case. The behaviour of solutions in neighbourhoods of the transition lines is studied in [**10, 33, 73**]. However, the complete asymptotic solution to the problem, unlike in the one-dimensional case, is not done. In Section 4.4, solutions localized in neighbourhoods of turning points and transition lines are constructed.

If $\Lambda > \Lambda^+$, there are no transition lines and it is possible to classify the free vibrations of a shell. We distinguish the *quasi-tangential* and the *quasi-transverse* vibrations according to the prevailing displacements (in the tangential or normal directions to the surface, respectively) (see [**73**]). If $\Lambda \gg 1$, the surface curvature does not essentially affect the vibrations and one can use simplified equations (similar to the equations of the tangential and transverse plate vibrations) to study these vibrations.

The classification of the free vibrations has been carried out by A. L. Radovinsky for shells containing a liquid [**168**], for shells in a magnetic field [**169**] and by N. N. Rogacheva for piezoceramic shells [**179**].

For shells oscillating in infinite liquid, the transition surfaces have been constructed by A. L. Popov and G. N. Chernychev [**164**].

4.3.4. Gol′denveizer's classification. A general method to classify the solutions of systems of partial differential equations was developed by A. L. Gol′denveizer *et al.* [**67, 73**] for thin shells as follows. The analysis of the problem leads us to pick out the unknowns and parameters which determine the type of solution. But it is desirable that the number of selected variables be as small as possible (for example, out of two variables with similar behaviour we select one). For example, for shell vibrations, we may pick out the variables

$$\Lambda, \quad w, \quad u, \quad \varepsilon, \quad t, \quad t_1, \quad \ldots, \tag{4.3.18}$$

which are respectively the frequency, the normal deflection, the tangential deflection, the tangential deformation, the indices of variation, etc. The orders of each of the variables (4.3.18) (except t and t_1) are compared with the main small parameter μ

$$\Lambda \sim \mu^a, \qquad w \sim \mu^b, \qquad u \sim \mu^c, \qquad \varepsilon \sim \mu^d, \qquad \ldots$$

Some of the indices (for example, a and b) are taken as arbitrary and we seek non-contradictory values for the remaining indices (c, d, t, t_1). The non-contradiction criterion is, as a rule, the equality of the orders of not less than two main terms in each of the equation of the system and the noncontradiction of the system itself when only the main terms are kept (see [**105**]). Depending upon which terms are the main ones, we get different simplified systems and correspondingly one or another character of the solution.

As a simple example we consider the system of equations (see (2.7.5))

$$\begin{aligned} &\mu^4\Delta^2 w - \Lambda w - \frac{\partial^2\Phi}{\partial x^2} = 0, \\ &\Delta^2\Phi + \frac{\partial^2 w}{\partial x^2} = 0, \qquad \Delta = \frac{\partial^2}{\partial x^2} + \frac{\partial^2}{\partial y^2}, \end{aligned} \tag{4.3.19}$$

which describes the quasi-transversal vibrations of a cylindrical shell. In deriving this equation, we have already neglected the tangential forces of inertia and some other terms; hence the analysis of this system gives only a partial representation of the vibrations of a cylindrical shell. In this case, however, it may be too difficult to consider the complete system (see [**73**]).

By eliminating Φ, we get an equation in w of the eighth order,

$$\mu^4\Delta^4 w - \Lambda\Delta^2 w + \frac{\partial^4 w}{\partial x^4} = 0. \tag{4.3.20}$$

We let

$$\Lambda \sim \mu^a, \qquad \frac{\partial}{\partial x} \sim \mu^{-t}, \qquad \frac{\partial}{\partial y} \sim \mu^{-t_1}, \tag{4.3.21}$$

and assume that $0 \le t < 2$ and $0 \le t_1 < 2$. If $t = 2$, the wave length of the deformation is commensurable with the shell thickness and system (4.3.19) is not valid any more.

When considering the orders of the terms in (4.3.20), it is convenient to emphasize two cases, namely $t_1 \ge t$ and $t_1 \le t$. In these cases the orders of the terms in (4.3.20) are, respectively, equal to

$$\begin{array}{llll} 4-8t_1 & a-4t_1 & -4t & (t_1 \ge t), \\ 4-8t & a-4t & -4t & (t_1 \le t). \end{array} \tag{4.3.22}$$

We provide the different values of the indices of variation t and t_1 and compare the orders (4.3.22) of the terms in (4.3.20).

As a result, the domain of the parameters t_1 and t splits into five subdomains (see Fig. 17), in each of which equation (4.3.20) may be replaced by a simplified equation,

$$\begin{aligned} &\text{(I)}\ -\Lambda\frac{\partial^4 w}{\partial y^4} + \frac{\partial^4 w}{\partial x^4} = 0,\ a = 4(t_1 - t), \\ &\text{(II)}\ \mu^4\frac{\partial^4 w}{\partial y^4} - \Lambda w = 0,\ a = 4(1 - t_1), \\ &\text{(III)}\ \mu^4\frac{\partial^4 w}{\partial x^4} - \Lambda w = 0,\ a = 4(1 - t), \\ &\text{(IV)}\ \mu^4\frac{\partial^8 w}{\partial x^8} - (\Lambda - 1)\frac{\partial^4 w}{\partial x^4} = 0,\ \Lambda - 1 \sim \mu^{a_1},\ a_1 = 4(1 - t), \end{aligned} \tag{4.3.23}$$

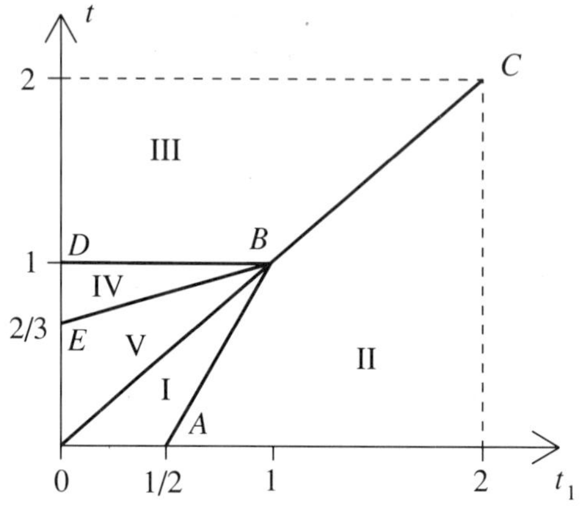

FIGURE 17. Splitting the (t, t_1)-plane.

$$\text{(V)} \quad -(\Lambda-1)\frac{\partial^4 w}{\partial x^4}+2\frac{\partial^4 w}{\partial x^2\partial y^2}=0,\ \Lambda-1\sim\mu^{a_1},\ a_1=2(t-t_1).$$

Equations (4.3.23) may be used in the open domains I–V. On the boundaries of the domains, the equations become more complex. For example, on the segment AB (excluding the point B)

$$\mu\frac{\partial^8 w}{\partial y^8}-\Lambda\frac{\partial^4 w}{\partial y^4}+\frac{\partial^4 w}{\partial x^4}=0, \qquad a=2-2t. \tag{4.3.24}$$

At the point B all terms of equation (4.3.20) are of the same order ($a=0$).

In the domains IV and V the values of Λ are close to 1. Indeed (see [**197, 223**]), the point $\Lambda=1$ is a point of accumulation of the vibration frequency of a cylindrical shell.

4.4. Local buckling and vibrations of shells

4.4.1. Convex shells. We use the system of equations (2.6.11) and (2.6.16) to study the buckling of the momentless initial stress state of a thin shell. If the coefficients of these equations are constant, the dents, under buckling, spread along the whole shell,

$$w=w_0e^{iz}, \qquad \Phi=\Phi_0e^{iz}, \qquad z=\frac{1}{\mu}(p_1\alpha_1+p_2\alpha_2), \tag{4.4.1}$$

where the wave numbers p_j satisfy the equation

$$\Lambda=H(p_j)=\frac{(q_1^2+q_2^2)^4+(k_2q_1^2+k_1q_2^2)^2}{(q_1^2+q_2^2)^2(t_1q_1^2+2t_3q_1q_2+t_2q_2^2)}, \tag{4.4.2}$$

with

$$q_j=p_jA_j^{-1}, \qquad k_j=R_j^{-1}.$$

We obtain the critical value Λ by minimizing the function H over all real p_j (see [**191**]).

If the coefficients k_i and t_i are variable, the buckling mode may be localized in neighbourhoods of some weakest points or lines on the neutral surface.

We, now, study a convex ($k_1k_2 > 0$) shells and seek its *weakest point*, $\alpha_i = \alpha_i^0$. We consider the function

$$\Lambda = H(p_i, \alpha_i), \tag{4.4.3}$$

where H is given by (4.4.2), and seek its minimum over the four variables p_i, α_i, $i = 1, 2$. Let its minimum, Λ_0, be at $p_i = p_i^0$, $\alpha_i = \alpha_i^0$.

If the point (α_1^0, α_2^0) is far from the shell edges and the 4×4 symmetric matrix

$$\begin{pmatrix} A & B \\ B^T & C \end{pmatrix}, \qquad A = \left\{ \frac{\partial^2 H}{\partial p_i^0 \partial p_j^0} \right\}, \; B = \left\{ \frac{\partial^2 H}{\partial p_i^0 \partial \alpha_j^0} \right\}, \; C = \left\{ \frac{\partial^2 H}{\partial \alpha_i^0 \partial \alpha_j^0} \right\}, \tag{4.4.4}$$

is positive definite, then the solution of system (2.7.2), which is localized in a neighbourhood of that point, has the form [**227, 238**]

$$\begin{aligned} w(\alpha_j, \mu) &\simeq \sum_{k=0}^{\infty} \mu^{k/2} w_k(\xi_j) \exp\left[i \left(\frac{\xi^T p^0}{\sqrt{\mu}} + \frac{1}{2} \xi^T S \xi \right) \right], \\ \Lambda &\simeq \sum_{k=0}^{\infty} \Lambda_k \mu^k, \end{aligned} \tag{4.4.5}$$

where

$$\xi_j = \frac{\alpha_j - \alpha_j^0}{\sqrt{\mu}}, \qquad \xi = \begin{bmatrix} \xi_1 \\ \xi_2 \end{bmatrix}, \qquad p^0 = \begin{bmatrix} p_1^0 \\ p_2^0 \end{bmatrix}.$$

The unknowns are the polynomials w_k in ξ_j, the corrections $\mu^k \Lambda_k$ to the critical load Λ_0 and the 2×2 constant matrix S with positive definite imaginary part.

Formula (4.4.5) generalizes formula (2.5.9) for the case of two independent variables. It is a sub-case of formula (4.1.2).

We do not discuss the determination of $w_k(\xi)$ and S, and note only that formula (4.4.5) determines a two-parameter family of close eigenvalues Λ, with common Λ_0, and

$$\begin{aligned} \Lambda_1 = \Lambda_1^{(m_1, m_2)} &= \frac{1}{d} \left[\left(\frac{1}{2} + m_1 \right) \omega_1 + \left(\frac{1}{2} + m_2 \right) \omega_2 \right], \\ & m_1, m_2 = 0, 1, 2, \ldots, \qquad \omega_i > 0, \qquad d > 0. \end{aligned} \tag{4.4.6}$$

Buckling occurs for the lowest eigenvalue corresponding to $m_1 = m_2 = 0$, and $w_0 \equiv 1$.

By separating the real and imaginary parts in (4.4.5), we find that each eigenvalue of (4.4.6) is asymptotically double. The buckling mode has the form

$$\begin{aligned} w &= \Re \left[\sum_{k=0}^{\infty} \mu^{k/2} w_k(\xi_j) e^{iz} \right] e^{-\xi^T S_2 \xi / 2}, \\ z &= \frac{\xi^T p^0}{\sqrt{\mu}} + \frac{\xi^T S_1 \xi}{2} + \theta, \qquad S = S_1 + i S_2. \end{aligned} \tag{4.4.7}$$

This method does not determine the phase θ, which takes one of the two values $0 \le \theta_1, \theta_2 < 2\pi$. The dents have an elongated form and fill the ellipse $\xi^T S_2 \xi = C$.

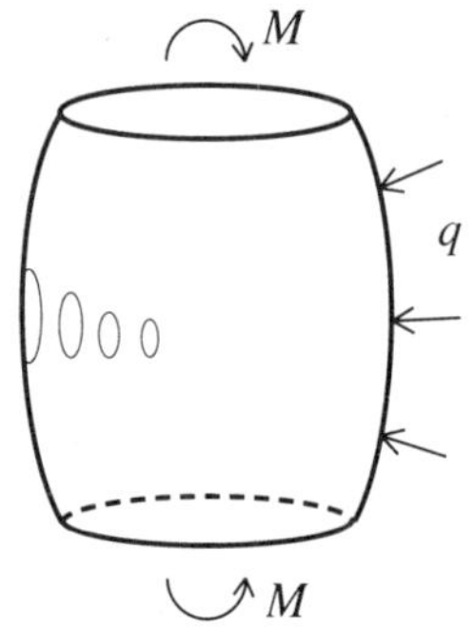

FIGURE 18. Buckling mode of an ellipsoid of rotation under combined loading.

This method has been used by G. I. Mikhasev [**132**] for the construction of the buckling mode of an ellipsoid of rotation under a combined loading (see Fig. 18).

4.4.2. Weakest point on the boundary. If the weakest point occurs on the boundary Γ of the domain occupied by a shell, the construction of the buckling mode becomes more complex, since the boundary conditions have to be satisfied. In this case, the buckling mode is represented as a sum of functions with different asymptotic expansions [**128**].

For shells of zero and negative Gaussian curvature, buckling modes localized at some points are not typical since perturbations of such shells have the tendency to spread along the characteristics. (Shells of negative, zero or positive Gaussian curvature have, respectively two, one or no systems of real characteristics.) The buckling problem of a cylindrical shell of varying thickness under non-uniform axial compression is an exception, for, in this problem, buckling modes are localized in the neighbourhoods of some points [**231**].

For shells of revolution of negative Gaussian curvature, localized buckling modes in a neighbourhood of the characteristics may occur only under non-axisymmetric loading. One such problem has been considered by A. L. Maiboroda in [**127**].

Localized buckling and vibration modes for shells of zero Gaussian curvature are considered below.

4.4.3. Conical and cylindrical shells. The free vibrations and buckling of a conical shell are described by the system of equations (2.6.11) and (2.6.16). We introduce coordinates, s and φ, on the neutral surface of a shell in such a way that

$$A_1 = 1, \qquad A_2 = s, \qquad k_1 = 0, \qquad k_2 = \frac{1}{s}k(\varphi),$$

where $s = s^0/R$ and s^0 is the distance between a point and the top of the cone.

We suppose that the shell is closed in the circumferential direction and four homogeneous boundary conditions are given at each of the two edges,

$$s = s_1(\varphi), \qquad s = s_2(\varphi), \qquad 0 \le \varphi \le \varphi_1, \tag{4.4.8}$$

which limit the shell.

We represent the approximate solution of system (2.6.11) as a sum of the main semi-momentless solution and the edge effect near the edges (4.4.8). The main solution is sought in the form [**228–230**]:

$$w(s,\varphi,\mu_1) = \sum_{k=0}^{\infty} \mu_1^{k/2} w_k(s,\xi) \exp\left[i\left(\frac{p_0\xi}{\sqrt{\mu_1}} + \frac{a\xi^2}{2}\right)\right], \tag{4.4.9}$$
$$\Lambda = \Lambda_0 + \mu_1\Lambda_1 + \mu_1^2\Lambda_2 + \ldots,$$

where $\mu_1^2 = \mu$ is a small parameter, $\xi = (\varphi - \varphi_0)/\sqrt{\mu_1}$, $w_k(s,\xi)$ are polynomials in ξ, $\Im p_0 = 0$ and $\Im a > 0$.

The coefficients, w_k, of the asymptotic expansion (4.4.9) depend on the parameter s. This is the only difference between this expansion and expansion (2.5.9). The solution (4.4.9) decreases away from the weakest generatrix $\varphi = \varphi_0$.

In the zeroth approximation, we get

$$\frac{k^2(\varphi)}{p_0^4}\frac{\partial^2}{\partial s^2}\left(s^3\frac{\partial^2 w_0}{\partial s^2}\right) + \frac{p_0^4}{s^3}w_0 + \Lambda_0 N w_0 = 0. \tag{4.4.10}$$

We have $Nw_0 = -t_2p_0^2s^{-1}w_0$ for the buckling problem and $Nw_0 = -sw_0$ for the free vibrations problem. The order of equation (4.4.10) makes it possible to satisfy only two boundary conditions at each edge (4.4.8). The choice of the boundary conditions, for equation (4.4.10), which provide a regular degeneracy of the boundary value problem, is considered in [**234**]. The remaining two boundary conditions at each edge are satisfied by the choice of the edge effect solutions. We note the analogy between this problem and the one-dimensional problem (see equation (3.3.12)).

The eigenvalue Λ_0 is a function of p_0 and φ_0, i.e. $\Lambda_0 = f(p_0,\varphi_0)$. The values of the parameters p_0 and φ_0 are obtained from the conditions

$$\frac{\partial f}{\partial p_0} = 0, \qquad \frac{\partial f}{\partial \varphi_0} = 0. \tag{4.4.11}$$

The parameter a is a root of the quadratic equation

$$a^2\frac{\partial^2 f}{\partial p_0^2} + 2a\frac{\partial^2 f}{\partial p_0\partial\varphi_0} + \frac{\partial^2 f}{\partial\varphi_0^2} = 0.$$

The compatibility conditions for the second approximation yields

$$w_0^{(m,n)}(s,\xi) = w_0^{(n)}(s)H_m(\xi),$$
$$\Lambda_1^{(m,n)} = -i\left(m+\frac{1}{2}\right)\left(a^{(n)}\frac{\partial^2 f^{(n)}}{\partial p_0^2} + \frac{\partial^2 f^{(n)}}{\partial p_0\partial\varphi_0}\right),$$
$$m = 0,1,2,\ldots, \qquad n = 1,2,3,\ldots,$$

where $H_m(\xi)$ is Hermite's polynomial of the m-th degree, $w_0^{(n)}$ is the eigenfunction of the boundary value problem for equation (4.4.10) corresponding to the eigenvalue $\Lambda_0^{(n)}$. The eigenvalues of the problem under consideration,

$$\Lambda^{(m,n)} = \Lambda_0^{(n)} + \mu_1\Lambda_1^{(m,n)} + \ldots,$$

are asymptotically double.

This asymptotic method may be used to study the vibrations and buckling of a cylindrical shell. In this case, the factor s in equation (4.4.10) has to be replaced by 1. The solutions of some problems of cylindrical shells may be obtained explicitly.

As an example, we consider the buckling of a freely supported circular cylindrical shell with a slanted edge under an uniform external pressure [**228**]. In this case,

$$A_1 = A_2 = k_2 = t_2 = 1.$$

The boundary conditions for equation (4.4.10) have to be taken as

$$w_0 = w_0'' = 0, \qquad \text{for } s = s_1,\ s = s_2. \tag{4.4.12}$$

Let $s_1(\varphi) = 0$ and $s_2(\varphi) = l + \tan\beta\cos\varphi$, where l is the shell length and $\beta \neq 0$ is the angle of the slope of the edge. If $\beta = 0$, both edges of the shell are straight and $s_2(\varphi) = l$. Then the solution of the boundary value problem (4.4.10), (4.4.12) is

$$w_0^{(n)} = \sin\frac{\pi n s}{s_2(\varphi)}, \qquad f^{(n)} = p_0^2 + \frac{\pi^4 n^4}{s_2^4(\varphi) p_0^6}, \qquad n = 1, 2, \ldots$$

In the applications, the lowest eigenvalue, corresponding to $n = 1$, of the buckling problem is especially important. Solving system (4.4.11) for $n = 1$, we get

$$\varphi_0 = 0, \qquad p_0^2 = 3^{1/4}\frac{\pi}{l_0}, \qquad l_0 = s_2(0).$$

Therefore, the weakest generatrix is the longest generatrix $\varphi = 0$. The zeroth approximation to the eigenvalue $\Lambda_0^{(1)} = 4\pi/(3^{3/4} l_0)$ gives the critical load by Southwell–Papkovich's formula [**201**] for a shell of length l_0 and with straight edges.

It is easy to show that

$$\frac{\partial^2 f^{(1)}}{\partial p_0^2} = 16, \qquad \frac{\partial^2 f^{(1)}}{\partial p_0 \partial q_0} = 0, \qquad \frac{\partial^2 f^{(1)}}{\partial \varphi_0^2} = \frac{4\pi\tan\beta}{3^{3/4} l_0}.$$

If the effect of the slanted edge is taken into account, then the correction to the first approximation is determined by the following formula:

$$\Lambda_1^{(m,1)} = \left(\frac{1}{2} + m\right)\sqrt{\frac{\partial^2 f^{(1)}}{\partial p_0^2}\,\frac{\partial^2 f^{(1)}}{\partial \varphi_0^2}} = 4\left(\frac{1}{2} + m\right)\sqrt{\frac{\Lambda_0^{(1)}\tan\beta}{l_0}},$$

$$m = 0, 1, 2, \ldots$$

To construct the next approximations it is necessary to use the system of equations (2.6.14) which is more precise than system (2.6.11). This question is considered in the papers by S. B. Filippov [**58, 59**] of the present collection.

In [**233**], a modification of this asymptotic method is presented which, in some cases, is more convenient than the method discussed above. The main solution is sought in the form

$$w(s, \varphi, \mu_1) = \sum_{k=0}^{\infty} \mu_1^k w_k(s, \varphi) \exp\left[\frac{i}{\mu_1}\int_{\varphi_0}^{\varphi} p(\varphi)d\varphi\right]. \tag{4.4.13}$$

Despite (4.4.9), in formula (4.4.13) we use series expansions in integer power of μ_1. This algorithm is described in details in the report [**19**] of this collection.

Localized vibrations and buckling modes near the generators of conical and cylindrical shells can be caused by changes in the geometrical and physical shell parameters, by a slanted edge and by a nonhomogeneous distribution of the initial stresses. The solutions of different buckling problems for shells of zero Gaussian curvature by means of the asymptotic methods described here may be found in [**128, 131–133, 207, 208**].

In [**136, 137**], vibrations and buckling of shells close to conical and cylindrical shells are analyzed. For such shells, the solutions may be represented in the form (4.4.13) if the deviation of the neutral surface from the ideal one is small (i.e. if it is of the order μ) and varies sufficiently smoothly.

4.5. Asymptotics in narrow domains for thin shells and beams

4.5.1. The equilibrium equation. There is an intensive development of asymptotic integration methods connected, to a marked degree, with the applications to thin beams and shells. Following G. Kirchhoff [**95**], the hypotheses on the distribution of the displacements and stresses as functions of the thickness have been used to derive the two-dimensional equilibrium equations for shells and plates and the one-dimensional equilibrium equations for beams. To estimate the errors in these hypotheses and in the ensuing equations, it is necessary to compare their solutions with the solutions of the equations of the three-dimensional theory of elasticity. However, only a few of the three-dimensional problems have exact solutions. Therefore, the construction of asymptotic expansions may render, here, an invaluable service.

A small parameter, which results from scaling, is used in the asymptotic expansion. This expansion is achieved, as in Section 4.2, by expressing the solution as a sum of two functions obtained by the main and the additional iterative processes, respectively. The main iterative process provides the solution far away from the edges of the domain and, for a shell or a plate, the three-dimensional problem reduces to a two-dimensional one. The additional iterative process allows us to satisfy the boundary conditions at the edge and to correctly formulate the boundary conditions for two-dimensional problems, which come as a consequence of the main process.

A complete consideration of the problems of asymptotic integration of three-dimensional equations for beams and shells is very complex. Therefore, we limit our consideration to two of the simplest problems to illustrate the method of asymptotic integration in narrow domains. In the last two Subsections 4.5.4 and 4.5.5, we briefly review the literature.

4.5.2. A plane elasticity problem. We consider a plane problem of the theory of elasticity in a narrow rectangular domain,

$$\begin{gathered}
\frac{\partial \sigma_x}{\partial x} + \frac{\partial \tau}{\partial y} = 0, \qquad \frac{\partial \tau}{\partial x} + \frac{\partial \sigma_y}{\partial y} = 0, \\
0 \le x \le l, \qquad 0 \le y \le h, \qquad \frac{h}{l} = \mu, \\
\sigma_x = K(\varepsilon_x + \nu\varepsilon_y), \qquad \sigma_y = K(\varepsilon_y + \nu\varepsilon_x), \\
\tau_{xy} = G\omega, \qquad K = \frac{E}{1-\nu^2}, \qquad G = \frac{E}{2(1+\nu)}, \\
\varepsilon_x = \frac{\partial u}{\partial x}, \qquad \varepsilon_y = \frac{\partial v}{\partial y}, \qquad \omega = \frac{\partial u}{\partial y} + \frac{\partial v}{\partial x},
\end{gathered} \tag{4.5.1}$$

where u and v are the displacement projections, E and G are the elastic moduli, ε_x, ε_y and ω are the deformations, σ_x, σ_y and τ are the stresses.

Let the edge $y = 0$ be clamped and the stresses on the remaining sides be given by

$$\begin{aligned} & u = v = 0, && \text{for } y = 0, \\ \sigma_y = \sigma_y^*, \quad & \tau = \tau^*, && \text{for } y = h, \\ \sigma_x = \sigma_x^{*0}, \quad & \tau = \tau^{*0}, && \text{for } x = 0, \\ \sigma_x = \sigma_x^{*l}, \quad & \tau = \tau^{*l}, && \text{for } x = l. \end{aligned} \tag{4.5.2}$$

We assume that the functions on the right-hand sides of (4.5.2) are infinitely differentiable.

We start by considering the main iterative process. If we make the substitutions

$$x = l\xi, \qquad y = h\eta, \tag{4.5.3}$$

then the terms of system (4.5.1) containing $\partial/\partial y$ are multiplied by μ^{-1}. We may seek the solution of the transformed system in the form

$$z = \mu^{\alpha_z} \sum_{k=0}^{\infty} \mu^k z_k(\xi, \eta), \tag{4.5.4}$$

where $z_k(\xi, \eta)$ are polynomials in η with coefficients expressed as derivatives of the functions $\sigma_y^*(\xi)$ and $\tau^*(\xi)$. Here the letter z replaces any of the unknown in (4.5.1) and the exponent α_z in the main term is equal to

$$\begin{aligned} \alpha_z &= 0, \qquad \text{if } z = (\sigma_x, \sigma_y, \tau, \varepsilon_y, \omega), \\ \alpha_z &= 1, \qquad \text{if } z = (u, v, \varepsilon_x). \end{aligned} \tag{4.5.5}$$

At every step of the main iterative process the unknown functions are uniquely determined.

In general, the solution (4.5.4) does not satisfy the boundary conditions (4.5.2) at $x = 0$ and $x = l$.

We now consider the additional iterative process in a neighbourhood of the edge $x = 0$. We make the substitutions

$$x = h\zeta, \qquad y = h\eta. \tag{4.5.6}$$

In constructing the solutions π of the boundary layer we have a problem of type (4.5.1) in the domain $0 \le \zeta \le \mu^{-1}$, $0 \le \eta \le 1$. Besides the homogeneous boundary conditions

$$u = v = \sigma_y = \tau = 0$$

on the edges $\eta = 0$ and $\eta = 1$, we require that the given series in powers of μ at the edge $\zeta = 0$,

$$\sigma_x = \sigma_x^\pi(\eta, \mu), \qquad \tau = \tau^\pi(\eta, \mu), \tag{4.5.7}$$

be such that the sum $z + \pi$ satisfies condition (4.5.2) at $\eta = 0$. We neglect the interference from the edges $x = 0$ and $x = l$ by assuming that the solution decreases at the opposite edge and we introduce the conditions

$$\sigma_x^\pi, \sigma_y^\pi, \tau^\pi, u^\pi, v^\pi \to 0, \qquad \text{as } \zeta \to \infty. \tag{4.5.8}$$

This problem has a unique solution for any functions σ_x^{*0} and τ^{*0}.

For plates made of *low-compressible* material, if the displacements are given on both edges, $y = 0$ and $y = h$, the asymptotics of the solutions differ from (4.5.4) [**123**].

4.5.3. Problem (4.5.1) **with different boundary conditions.** We consider the same problem (4.5.1), but with different boundary conditions. We suppose that the stresses are given at all four edges. This problem splits into the sum of a symmetric and an antisymmetric problems, the symmetry and antisymmetry being with respect to the neutral surface.

We consider an antisymmetric problem of beam bending. We take the boundary conditions in the form

$$\sigma_y = \pm\sigma_y^*, \qquad \tau = \tau^*, \qquad \text{at } y = \pm\frac{h}{2}. \tag{4.5.9}$$

We require that the functions in the boundary conditions at $x = 0$ and $x = l$ satisfy the corresponding evenness condition. The forces acting on the beam are assumed to be self-balanced.

We consider the main iterative process. We scale the problem (4.5.3) and seek the solution in the form (4.5.4); however the orders α_z of the unknown functions take different values:

$$\begin{aligned} \alpha_z &= -3, && \text{if } z = (v), \\ \alpha_z &= -2, && \text{if } z = (\sigma_x, u, \varepsilon_x, \varepsilon_y), \\ \alpha_z &= -1, && \text{if } z = (\tau, \omega), \\ \alpha_z &= \ \ 0, && \text{if } z = (\sigma_y). \end{aligned} \tag{4.5.10}$$

Four iterations are required only to construct the main terms in (4.5.4). To have easier transformations, the stresses are found first. To reach this aim, we introduce the equations of stress compatibility, which, together with the equilibrium equations, form a complete system,

$$\begin{aligned} \frac{\partial^2\sigma_x}{\partial\eta^2} - 2\mu\frac{\partial^2\tau}{\partial\xi\partial\eta} + \mu^2\frac{\partial^2\sigma_y}{\partial\xi^2} &= 0, \\ \frac{\partial\sigma_y}{\partial\eta} + \mu\frac{\partial\tau}{\partial\xi} &= 0, \\ \frac{\partial\tau}{\partial\eta} + \mu\frac{\partial\sigma_x}{\partial\xi} &= 0, \end{aligned} \tag{4.5.11}$$

which may be integrated without determining the displacements. We find the first terms in (4.5.4),

$$\begin{aligned} &\sigma_x = c(\xi)\eta, && \tau = \mu\left(\frac{1}{8} - \frac{\eta^2}{2}\right)\frac{\partial c}{\partial\xi}, \\ &\sigma_y = \mu^2\left(\frac{\eta^3}{6} - \frac{\eta}{8}\right)\frac{\partial^2 c}{\partial\xi^2}, && \frac{\partial^2 c}{\partial\xi^2} = -\frac{24}{\mu^2}\sigma_y^*. \end{aligned} \tag{4.5.12}$$

Now, from system (4.5.1) we find the top order terms for the displacements

$$\frac{\partial u}{\partial\xi} = \frac{l}{E}c(\xi)\eta, \qquad \frac{\partial^2 v}{\partial\xi^2} = -\frac{1}{\mu}\frac{\partial^2 u}{\partial\eta\partial\xi} = -\frac{lc(\xi)}{E\mu}, \tag{4.5.13}$$

from which we get the equation for the beam deflection,

$$(4.5.14)\qquad \frac{\partial^4 v}{\partial \xi^4} = \frac{24l}{E\mu^3}\sigma_y^* \qquad \text{or} \qquad \frac{Eh^3}{12}\frac{\partial^4 v}{\partial x^4} = 2\sigma_y^*.$$

The tangential forces τ^* from (4.5.9) do not enter the constructed approximation. They may be taken into account only in the next approximation. For equation (4.5.14) two boundary conditions have to be introduced at each of the edges $\xi = 0$ and $\xi = 1$. These conditions may be found from the additional iterative process.

After scaling (4.5.6), in the zeroth approximation for the additional iterative process, we have the following problem (at the edge $\zeta = 0$):

$$(4.5.15)\qquad \begin{gathered} \frac{\partial \sigma_x}{\partial \zeta} + \frac{\partial \tau}{\partial \eta} = 0, \qquad \frac{\partial \tau}{\partial \zeta} + \frac{\partial \sigma_y}{\partial \eta} = 0, \\ \sigma_x = \sigma_x^* - c(0)\eta, \qquad \tau = \tau^* - \mu\left(\frac{1}{8} - \frac{\eta^2}{2}\right)\left(\frac{\partial c}{\partial \xi}\right)_0, \qquad \zeta = 0, \\ \sigma_x, \sigma_y, \tau \to 0, \qquad \text{as } \zeta \to \infty. \end{gathered}$$

The solution of this problem is damped as $\zeta \to \infty$ only if the load at the edge $\zeta = 0$ is self-balanced. Load self-balance is achieved by an appropriate choice of $c(0)$ and $(\frac{\partial c}{\partial \zeta})_0$, i.e. by introducing appropriate boundary conditions for equation (4.5.14).

4.5.4. A brief review of beams. We briefly review the works on beams and shells by starting with thin beams. The foundations of the engineering one-dimensional theories of thin beams have been laid by B. de Saint-Venant [**184**], G. Kirchhoff [**94, 95**] and A. Clebsch [**36**]. In these and in most of the later works devoted to the derivation of the one-dimensional equations, the same hypothesis about the distributions of the displacements and stresses along the beam thickness has been used. The easiest model, due to Kirchhoff and Clebsch, considers only bending and twisting deformations. In more complex models (S. P. Timoshenko [**210**]) shift deformations and elongation of the neutral axis are also taken into account.

In very few works (compared with the two-dimensional shell theories), for example [**31, 35, 44, 63, 82, 153, 161**], the one-dimensional beam equations are derived from the three-dimensional theory of elasticity. In the first approximation one gets the Kirchhoff–Clebsch equations. The next approximations refine the models.

Thin-wall beams are interesting objects for asymptotic analysis. They occupy an intermediate position between regular beams and shells. To develop the engineering theory of thin-wall beams, an additional degree of freedom, i.e. the deplanation of the cross-section, is introduced. In deriving the one-dimensional equation for thin-wall beams by the asymptotic approach, one has to solve the problem with two small geometric parameters. One of them is equal to the ratio of the largest diameter of the cross-section to the beam length and the second is equal to the ratio of the largest to the smallest diameters of the cross-section. The papers [**45, 162**] are devoted to the asymptotic studies of thin-wall beams.

4.5.5. A brief review of shell theory. The study of the equations of the theory of shells and plates, based on the asymptotic integration of the three-dimensional equations of elasticity theory, are developed by A. L. Gol′denveizer [**67**] and his pupils [**78, 98, 178**]. We also note the references [**3, 21, 60**]. A detailed review of this area may be found in [**69**]. We discuss some results.

In problems of static shell theory, the strain-stress state is split into the main state and the boundary layer. The solution of the three-dimensional equations of the theory of elasticity in the domain containing the shell with appropriate boundary conditions is called *the strain-stress state.* The main strain-stress state spreads through the interior of the shell and is determined by the main iterative process. In the zeroth approximation, this process leads to the two-dimensional Kirchhoff–Love equations. The boundary layer is localized near the boundary Γ of the domain G containing the shell (and also possibly near other lines of distortion, for example, concentrated loads and breaks in the surface geometry). At distances of the order of the shell thickness, h, the boundary layer disappears. The boundary layer is determined by the additional iterative process.

The asymptotic integration of the equations of the theory of elasticity allows one to answer the question about the order of the error, Δ, of two-dimensional shell theories. The first result in this direction has been obtained by V. V. Novozhilov and R. M. Finkel′shtein [**145**], i.e.

$$\Delta \sim h_*, \tag{4.5.16}$$

where h_* is the relative shell thickness ($h_* = h/R$). Later A. L. Gol′denveizer [**67**] (see also [**96**]) has shown that Δ depends on the index of variation of the stress state. If t is the index of variation calculated with respect to the parameter h_* (i.e. $\partial\Phi/\partial\alpha \sim h_*^{-t}\Phi$), then

$$\Delta \sim \max\left\{h_*, h_*^{2-2t}\right\}, \tag{4.5.17}$$

i.e., for $t = 1$, the possibilities of the two-dimensional theory to describe the stress state appear to be exhausted.

The estimate (4.5.17) is concerned with the internal stress state. Two facts may increase the error Δ, that is, the boundary layer and the small difference between the deformation and the bending of the neutral surface (for details, see [**69**]).

For stationary dynamical problems of plates and shells, the estimate (4.5.17) needs to be made more precise by taking into account the effect of the vibration frequency. The papers [**72, 85, 86**] are devoted to this problem. For plates, the question about the two-dimensional dynamical equations is discussed by G. I. Petrashen′ in [**154**]. The question of the error of the two-dimensional equations for shells of simple geometry may be solved by comparison with the exact solutions of the three-dimensional problems of elasticity theory [**130**].

We note the recently published works [**4, 71, 159, 250, 270**] in which the area of application of different two-dimensional plates and shells theories is discussed.

CHAPTER 5

Asymptotic Solutions of Nonlinear Differential Equations

There is a large variety of asymptotic expansions of solutions of nonlinear differential equations; see, for example, the monographs by N. N. Bogolyubov and Yu. A. Mitropolsky [**25, 26**], J. D. Cole and J. Kevorkian [**37**], K. V. Kholshevnikov [**92**], N. M. Krylov and N. N. Bogoluybov [**104**], I. G. Malkin [**122**], V. P. Maslov [**124, 126**], E. F. Mishchenko and N. Kh. Rosov [**134**], A. H. Nayfeh [**139, 140**], I. M. Rapoport [**170**], M. Van Dyke [**249**], A. B. Vasil'eva and V. F. Butuzov [**253, 254**] and W. Wasov [**266**]. We shall limit ourselves to singular perturbation and eigenvalue problems which have applications to nonlinear problems of the theory of shells.

5.1. The Cauchy problem for ordinary differential equations

5.1.1. A degenerate differential-algebraic system. We consider the following Cauchy problem for a system of $m+n$ nonlinear differential equations:

$$\begin{aligned} &\frac{dy}{dt} = f(y,z,t), \qquad y = (y_1,\dots,y_m)^T, \\ &\mu\frac{dz}{dt} = F(y,z,t), \qquad z = (z_1,\dots,z_n)^T, \\ &y(0) = y^0, \qquad z(0) = z^0, \qquad 0 \le t \le T, \end{aligned} \tag{5.1.1}$$

which, as $\mu \to 0$, degenerates into a system of m differential and n algebraic equations,

$$\frac{dy_0}{dt} = f(y_0, z_0, t), \qquad F(y_0,z_0,t) = 0, \qquad y_0(0) = y^0. \tag{5.1.2}$$

Here y and f are m-vectors and z and F are n-vectors.

Suppose that the algebraic system

$$F(y_0,z_0,t) = 0 \tag{5.1.3}$$

of (5.1.2) has an isolated solution,

$$z_0 = z_0(y_0,t). \tag{5.1.4}$$

If we substitute this solution into the differential system of (5.1.2), we obtain a degenerate problem,

$$\frac{dy_0}{dt} = f\left(y_0, z_0(y_0,t), t\right), \qquad y_0(0) = y^0. \tag{5.1.5}$$

The degeneracy of problem (5.1.1) into problem (5.1.5) raises two questions:

(a) What is the relation between the solutions of these problems as $\mu \to 0$?
(b) What is the analytical structure of the solution of problem (5.1.1) for $\mu > 0$?

By Tikhonov's theorem [**206**] the answer to question (a), under some conditions (mainly on the functions f and F in (5.1.1)), is as follows:

$$\begin{aligned} &y(t,\mu) \to y_0(t), && \text{as } \mu \to 0, && 0 \le t \le T, \\ &z(t,\mu) \to z_0(t) = z_0(y_0(t),t), && \text{as } \mu \to 0, && \varepsilon \le t \le T. \end{aligned} \tag{5.1.6}$$

Naturally, in general, the function $z_0(t)$ does not satisfy the boundary condition of (5.1.1).

The solution of problem (5.1.1) in the form of asymptotic series in powers of μ is constructed in the monograph by A. B. Vasil′eva and V. F. Butusov [**253, 254**] and is briefly described below.

Let x replace any of the letters y or z. We seek the solution $x(t,\mu)$ of problem (5.1.1) in the form

$$x(t,\mu) \simeq \sum_{k=0}^{\infty} x_k(t)\mu^k + \sum_{k=0}^{\infty} X_k(\tau)\mu^k, \tag{5.1.7}$$

where $\tau = t/\mu$ is a fast variable and the functions $X_k(\tau)$ satisfy the condition

$$X_k(\tau) \to 0, \qquad \text{as } \tau \to \infty, \tag{5.1.8}$$

i.e. $X_k(\tau)$ is a function of the boundary layer.

If the series (5.1.7) is to be an asymptotic solution of problem (5.1.1), it needs to satisfy the following conditions:

(1) the function (5.1.4) is an isolated solution of system (5.1.3);
(2) the right-hand sides $F(y,z,t)$ and $f(y,z,t)$ are analytic functions of their arguments;
(3) the solution (5.1.4) is an asymptotically stable stationary point of the adjoint system:

$$\frac{dz}{d\tau} = F(y,z,t), \tag{5.1.9}$$

and all the roots of the additional characteristic equation,

$$\det(A - \lambda I) = 0, \qquad A = \left(\frac{\partial F}{\partial z}\right)\Big|_{z=z_0(y_0,t)}, \tag{5.1.10}$$

have negative real parts;
(4) the initial point z^0 belongs to the domain of attraction of the stable solution (5.1.4).

Note that in (5.1.9) y and t are considered as parameters.

Condition (3) is called the stability condition from the right and, in this case, the root $z(y,t)$ is called stable from the right. Similarly, if all roots of equation (5.1.10) have positive real parts, we speak of stability from the left.

Conditions (1)–(4) may be weakened, if we wish to obtain only the relation (5.1.6).

The construction of the subsequent terms in series (5.1.7) is remarkably similar to Vishik–Lyusternik's algorithm (see Section 3.1). The zeroth approximation to

$y_0(t)$ is determined by (5.1.5) and $Y_0(\tau) \equiv 0$. The function $z_0(t)$ is determined by formula (5.1.4). The construction of $Z_0(\tau)$ leads to the Cauchy problem

$$\frac{dZ_0}{d\tau} = F\left(y^0, z_0(0) + Z_0(\tau), 0\right) - F\left(y^0, z_0(0), 0\right), \qquad Z_0(0) = z^0 - z_0(0), \tag{5.1.11}$$

the solution of which, by conditions (3) and (4) above, satisfies (5.1.8).

The subsequent terms of the series (5.1.7) are determined from linear equations in the following order, Y_1, y_1, z_1, Z_1, Y_2, ... The functions Y_1 are determined by quadratures,

$$Y_1(\tau) = \int_\infty^\tau \left[f\left(y^0, z_0(0) + Z_0(\widetilde{\tau}), 0\right) - f\left(y^0, z_0(0), 0\right)\right] d\widetilde{\tau}. \tag{5.1.12}$$

To find y_1, z_1 we solve the Cauchy problem

$$\frac{dy_1}{dt} = \left(\frac{\partial f}{\partial y}\right)_0 y_1 + \left(\frac{\partial f}{\partial z}\right)_0 z_1, \qquad y_1(0) = -Y_1, \tag{5.1.13}$$

$$\frac{dz_0}{dt} = \left(\frac{\partial F}{\partial y}\right)_0 y_1 + \left(\frac{\partial F}{\partial z}\right)_0 z_1, \tag{5.1.14}$$

where the subscript $(\cdot)_0$ means that the corresponding derivatives are evaluated on the solution (5.1.4). By condition (4), the determinant of the matrix $A = (\partial F/\partial z)_0$ is nonzero and, from (5.1.14), we find z_1, which we then substitute into (5.1.13).

To construct the subsequent approximations we transform the right-hand sides of equations (5.1.1) into

$$F(x, t) = F(\hat{x}, t) + F^*(\widehat{X}, \tau), \tag{5.1.15}$$

where $\hat{x}$ and $\widehat{X}$ denote the first and the second sums, respectively, on the right-hand side of (5.1.7). The expression (5.1.15), which determines the function F^*,

$$F^*(\widehat{X}, \tau) = F\left(\hat{x}(\mu\tau) + \widehat{X}(\tau), \mu\tau\right) - F\left(\hat{x}(\mu\tau), \mu\tau\right), \tag{5.1.16}$$

has already been used in the derivation of (5.1.11) and (5.1.12).

The equation for $Z_1(\tau)$ is rather complex:

$$\begin{aligned} \frac{dZ_1}{d\tau} &= \left(\frac{\partial F}{\partial y}\right)_{00} \left(Y_1(\tau) + y_1(0) + \tau y_0'(0)\right) \\ &\quad + \left(\frac{\partial F}{\partial z}\right)_{00} \left(Z_1(\tau) + z_1(0) + \tau z_0'(0)\right) \\ &\quad + \tau \left(\frac{\partial F}{\partial \tau}\right)_{00} - \left(\frac{\partial F}{\partial y}\right)_0 \left(y_1(0) + \tau y_0'(0)\right) \\ &\quad - \left(\frac{\partial F}{\partial z}\right) \left(z_1(0) + \tau z_0'(0)\right) - \tau \left(\frac{\partial F}{\partial \tau}\right)_0, \\ Z_1(0) &= -\, z_1(0), \end{aligned} \tag{5.1.17}$$

where the double subscript $(\cdot)_{00}$ means that the derivative is calculated at $y = y^0$, $z = z_0(0) + Z_0(\tau)$ and $t = 0$.

5.1.2. Exact solution. It is proved, in [**253, 254**], that the formal series (5.1.7) are the asymptotic expansions of the exact solutions as $\mu \to 0$ ($\mu > 0$).

Problem (5.1.1) can also be generalized to the case where the right-hand sides of system (5.1.1) depend regularly on μ.

5.2. Degeneracy of nonlinear boundary value problems

5.2.1. The formulation of nonlinear boundary value problems. When considering boundary value problems, we shall limit ourselves to the cases where the solution can be represented in the form of a sum of slowly varying functions and functions of boundary layer type. Besides the boundary layer in the neighbourhoods of the endpoints of the interval of integration, an *interior boundary layer* may appear.

We consider the boundary value problem

$$\frac{dy}{dt} = f(y, z, t), \qquad \mu \frac{dz}{dt} = F(y, z, t), \qquad t \in [0, 1], \tag{5.2.1}$$

where y and z are m- and n-dimensional vectors, respectively. Suppose that $m + n$ general type boundary conditions are given,

$$l_i\left(y(0), y(1), z(0), z(1)\right) = 0, \qquad i = 1, \dots, m + n. \tag{5.2.2}$$

Although the degenerate system is of the same form as (5.1.2), the question of the corresponding boundary conditions is non trivial.

Let $z_0(y_0, t)$ be an isolated solution of equation (5.1.3). In a neighbourhood of this solution, we consider the adjoint system (5.1.9), in which y and t are parameters. For any t and y_0 in some domain, suppose that equation (5.1.10) has n_1 roots with negative real parts and n_2 roots with positive real parts ($n_1 + n_2 = n$). In this case, we seek the solution of problem (5.2.1) in the form (5.1.7), but with two boundary layers,

$$x(t, \mu) = \sum_{k=0}^{\infty} x_k(t)\mu^k + \sum_{k=0}^{\infty} X_k^{(0)}(\tau_0)\mu^k + \sum_{k=0}^{\infty} X_k^{(1)}(\tau_1)\mu^k, \tag{5.2.3}$$

$$\begin{gathered} X_k^{(0)}(\tau_0) \to 0, \text{ as } \tau_0 \to \infty, \qquad X_k^{(1)}(\tau_1) \to 0, \text{ as } \tau_1 \to -\infty, \\ \tau_0 = \frac{t}{\mu}, \qquad \tau_1 = \frac{t-1}{\mu}. \end{gathered} \tag{5.2.4}$$

We limit our discussion to the questions of the boundary conditions for the degenerate problem (5.1.2). In the zeroth approximation, conditions (5.2.2) can be written in the form

$$l_i\left(y_0(0), y_0(1), z_0(0) + Z_0^{(0)}(0), z_0(1) + Z_0^{(1)}(0)\right) = 0. \tag{5.2.5}$$

By (5.2.4), the family of functions Z_0^0 $\left(Z_0^{(1)}\right)$ contains n_1 (n_2) constants. The elimination of these functions from conditions (5.2.5) leads to m relations, which do not contain functions of the boundary layer and which are the boundary conditions for the degenerate problem (5.1.2).

A rather complex example has been considered by P. E. Tovstik [**240**] in a paper in this volume. In Section 5.2.3 below, we shall consider pedagogical examples, as discussed in the book by A. B. Vasil′eva and V. F. Butusov [**253**].

5.2.2. A boundary value problem with non-unique solution. We consider the boundary value problem

$$(5.2.6)\qquad \begin{aligned} &\frac{dy}{dt} = z,\\ &\mu\frac{dz}{dt} = -(z-y-1)(z-y)(z-y+1) =: F(y,z,t),\\ &y(1) = 0, \qquad z(0) = z^0, \end{aligned}$$

where y and z are scalar functions. The equation $F(y,z,t) = 0$ has, in this case, three isolated solutions,

$$(1): \quad z = y+1, \qquad (2): \quad z = y, \qquad (3): \quad z = y-1,$$

and each of these may give a solution of the boundary value problem (5.2.6). These solutions correspond to the slowly varying functions of the zeroth approximation and to the adjoint equations (5.1.9), i.e.

$$(5.2.7)\qquad \begin{array}{llll} (1) & y_0 = e^{t-1} - 1, & z_0 = e^{t-1}, & \dfrac{dZ}{d\tau} = -Z(Z+1)(Z+2),\\ (2) & y_0 = 0, & z_0 = 0, & \dfrac{dZ}{d\tau} = -(Z-1)Z(Z+1),\\ (3) & y_0 = 1 - e^{t-1}, & z_0 = -e^{t-1}, & \dfrac{dZ}{d\tau} = -(Z-2)(Z-1)Z. \end{array}$$

More specifically, we have:

(1) The root $z = y+1$ of equation $F = 0$ is stable from the right and the domain of attraction of the root $Z = 0$ of system (5.2.7) is $-1 < Z(0) < \infty$ or, by (5.1.11), z^0 satisfies the inequalities

$$(5.2.8)\qquad -1 + e^{-1} < z^0 < \infty,$$

and the solution of problem (5.2.6) converges to y_0 and z_0 (the case (a)), as $\mu \to 0$, if $t \geq \varepsilon > 0$.

(2) The root $z = y$ is stable from the left; thus the boundary value problem (5.2.6) has no solution of type (5.2.3), convergent to $y_0 = z_0 \equiv 0$ as $\mu \to 0$ (if $z^0 \neq 0$).

(3) The root $z = y - 1$ is stable from the right with the following domain of attraction:

$$(5.2.9)\qquad -\infty < z^0 < 1 - e^{-1}.$$

Summing up the different cases, we see that, for $-\infty < z^0 < -1+e^{-1}$, problem (5.2.6) has only one solution convergent to the functions y_0 and z_0 as $\mu \to 0$ (see (5.2.7), case (3) and Fig. 19a). For $-1+e^{-1} < z^0 < 1-e^{-1}$, problem (5.2.6) has two solutions, which correspond to the cases (1) and (3), respectively (see Fig. 19b). For $1 - e^{-1} < z^0 < \infty$, we again get only one solution which corresponds to the case (1) (see Fig. 19c).

5.2.3. An example of interior boundary layer. We now consider an example of interior boundary layer (a more detailed discussion is found in [**253**]),

$$(5.2.10)\qquad \frac{dy}{dx} = z, \qquad \mu\frac{dz}{dt} = 1 - z^2, \qquad y(0) = y(1) = 0.$$

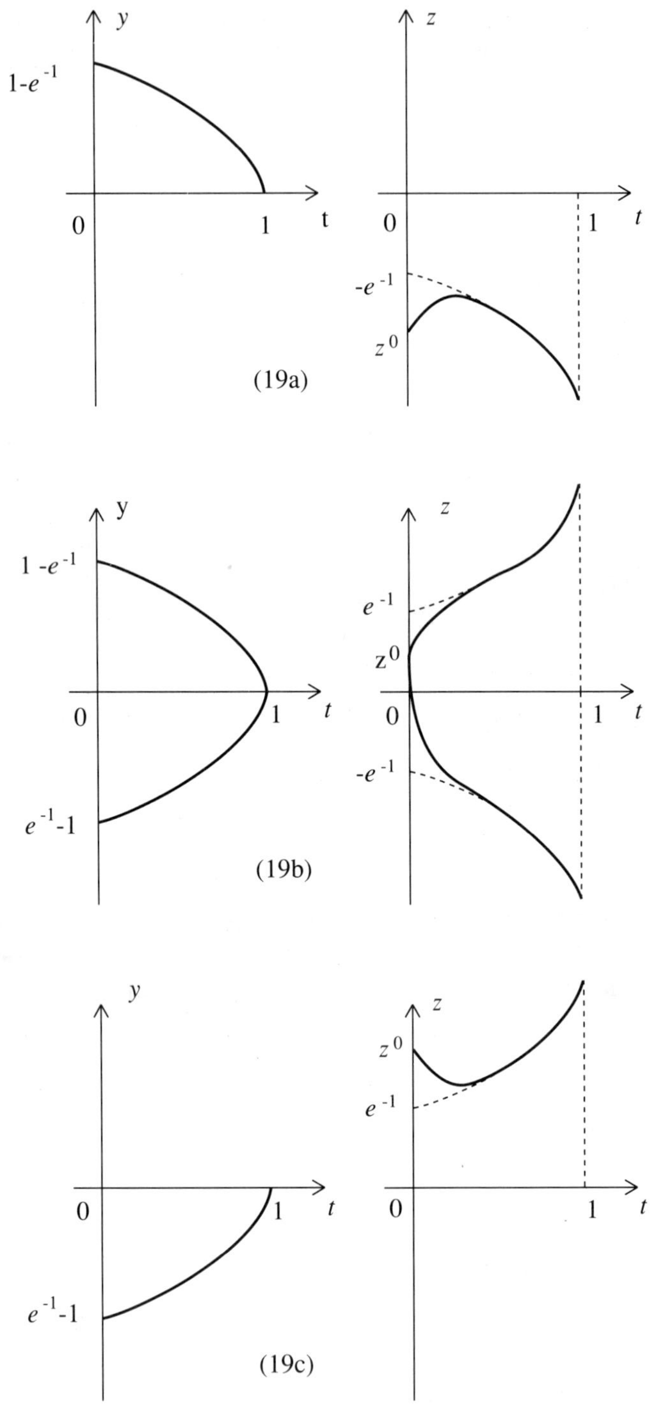

FIGURE 19. The effect of the initial conditions on the solutions.

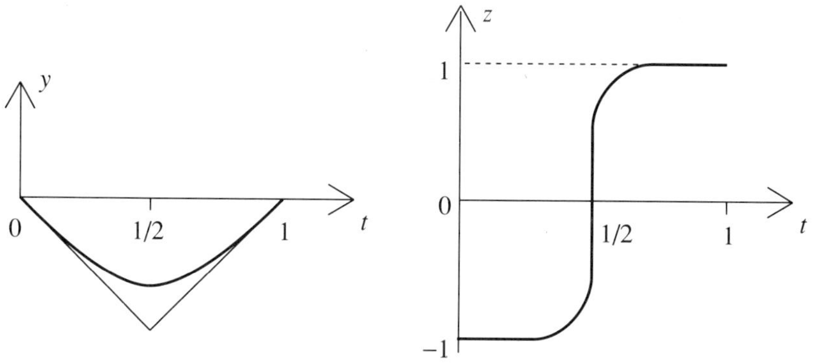

FIGURE 20. Example of interior boundary layer.

The exact solution of (5.2.10) is easily written as

$$y = \ln\left[\cosh\left(\frac{2t-1}{2\mu}\right)\Big/\cosh\left(\frac{1}{2\mu}\right)\right], \qquad z = \tanh\left(\frac{2t-1}{2\mu}\right). \tag{5.2.11}$$

The asymptotic approach considered here shows that the equation $F = 0$ has two solutions: (1) $z = -1$ and (2) $z = 1$; the first one is stable from the left and the second is stable from the right. Since both boundary conditions in (5.2.10) are introduced for the function y, both solutions of the equation $F = 0$,

$$\begin{aligned} &(1) \quad z = -1, \quad y = -t + C, \quad C = 0, \quad t < \frac{1}{2}, \\ &(2) \quad z = 1, \quad y = t + C, \quad C = -1, \quad t > \frac{1}{2}, \end{aligned} \tag{5.2.12}$$

enter in the construction of the solution of problem (5.2.10). The straight lines $y = -t$ and $y = t - 1$ intersect at the interior point

$$(t, y) = \left(\frac{1}{2}, -\frac{1}{2}\right).$$

In a neighbourhood of this point, there is an internal boundary layer which smoothens the jumps of the function (5.2.12) at $t = 1/2$ (compare with the exact solution (5.2.11) and see Fig. 20).

5.3. The Liapunov–Schmidt perturbation method

5.3.1. The mathematical bifurcation problem. At the beginning of this century, A. M. Liapunov [**110**] and E. Schmidt [**186**] laid the foundation of the bifurcation theory for nonlinear equations. They reduced the problem of determining the number and behaviour of solutions of integral or differential equations to the problem of determining the number and behaviour of roots for systems of implicit analytic functions.

Liapunov–Schmidt's method, its different versions and its development in recent work is considered in detail in the paper by M. S. Berger [**23**] and in the monograph by M. M. Vainberg and V. A. Trenogin [**248**].

The idea of this method is to split the system of equations into two, one in a finite n-dimensional subspace, and the other in its infinite-dimensional orthogonal complement. Thus, the initial problem reduces to a system of n equations in n unknowns (the so called *bifurcation equations*).

Mathematically, the bifurcation problem for the solutions of a nonlinear equation may be represented in the following form. Let $G(U, \Lambda)$ be a nonlinear operator which maps the direct sum, $E \oplus E_0$, of two linear spaces, E and E_0 ($U \in E$, $\Lambda \in E_0$), to a linear space, E_1. The problem can be split into three parts [**23**]:

(1) the existence of a Λ (for example, λ_0), at which there is a jump in the number of solutions of the equation $G(U, \Lambda) = 0$;
(2) the determination of the number of *small-norm* solutions in a neighbourhood of $\Lambda = \lambda_0$;
(3) the study of the behaviour of the small-norm solutions in a neighbourhood of $\Lambda = \lambda_0$.

If it is known that, for $\Lambda = \lambda_0$, the equation $G(U, \lambda_0) = 0$ has the solution $U = U_0$, then, by the shifts $\lambda_0 + \lambda = \Lambda$ and $U_0 + u = U$ and the substitution

$$\Phi(u, \lambda) = G(U_0 + u, \lambda_0 + \lambda),$$

the problem is reduced to the determination of the solutions which branch off the zero solution (since $\Phi(0, 0) = 0$) for

$$\Phi(u, \lambda) = 0. \tag{5.3.1}$$

5.3.2. The bifurcation equations. Let the operator $\Phi(u, \lambda)$ be continuously differentiable with respect to u in a neighbourhood of the origin and this derivative defines the linear operator

$$B = -\Phi_u(0, 0).$$

Then equation (5.3.1) can be written in the form

$$Bu = R(u, \lambda), \tag{5.3.2}$$

where

$$R(u, \lambda) := \Phi(u, \lambda) - \Phi_u(0, 0)u. \tag{5.3.3}$$

An element $\varphi \in E$ is called a zero of the linear operator B if it is a solution of the equation $Bu = 0$.

The zeros of B form a closed linear set, i.e. a subspace in E, called the *null space*, or *kernel*, of B and is denoted by $N(B)$. The dimension of $N(B)$, called the *nullity* of B, is equal to the number of independent zeros of B.

If B has no zeros, then the operator B^{-1} exists and is bounded ([**248**]), and the equation

$$Bu = h, \tag{5.3.4}$$

has a unique solution for every $h \in E_1$.

We recall some definitions.

The space E^*, *dual* to E, is the space of continuous linear functionals on E.

The linear operator B is called a *normally solvable operator*, if equation (5.3.4) is solvable for any $h \in E_1$, or if in the space E_1^* there is a set $N^*(B)$ of functionals ψ such that, in order that equation (5.3.4) be solvable, it is necessary and sufficient that, for every ψ from $N^*(B)$,

$$(h, \psi) = 0,$$

where (h, ψ) is the value of the linear functional ψ on the element h.

A normally solvable operator is called a *Fredholm operator* if the cardinality of the sets $N(B)$ and $N^*(B)$ is finite and equal.

We now assume that the operator B is a Fredholm operator and the nullity, n, of B is at least one, $n \geq 1$. Let $\{\varphi_1, \dots, \varphi_n\}$ be a basis of the subspace $N(B)$ and $\{\psi_1, \dots, \psi_n\}$ be a basis of the subspace $N^*(B)$.

Then the general solution of the homogeneous equation $Bu = 0$ can be written in the form $u = \sum_{i=1}^n \xi_i \varphi_i$, were ξ_i are arbitrary numbers and the solvability conditions of the nonhomogeneous equation (5.3.4) are

$$(h, \psi_i) = 0, \qquad i = 1, \dots, n.$$

We introduce the *projector*, P, of E on $N(B)$,

$$Pu = \sum_{i=1}^{n} (u, \gamma_i)\varphi_i, \qquad \forall x \in E, \tag{5.3.5}$$

where $\gamma_1, \dots, \gamma_n$ are functionals in E^* such that

$$(\varphi_i, \gamma_j) = \delta_{ij}, \qquad i, j = 1, \dots, n.$$

(In the general case, the $\gamma_1, \dots, \gamma_n$ may distinct from the $\psi_1, \dots, \psi_n$.)

The operator P reduces the space E into the direct sum

$$E = N(B) \oplus N(B)^{\perp},$$

where $N(B)^{\perp}$ consists of the elements orthogonal to $N(B)$, i.e.

$$(u, \gamma_i) = 0, \qquad i = 1, \dots, n, \qquad \forall u \in E.$$

We introduce the projector Q into E_1, defined by

$$Qh = \sum_{k=1}^{n} (h, \psi_k) z_k, \qquad \forall h \in E_1, \tag{5.3.6}$$

where the $z_1, \dots, z_n$ are elements of E_1 such that

$$(z_k, \psi_l) = \delta_{kl}, \qquad k, l = 1, \dots, n.$$

The operator Q reduces the space E_1 into the direct sum of subspaces,

$$E_1 = E_{1,n} \oplus E_{1,n}^{\perp}.$$

If $E = E_1$, we note that, the equality $N(B) = E_{1,n}$ is true only if the determinant of the matrix $\| (\varphi_i, \varphi_j) \|$ does not vanish, det $\| (\varphi_i, \psi_j) \| \neq 0$. In this case,

$$(\varphi_i, \psi_j) = \delta_{ij}, \qquad i, j = 1, \dots, n,$$

and

$$Pu = Qu = \sum_{k=1}^{n} (u, \psi_k) \varphi_k.$$

Returning to equation (5.3.2), we assume that $u = x + y$, where

$$x = (I - P)u \in N(B)^{\perp},$$
$$y = Pu \in N(B),$$

and project equation (5.3.2) onto $E_{1,n}$ and $E_{1,n}^{\perp}$:

$$\begin{aligned} \widehat{B}x &= (I - Q)R(x + y, \lambda) \\ 0 &= QR(x + y, \lambda). \end{aligned} \tag{5.3.7}$$

This system is equivalent to equation (5.3.2). Here $\widehat{B}$ is the restriction of B to $N(B)^{\perp}$ and Q is the projector defined by (5.3.6).

The first equation of (5.3.7) can be simplified by taking into account the second equation, i.e.

$$\widehat{B}x = R(x + y, \lambda). \tag{5.3.8}$$

Since $\widehat{B}^{-1}$ exists in a sufficiently small neighbourhood of the origin, $(x, y, \lambda) = (0, 0, 0)$, then, by the implicit function theorem (see [**248**]), there exists a unique continuous solution of equation (5.3.8), $x = x(y, \lambda)$, such that

$$x(0, 0) = 0.$$

Substituting this solution into the second equation of (5.3.7) we get

$$QR(x(y, \lambda) + y, \lambda) = 0. \tag{5.3.9}$$

This equation, which is usually called *Liapunov–Schmidt's equation*, can be written as a system of n scalar equations in n scalar unknowns.

Since $y \in N(B)$ is uniquely represented in the form

$$y = \sum_{i=1}^{n} \xi_i \varphi_i,$$

i.e.

$$x(y, \lambda) \equiv x(\xi_1, \dots, \xi_n, \lambda),$$

then, taking into account the relations (5.3.6), we can write equation (5.3.9) in the following form:

$$\begin{aligned} L^{(k)}(\xi_1, \dots, \xi_n, \lambda) &:= \left(R\left(x(\xi_1, \dots, \xi_n, \lambda) + \sum_{i=1}^{n} \xi_i \varphi_i, \lambda \right), \psi_k \right) \\ &= 0, \qquad k = 1, \dots, n. \end{aligned} \tag{5.3.10}$$

Equations (5.3.10) are called the *bifurcation equations*. In this case, the functions $L^{(k)}(\xi_1, \dots, \xi_n, \lambda)$ are continuous for all sufficiently small $\xi_1, \dots, \xi_n, \lambda$, and

$$L^{(k)}(0, \dots, 0, 0) = 0, \qquad k = 1, \dots, n.$$

The advantage of the bifurcation equation (5.3.10) is that the number of their small solutions is equal to the number of small solutions of equation (5.3.2) [**248**].

5.3.3. Schmidt's operator. To obtain the bifurcation equations (5.3.10) in another way, we introduce the operator $\widetilde{B}$, connected with the operator B by the formula

$$\widetilde{B}u = Bu + \sum_{i=1}^{n} (u, \gamma_i) z_i. \tag{5.3.11}$$

The operator $\widetilde{B}$ is called a *Schmidt operator* and it extends the operator $\hat{B}$ to E.

By the generalized Schmidt lemma [**248**], the operator $\widetilde{B}^{-1}$ exists and is bounded. We write $\widetilde{B}^{-1} = \Gamma$. If we take relations (5.3.11) into account, equation (5.3.2) may be represented in the form

$$\widetilde{B}u = R(u, \lambda) + \sum_{i=1}^{n} \xi_i z_i, \tag{5.3.12}$$

where

$$\xi_k = (u, \gamma_k), \qquad k = 1, \dots, n. \tag{5.3.13}$$

After the change of variable

$$u = x + \sum_{i=1}^{n} \xi_i \varphi_i,$$

x can be determined from

$$\widetilde{B}x = R(x + \sum_{i=1}^{n} \xi_i \varphi_i, \lambda).$$

(From (5.3.11), $\widetilde{B}\varphi_i = z_i$, and $\Gamma z_i = \varphi_i$, $i = 1, \dots, n$). The last equation has a unique solution x for all small $\lambda, \xi_1, \dots, \xi_n$, and, hence, equation (5.3.12) has a unique small solution

$$g = x(\xi_1, \dots, \xi_n, \lambda) + \sum_{i=1}^{n} \xi_i \varphi_i.$$

But this solution must satisfy equation (5.3.13).

Since $(\varphi_i, \gamma_j) = \delta_{ij}$, we obtain

$$(x(\xi_1, \dots, \xi_n, \lambda), \gamma_k) = 0, \qquad k = 1, \dots, n, \tag{5.3.14}$$

for the determination of $\xi_1, \dots, \xi_n$. This system is also the bifurcation equations. Since, by definition, the dual operator B^* maps E^* to E_1^*, i.e. $(x, B^*\psi) = (Bx, \psi)$, then we can get $\gamma_k = (\widetilde{B})^* \psi_k$. Hence, equations (5.3.14) are of the form

$$\left(\widetilde{B}x(\xi_1, \dots, \xi_n, \lambda), \psi_k\right) = 0.$$

If we take into account (5.3.12), we see that this equation is the same as equation (5.3.10).

5.3.4. Analytic bifurcation equations. If the operator $\Phi(u, \lambda)$ is analytic in some neighbourhood of the origin, then the functions $L^{(k)}(\xi_1, \dots, \xi_n, \lambda)$ are analytic in all their variables and the bifurcation equation can be written in the form [**248**]

$$\sum_{i_1 + \cdots + i_n \geq 2} L^{(k)}_{i_1 \dots i_n l 0}(\lambda) \xi_1^{i_1} \dots \xi_n^{i_n} + \sum_{i_1 + \cdots + i_n \geq 0} \sum_{l=1}^{\infty} L^{(k)}_{i_1 \dots i_n l}(\lambda) \xi_1^{i_1} \dots \xi_n^{i_n} = 0, \qquad k = 1, \dots, n, \tag{5.3.15}$$

where the coefficients, $L^{(k)}_{i_1 \dots i_n l}(\lambda)$, of the bifurcation equation are power functionals in λ of the order l, for $l \geq 1$, and scalars for $l = 0$.

If λ is a scalar parameter, then

$$L_{i_1\dots i_{nl}}(\lambda) = L_{i_1\dots i_{nl}}\lambda^l$$

and the bifurcation equation (5.4.9) has the form

$$\sum_{i_1+\dots+i_n\geq 2} L^{(k)}_{i_1\dots i_{n0}}\xi_1^{i_1}\dots\xi_n^{i_n} + \sum_{i_1+\dots+i_n\geq 0}\xi_1^{i_1}\dots\xi_n^{i_n}\sum_{l=1}^{\infty} L^{(k)}_{i_1\dots i_{nl}}\lambda^l = 0, \qquad k=1,\dots,n. \tag{5.3.16}$$

If the nullity of B is equal to one, we have a one-dimensional bifurcation. In this case, equation (5.3.2) may be written as

$$Bu = F_{01}\lambda + \sum_{i+j\geq 2} F_{ij}u^i\lambda^j, \tag{5.3.17}$$

and the system of bifurcation equations (5.3.16) transforms into the equation

$$\sum_{k=2}^{\infty} L_{k0}\xi^k + \sum_{k=0}^{\infty}\xi^k\sum_{l=1}^{\infty} L_{kl}\lambda^l = 0. \tag{5.3.18}$$

5.3.5. Recursive construction of the solution. The Liapunov–Schmidt method allows us not only to determine the number of solutions which branch off the main one, but also to construct these solutions in a neighbourhood of the bifurcation point.

It is easiest to determine the solution in the case of a one-dimensional bifurcation, for then the initial equation can be written as

$$\widetilde{B}u = F_{01}\lambda + \sum_{i+j\geq 2} F_{ij}u^i\lambda^j + \xi z, \tag{5.3.19}$$

where

$$\xi = (u, \gamma)$$

and

$$F_{ij} = \frac{1}{i!j!}\,\frac{\partial^{i+j}\Phi(0,0)}{\partial u^i\partial\lambda^j}.$$

We look for a small solution, $u = u(\xi,\lambda)$, of equation (5.3.19) in the form of a series,

$$u = \sum_{i=1}^{\infty} x_{i0}\xi^i + \sum_{i=0}^{\infty}\xi^i\sum_{j=1}^{\infty} x_{ij}\lambda^j. \tag{5.3.20}$$

The substitution of this series into (5.3.19) yields the bifurcation equation. Thus

$$L_{ij} = (x_{ij}, \gamma). \tag{5.3.21}$$

Substituting the series (5.3.20) into (5.3.19) and equating the coefficients of identical powers of λ and ξ we get a recurrent system for the determination of x_{ij}:

$$\begin{aligned}
\widetilde{B}x_{01} &= F_{01},\\
\widetilde{B}x_{02} &= F_{02} + 2F_{11}x_{01} + F_{20}x_{01}^2,\\
&\vdots\\
\widetilde{B}x_{10} &= z,\\
\widetilde{B}x_{20} &= F_{20}x_{10}^2,\\
&\vdots\\
\widetilde{B}x_{11} &= 2F_{11}x_{10} + 2F_{20}x_{10}x_{01},\\
\widetilde{B}x_{21} &= 2F_{11}x_{20} + 2F_{20}x_{01}x_{20} + 3F_{21}x_{10}^2 + 3F_{30}x_{01}x_{10}^2,\\
&\vdots
\end{aligned} \tag{5.3.22}$$

From (5.3.22) we can subsequently determine the x_{ij}:

$$x_{ij} = \Gamma\widetilde{F}_{ij}, \tag{5.3.23}$$

where $\widetilde{F}_{ij}$ are the right-hand sides in (5.3.22). By relation (5.3.21) and the fact that $\Gamma^*\gamma = \psi$, we can obtain the coefficients of the bifurcation equation:

$$\begin{aligned}
L_{01} &= (F_{01}, \psi),\\
L_{02} &= (F_{02} + 2F_{11}(\Gamma F_{01}) + F_{20}(\Gamma F_{01})^2, \psi),\\
&\vdots\\
L_{20} &= (F_{20}\varphi^2, \psi),\\
&\vdots\\
L_{11} &= (2F_{11}\varphi + 2F_{20}\varphi(\Gamma F_{01}), \psi),\\
&\vdots
\end{aligned} \tag{5.3.24}$$

If the nullity, n, of B is at least 2 ($n \geq 2$), then similar transformations can be made. In this case, equation (5.3.12) is equivalent to the system

$$\begin{aligned}
u &= a_{0\ldots01}\lambda + \sum_{i=1}^{n}\xi_i\varphi_i + \sum_{k_1+\cdots+k_n+k\geq 2} a_{k_1\ldots k_n k}\xi_1^{k_1}\ldots\xi_n^{k_n}\lambda^k,\\
L_i(\xi_1,\ldots,\xi_n,\lambda) &:= \sum_{k_1+\cdots+k_n\geq 2} L^{(i)}_{k_1\ldots k_n 0}\xi_1^{k_1}\ldots\xi_n^{k_n}\\
&\quad + \sum_{k_1+\cdots+k_n\geq 0}\xi_1^{k}\ldots\xi_n^{k_n}\sum_{k\geq 1} L^{(i)}_{k_1\ldots k_n k}\lambda^k\\
&= 0,
\end{aligned} \tag{5.3.25}$$

where $\{\varphi_1,\ldots,\varphi_n\}$ is a basis of the null subspace $N(B)$.

5.3.6. An example. Let us apply the above theory to the bending of a uniform bar, with ends fixed, under an axial load. The angle $\theta(x)$ of the tangent to the bar satisfies the following equilibrium equation:

$$\theta_{xx} + \Lambda \sin\theta = 0, \qquad 0 \le x \le 1, \tag{5.3.26}$$

with the boundary conditions:

$$\theta_x(0) = \theta_x(1) = 0.$$

The value of Λ in (5.3.26) is proportional to the axial load.

The function $\theta_0(x) \equiv 0$ is a solution of equation (5.3.26) for every Λ. The linearized problem near $\theta_0(x)$ has the form

$$\theta_{xx} + \Lambda\theta = 0, \qquad \theta_x(0) = \theta_x(1) = 0. \tag{5.3.27}$$

This problem has the eigenvalues $\Lambda_i = (\pi n)^2$, and the corresponding eigenfunctions $\theta(x) = \cos(\pi n x)$. We consider the solution of the problem in a neighbourhood of the first eigenvalue $\Lambda = \pi^2$. Let $\Lambda = \pi^2 + \lambda$. We represent equation (5.3.26) (similar to (5.3.17)) in the form

$$\begin{aligned} B\theta &= \theta_{xx} + \pi^2\theta \\ &= -\theta\lambda - (\pi^2 + \lambda)\sum_{i=2}^{\infty} \theta^{2i-1}\frac{(-1)^{i+1}}{(2i-1)!} := h(x). \end{aligned} \tag{5.3.28}$$

This equation has a solution if

$$\int_0^1 h(x)\cos(\pi x)\,dx = 0.$$

In this case, $E = E_1$, $n = 1$, $\varphi = \cos(\pi x)$, and the functional γ coincide with the functional ψ. It then follows from the condition $(\varphi, \gamma) = 1$ that

$$(\theta, \gamma) = \int_0^1 2\theta\cos(\pi x)\,dx.$$

We look for the solution θ of equation (5.3.28) in the form

$$\theta = \xi\cos(\pi x) + \sum_{i=1}^{\infty}\xi^{2i+1}\theta_{2i+1,0} + \sum_{i=0}^{\infty}\xi^{2i+1}\sum_{j=1}^{\infty}\theta_{2i+1,j}\lambda^j, \qquad \xi \ll 1. \tag{5.3.29}$$

This solution will be expressed as a series in odd power of ξ, because the operator B is an odd function of ψ. We substitute the series (5.3.29) into system (5.3.30) which is equivalent to (5.3.28),

$$\begin{aligned} \widetilde{B}\theta &= \theta_{xx} + \pi^2\theta + \left(\int_0^1 2\theta\cos(\pi x)\,dx\right)\cos(\pi x) \\ &= -\theta\lambda - (\pi^2 + \lambda)\sum_{i=2}^{\infty}\theta^{2i-1}\frac{(-1)^{i+1}}{(2i-1)!} + \xi\cos(\pi x), \end{aligned} \tag{5.3.30}$$

where

$$\xi = \int_0^1 2\theta\cos(\pi x)\,dx.$$

Finally, we equate the coefficients with equal powers of λ and ξ.

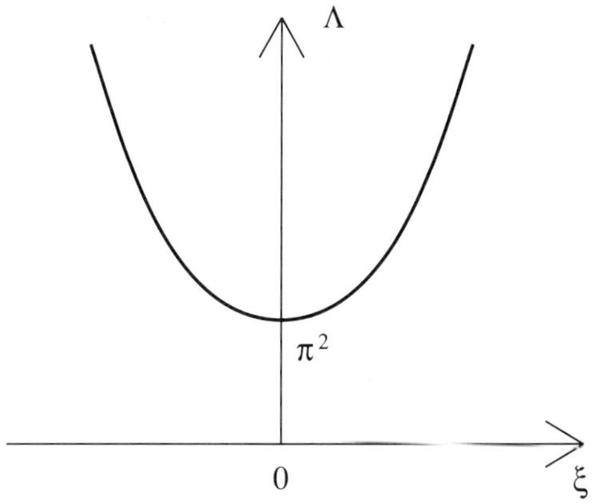

FIGURE 21. A bifurcation phenomenon.

The first equations of system (5.3.22) for the determination θ_{ij} is

$$\widetilde{B}\theta_{3,0} = \frac{\pi^2}{6}\cos^3(\pi x), \qquad \widetilde{B}\theta_{1,1} = -\cos(\pi x).$$

The solution of this system is

$$\theta_{3,0} = -\frac{\cos(3\pi x)}{24\times 8} + \frac{\pi^2}{8}\cos(\pi x), \qquad \theta_{1,1} = -\cos(\pi x).$$

We list the first coefficients of the bifurcation equations (like (5.3.24)):

$$L_{30} = (\theta_{3,0}, \gamma) = \int_0^1 2\theta_{3,0}\cos(\pi x)\,dx = \frac{\pi^2}{8},$$

$$\vdots$$

$$L_{11} = (\theta_{1,1}, \gamma) = \int_0^1 2\theta_{1,1}\cos(\pi x)\,dx = -1.$$

$$\vdots$$

Hence, we see that the bifurcation equations are

$$\frac{\pi^2}{8}\xi^3 - \lambda\xi + \lambda^2\xi + \lambda\xi^3\left(\frac{1}{8} - \frac{\pi^2}{2}\right) + \xi^5\frac{3\pi^2}{64}\left(\pi^2 - \frac{1}{8}\right) + \cdots = 0. \tag{5.3.31}$$

The turning angle of the bar under a load closed to the critical load is

$$\theta \approx \xi\cos(\pi x) - \frac{\xi^3}{24\times 8}\cos(3\pi x) + \left(\xi^3\frac{\pi^2}{8} - \xi\lambda\right)\cos(\pi x) + \dots,$$

or, by taking into account the bifurcation equation, we have

$$\theta \approx \xi\cos(\pi x) - \frac{\xi^3}{24\times 8}\cos(3\pi x) + \dots$$

The dependence of ξ on Λ in a neighbourhood of the critical value $\Lambda = \pi^2$ is represented in Fig. 21.

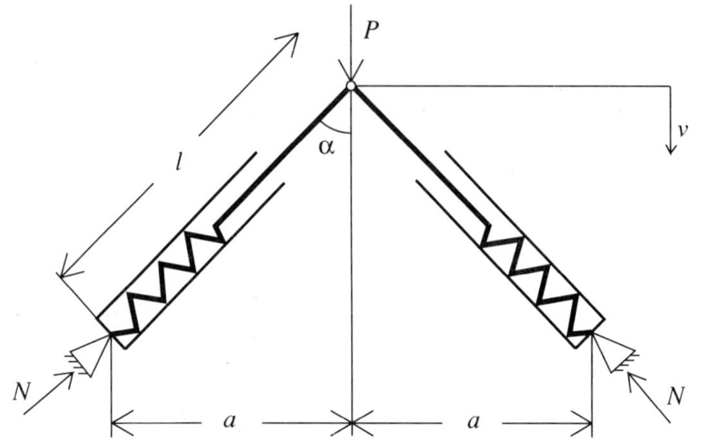

FIGURE 22. A von Mises truss.

It follows from equation (5.3.31), that, for small λ, there are three solutions of equation (5.3.26), i.e. three equilibrium states.

5.4 Nonlinear shell buckling

5.4.1. Limit and bifurcation points. In the stability theory of elastic conservative systems, we distinguish two types of loss of stability (see [**205**]).

The first type of loss of stability, according to Poincaré's classification, is connected with the notion of *limit points*, i.e. points of strict maximum on the load-deflection curves (or equilibrium state curves). These curves are the locus of points each characterizing a particular state of equilibrium in the (f, P)-plane, where f is the deflection of a representative point of a system and P is the force that causes the deflection. This loss of stability is usually called *clicking.*

An example of a construction which looses stability by clicking is the von Mises truss. A simplified version of this structure is represented in Fig. 22.

We suppose that the bars are absolutely rigid and the springs attached to the bars have stiffness c. When $P = 0$, the angle $\alpha = \alpha_0$ and $l = l_0$. We consider only the symmetric state equilibrium. From the equilibrium condition of the hinge,

$$N = P/2\cos\alpha,$$

and the deformation of any one of the springs satisfies the following relation:

$$N = c\frac{\Delta l}{l_0} = c\left(1 - \frac{\sin\alpha_0}{\sin\alpha}\right).$$

Equating the right-hand sides of these two relations, we can find the dependence of the load P on the value of the angle α,

$$P = 2c\left(1 - \frac{\sin\alpha_0}{\sin\alpha}\right)\cos\alpha.$$

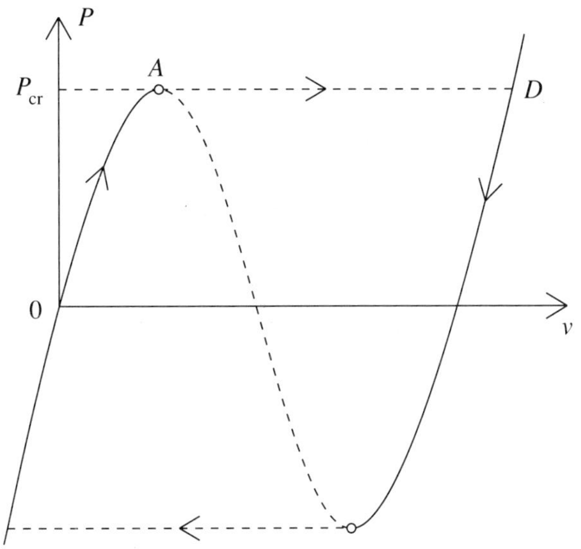

FIGURE 23. Effect of loading on the displacement for von Mises truss.

If we replace the angle α by a linear displacement v, then

$$P = 2c\left(1 - \frac{v}{a}\tan\alpha_0\right)\left[\left(\sqrt{\tan^2\alpha_0 + \left(1 - \frac{v}{a}\tan\alpha_0\right)^2}\right)^{-1} - \cos\alpha_0\right]. \tag{5.4.1}$$

The dependence (5.4.1) is represented in Fig. 23.

If the load P increases from zero, then the system jumps from point A to point D. Thus, the limit point A corresponds to the critical load.

The second type of loss of stability is connected with bifurcation points, i.e. points where two or more equilibrium states of the structure intersect (this loss of stability is usually called *bifurcation*).

The simplest example of such construction is represented in Fig. 24.

One end of an absolutely rigid bar is fixed to an elastic hinge of stiffness k. The other end is acted upon by load P, which keeps its direction fixed under deformation.

In the deflected state we have the equilibrium condition

$$P\,l\sin\psi = k\psi. \tag{5.4.2}$$

At point B, in Fig. 25, the equilibrium state $P = k\psi/(l\sin\psi)$ branches from the initial equilibrium state $\psi = 0$ and B is a bifurcation point.

We note that, from a mathematical point of view, it is usually easier to evaluate bifurcation points than to determine limit points since the former are usually determined as eigenvalues of linearized problems for the main pre-buckling equilibrium state. In other words, one can determine the load for which, together with the initial equilibrium mode, there exist modes infinitely close to the initial mode. (For the example considered here, linearization near the main equilibrium state $\psi = 0$ from (5.4.2) gives $(Pl - k)\psi = 0$, and, if $\psi \neq 0$ then $P = k/l$.)

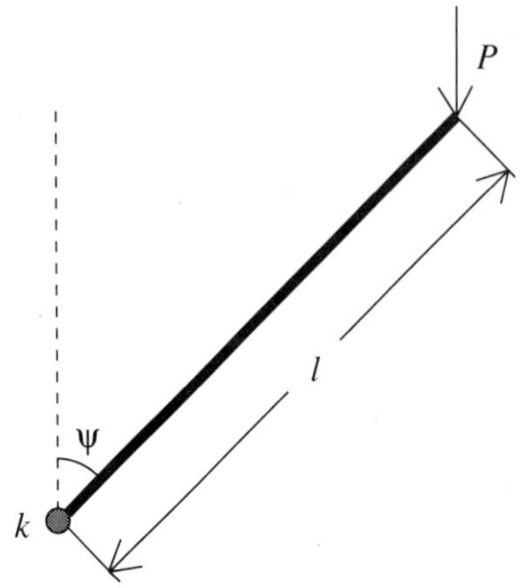

FIGURE 24. Hinge cantilever.

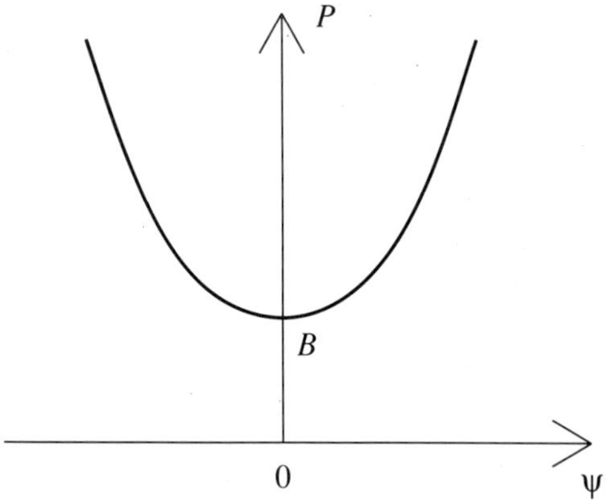

FIGURE 25. Effect of loading on the displacement of hinged cantilever.

5.4.2. The stability of shells. The first (classical) critical loads for shells have been obtained by the above method in the beginning of this century (S. P. Timoshenko [**210**] and R. Lorenz [**118**]). But a comparison of experimental and theoretical critical loads showed that experimental loads (having large dispersion) are one-third smaller than classical critical loads (obtained from the linearized problem).

In 1939, T. Karman and H. S. Tsien [**89, 90**] introduced an essentially new method for the study of shell buckling by considering finite displacements instead of small deviations from the pre-buckling state. They showed that the nonlinear effects played an essential role in buckling. It was found that new equilibrium forms, not close to the momentless form, exist for loads which are essentially smaller than the classical critical loads.

In [**89**], for the first time the notion of top and bottom critical loads has been introduced. The *top critical load* (point B in Fig. 26) is the largest critical load, before which the initial state has no close equilibrium forms. The *lower-bound*

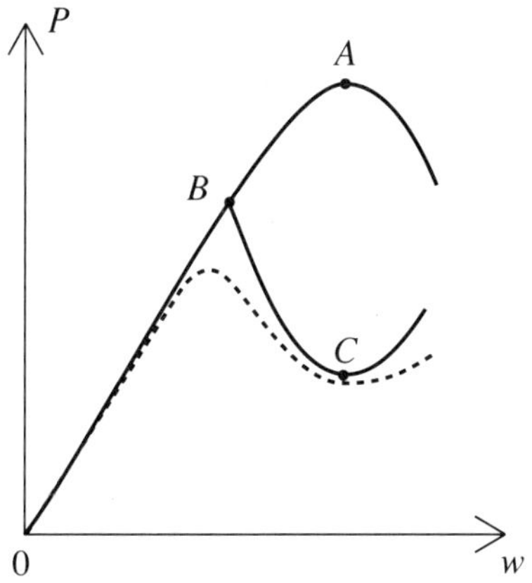

FIGURE 26. Top and lower-bound critical loads, B and C, respectively.

critical load (point C in Fig. 26) is the smallest critical load, before which the initial state is unique. The dotted line corresponds to a non-ideal structure (i.e. with initial deflections).

5.4.3. Shallow shells. In [**89, 90**], T. Karman and H. S. Tsien use, as initial equations, the nonlinear equations of shallow shells,

$$
\begin{aligned}
D\Delta^2 w &= \Delta_k \Phi + L(\Phi, w),\\
\frac{1}{Eh}\Delta^2 \Phi &= -\Delta_k w - \frac{1}{2}L(w, w),
\end{aligned}
\tag{5.4.3}
$$

which differ from the linear equations (2.6.11), by the second terms on the right-hand sides of (5.4.3). The operators Δ and Δ_k and the stress function Φ are determined in Section 2.6 and

$$L(w, \Phi) = L_1 w\, L_2 \Phi + L_2 w\, L_1 \Phi - 2L_3 w\, L_3 \Phi,$$

where

$$
\begin{aligned}
L_1 w &= \frac{1}{A}\frac{\partial}{\partial \alpha_1}\left(\frac{1}{A}\frac{\partial w}{\partial \alpha_1}\right) + \frac{1}{A_1 A_2^2}\frac{\partial A_1}{\partial \alpha_2}\frac{\partial w}{\partial \alpha_2}, \qquad (1 \leftrightarrow 2),\\
L_3 w &= -\frac{1}{A_1 A_2}\frac{\partial^2 w}{\partial \alpha_1 \partial \alpha_2} + \frac{1}{A_1^2 A_2}\frac{\partial A_1}{\partial \alpha_2}\frac{\partial w}{\partial \alpha_1} + \frac{1}{A_1 A_2^2}\frac{\partial A_2}{\partial \alpha_1}\frac{\partial w}{\partial \alpha_2},
\end{aligned}
$$

and similarly for $L_i \Phi$.

As it was noted in [**89, 90**], for complete information about the shell behaviour after buckling, one has to solve both equations of system (5.4.3). But, because the problem is very complex, the approximate Ritz's variational method has been applied.

For example, in the case of a cylindrical shell under axial compression, the deflection function, w, was represented in the form

$$w = W_0 + W_1 \cos\frac{mx}{R}\cos\frac{ny}{R} + W_2 \cos^2\frac{mx + ny}{2R}\cos^2\frac{mx - ny}{2R}. \tag{5.4.4}$$

The first term, W_0, characterizes the shell radial extension. The second term, containing the coefficient W_1, corresponds to the classical linear theory and the last term, which provides the rhombic dents directed inside the shell, agrees with experimental results.

The stress function may be found from relation (5.4.4) and the second equation of (5.4.3).

The unknown scalars, W_0, W_1 and W_2, are determined from the minimum condition for the potential energy,

$$\Pi = U_\varepsilon + U_\varkappa - U,$$

where

$$U_\varepsilon = \frac{1}{2Eh}\iint_G \left[(\Delta\Phi)^2 - (1+\gamma)L(\Phi,\Phi)\right] dG$$

is the elongation-compression energy of the neutral surface,

$$U_\varkappa = \frac{D}{2}\iint_G \left[(\Delta w)^2 - (1-\gamma)L(w,w)\right] dG$$

is the bending energy and U is the work done by the external forces.

In the case of a cylindrical shell under axial compression,

$$\begin{aligned} U &= h\int_0^{2\pi R} \sigma_x\Big|_{x=0}^{x=L} dy \int_0^L \frac{\partial U}{\partial x}\,dx \\ &= \frac{1}{Eh}\int_0^L\int_0^{2\pi R}\left(\frac{\partial^2\Phi}{\partial y^2}\right)\Big|_{x=0}^{x=L}\left[\frac{\partial^2\Phi}{\partial y^2} - \gamma\frac{\partial^2\Phi}{\partial x^2} - \frac{Eh}{2}\left(\frac{\partial w}{\partial x}\right)^2\right] dxdy. \end{aligned}$$

By eliminating the parameters W_0 and W_1 from the equations

$$\frac{\partial\Pi}{\partial W_i} = \Phi_i(W_0, W_1, W_2, \sigma_x) = 0, \qquad i = 0,1,2, \tag{5.4.5}$$

we obtain a relation between the compression stress and the buckling amplitude,

$$\frac{\sigma_x R}{Eh} = g_0(n,m) + \zeta g_1(n,m) + \zeta^2 g_2(n,m), \tag{5.4.6}$$

where $\zeta = W_2R/h$ and g_i are functions of the wave numbers in the circumferential and generator directions, m and n, respectively.

We see that

$$\frac{\sigma_x R}{Eh} = g_0(n,m), \qquad \text{as } \zeta \to 0, \tag{5.4.7}$$

and the function g_0 corresponds to the function f in (2.7.5). By minimizing $g_0(m,n)$ over n and m, one can get the classical critical load, which is achieved, as it is known, for infinitely many pairs (n,m), tied by the relations:

$$\frac{\eta(1+\mu^2)^2}{\mu^2} = 2\sqrt{3(1-\nu^2)}, \qquad \eta = n^2\frac{h}{R}, \qquad \mu = \frac{m}{n}. \tag{5.4.8}$$

The function $\min_{mn}\sigma_x(\zeta)$ is graphed for different values of m and n in Fig. 26 (see the curve OBC). Thus it was shown in [**89, 90**] that the load, which is resisted by a shell, decreases as post-buckling deformations increase.

Following the pioneering work of Karman and Tsien, similar research has been extensively conducted, distinguishing itself only by the form of relation (5.4.4) for the deflection function. The values of lower-bound critical loads obtained in these works agreed better with experimental data than the classical critical load. As a result, it has been proposed to consider the lower-bound load as a critical load. But later, with the help of numerical methods, it was found that the value of the lower-bound load decreases when more terms are kept in the formula for the form of deflection (similar to (5.4.4)). Some authors even obtained a negative value for the critical load.

L. Donnell [**40**] attributed the existence of a low critical load, found in experiments, to the initial imperfections in the form of the neutral surface. This fact was confirmed by Tennyson's experiments [**203**] for very precise shell models produced by vacuum technics. These results agree very well with the classical linear theory. However, during the following years, little attention has been given to the general nonlinear shell theory because of its complexity.

5.4.4. Koiter's theory. W. T. Koiter has investigated the behaviour of different elastic systems near a bifurcation point for small finite deflections from the main state by means of the nonlinear theory. The basic results of Koiter's theory are briefly reported in [**96**] and a survey of works based on Koiter's theory can be found in [**97**]. We summarize below the basic propositions of this theory.

Let external loads be proportional to the single parameter Λ. We assume that for the main state I the deflections $W(\Lambda)$ are known functions of Λ and if the system changes to another state then additional small (but not infinitesimal) deflections u arise. Koiter used the energy method. The change of energy in the transition from state I to state II may be represented by a series in the function u and its derivatives:

$$\Pi_{II} - \Pi_I = \sum_{i=1}^{\infty} \Pi_i[u; \Lambda], \tag{5.4.9}$$

where $\Pi_i[u; \Lambda]$ contain the i-th power of the function u and derivatives of u.

The equilibrium condition for the main state is

$$\Pi_1[u; \Lambda] = 0.$$

This state is stable if the following condition holds:

$$\Pi_2[u; \Lambda] \geq 0. \tag{5.4.10}$$

If equality in this formula holds only for the $u = 0$, this condition is also sufficiently. If the parameter Λ reaches a critical value, the relation (5.4.10) is valid, but now equality is possible not only for $u = 0$ but also for $u = \xi u_1$, where u_1 is the function describing the new buckling mode, and ξ is the parameter characterizing the buckling amplitude. The critical value Λ_1 and the buckling mode u_1 are determined from the solution of the eigenvalues problem:

$$\min_u \Pi_2[u; \Lambda] = 0.$$

We study a new state of equilibrium with the load parameter λ close to Λ_1. This state is characterized by a small additional deflection u. We represent u in form:

$$u = \xi u_1 + \bar{u},$$

and the functionals $\Pi_i[u;\Lambda]$ by a Taylor series in a neighbourhood of Λ_1:

$$\Pi_i[u;\Lambda] = \sum_{i=1}^{\infty} \Pi^{(i)}[u]\lambda^i, \tag{5.4.11}$$

where $\lambda = \Lambda - \Lambda_1$. Thus, relation (5.4.9) may be written as

$$\begin{aligned}\Pi_{II} - \Pi_I = &\sum_{i=2}^{\infty}\sum_{j=0}^{\infty} \xi^i \lambda^j \Pi_i^{(j)}[u_1] \\ &+ \sum_{k=1}^{\infty}\sum_{j=0}^{\infty} \Pi_k^{(j)}[\bar{u}]\lambda^j + \sum_{k=1}^{\infty}\sum_{i=1}^{\infty}\sum_{j=0}^{\infty} \xi^i \lambda^j \Pi_{ik}^{(j)}[u_1,\bar{u}],\end{aligned} \tag{5.4.12}$$

where

$$\Pi_{11}[u_1,\bar{u}] = 0, \qquad \Pi_2[u_1] = 0.$$

Here $\Pi_{ik}^{(j)}[u_1,\bar{u}]$ are homogeneousfunctionals of the i-th power in u and its derivatives and k-th power in $\bar{u}$ and its derivatives, i.e.

$$\Pi_{ik}^{(j)}[au_1, b\bar{u}] \equiv a^i b^k \Pi_{ik}^{(j)}[u_1,\bar{u}].$$

The minimum condition for expression (5.4.12) for a constant value of the buckling amplitude ξ, in the first approximation, is

$$\min_{\bar{u}}(\Pi_{II} - \Pi_I) = \min_{\bar{u}}(\Pi_2[\bar{u}] + \xi\lambda\Pi'_{11}[u_1,\bar{u}] + \xi^2\Pi_{21}[u_1,\bar{u}]). \tag{5.4.13}$$

From the solution of this problem, it follows that the function $\bar{u}$ is a small quantity of the second order in ξ and λ and hence can be represented as

$$\bar{u} = \xi\lambda\varphi + \xi^2\psi.$$

If this fact is taken into account, then the minimum of the function (5.4.12), in the first approximation, may be represented as

$$\Pi(\xi,\lambda) = A_2\xi^2\lambda + A_n\xi^n, \tag{5.4.14}$$

where

$$A_2 = \Pi'_2[u_1],$$

$$n = 3, \qquad \text{if} \qquad A_3 = \Pi_3[u_1] \neq 0;$$

$$n = 4 \qquad \text{and} \qquad A_4 = \Pi_4[u_1] - \Pi_2[\psi], \qquad \text{if} \qquad A_3 = 0.$$

From the equilibrium condition for the buckling mode, we have

$$\frac{\partial}{\partial\xi}\Pi(\xi;\lambda) = 2A_2\xi\lambda + nA_n\xi^{(n-1)}. \tag{5.4.15}$$

In order to determine whether the equilibrium state is stable or not, we evaluate the second derivative of Π with respect to ξ

$$\frac{\partial^2}{\partial\xi^2}\Pi(\xi;\lambda) = 2A_2\lambda + n(n-1)A_n\xi^{(n-2)}. \tag{5.4.16}$$

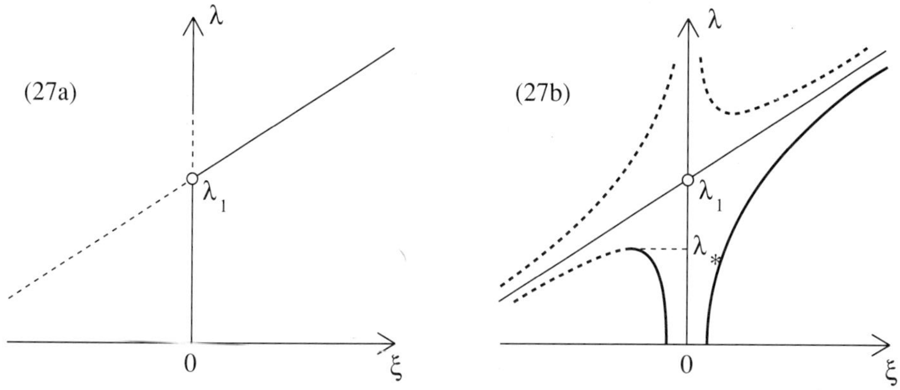

FIGURE 27. Load-displacement for $n = 3$.

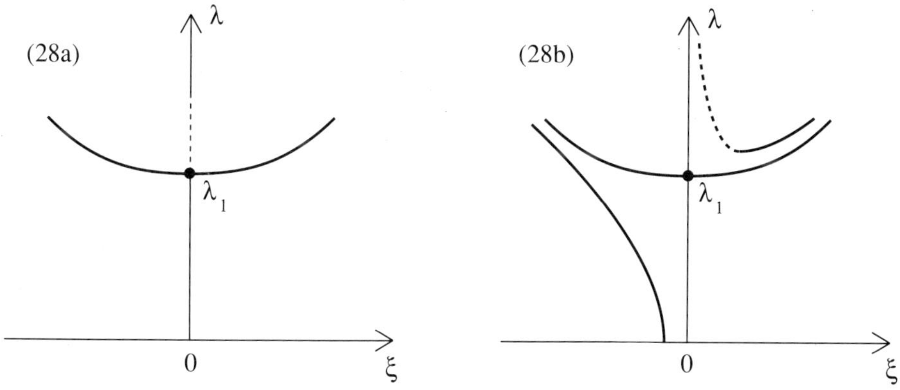

FIGURE 28. Load-displacement for $n = 4$ and $A_4 > 0$.

Equation (5.4.15) is satisfied for $\xi = 0$ and in this case $\bar{u} = 0$. This fact corresponds to the main equilibrium state, which is stable for $\Lambda < \Lambda_1$ and unstable for $\Lambda > \Lambda_1$. The rest of the solutions of equation (5.4.15) is given by

$$\xi^{(n-2)} = -\frac{2A_2}{nA_n}\lambda \tag{5.4.17}$$

and the corresponding stability condition is

$$\frac{\partial^2}{\partial \xi^2}\Pi(\xi;\lambda) = -2A_2(n-2)\lambda. \tag{5.4.18}$$

For $n = 3$ the equilibrium state under the critical load is unstable; the adjacent equilibrium states described by (5.4.17) are stable under loads greater than the critical load and unstable if the load is smaller than the critical value (see Fig. 27a).

For $n = 4$, if $A_4 > 0$, the equilibrium state under the critical load is stable and stable adjacent equilibrium states exist for loads above the critical load (see Fig. 28a). This case illustrates the behaviour of compressed bars or plates.

If $A_4 < 0$, the equilibrium state under the critical load is unstable and unstable equilibrium states exist for loads below the critical value (see Fig. 29a). This case

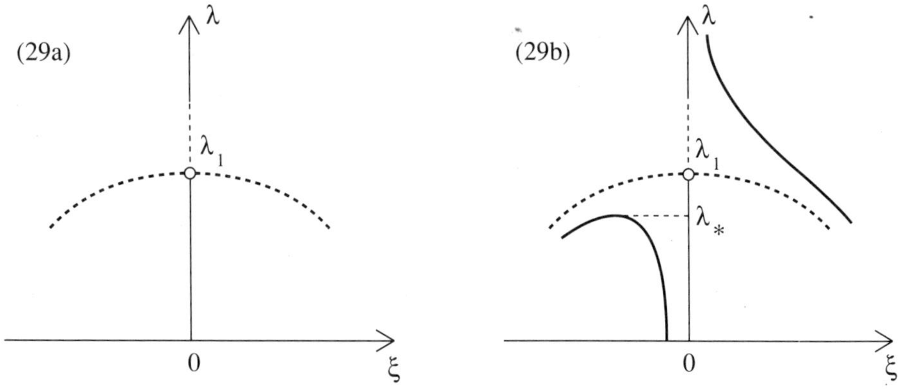

FIGURE 29. Load-displacement for $n = 4$ and $A_4 < 0$.

illustrates many problems of shell buckling.

If the load is less than the critical value, all adjacent equilibrium states are unstable (see the dashed lines in Figs. 27a, 28a and 29a).

5.4.5. Constructions with initial imperfections. If the construction has initial imperfections, the main equilibrium state corresponding to the equilibrium form of the simplified models differs from that of the real structure. The formula for the energy increment under the transition from the main state to an adjacent state has an additional term which is linear in the deflection u. If the initial imperfections are small (i.e. are of the order of the relative thickness), this additional term is also linear in the initial imperfections. In this case, in the first approximation, expression (5.5.12) is written as

$$\Pi_{II} - \Pi_I = \Pi_2[u; \lambda] + \Pi_3[u; \lambda] + \Pi_4[u; \lambda] + \bar{u}_0 \Lambda Q_1[u], \tag{5.4.19}$$

where $\bar{u}_0$ is the parameter determining the depth of the initial imperfections of the form and $Q_1[u]$ is a linear functional, depending on the type of initial imperfections. Using the method described above, we get

$$\Pi^*(\xi; \lambda) = A_2 \xi^2 \lambda + A_n \xi^n + \bar{u}_0 \Lambda B \xi,$$

where $B = Q_1[u_1]$.

In this case, the equilibrium condition is

$$\frac{\partial}{\partial \xi} \Pi^*(\xi; \Lambda) = 2A_2 \xi \lambda + n A_n \xi^{(n-1)} + \bar{u}_0 \Lambda B = 0, \tag{5.4.20}$$

and the buckling criterion is the same as (5.4.18). By considering equation (5.4.20) together with the buckling condition

$$\frac{\partial^2}{\partial \xi^2} \Pi(\xi; \Lambda) = 0$$

for the critical value Λ^*, we can obtain an asymptotic formula which connects the deviation of the critical load, obtained from the classical load for the "ideal" structure, with the value of the initial imperfections.

For $n = 3$,

$$(\Lambda_1 - \Lambda^*)^2 = 3\frac{BA_3}{A_2^2}\bar{u}_0\Lambda^*. \tag{5.4.21}$$

For $n = 4$ and $A_4 < 0$,

$$(\Lambda_1 - \Lambda^*)^3 = \frac{27B^2A_4}{8A_2^3}\bar{u}_0^2\Lambda^{*2}. \tag{5.4.22}$$

These loads are shown in Figs. 27b, 28b and 29b. The most important result is the appearance of limit points, corresponding to Λ^* in the case of an unstable critical state for an "ideal" simplified model. In these cases, the value of the real critical load Λ^* happens to be very sensitive to the value of the initial imperfections.

The simplest case, for which formulae of type (5.4.21) and (5.4.22) may be derived, is the case where the initial imperfections have the form of classical buckling. Some of these formulae were obtained by W. T. Koiter [**96, 97**].

In [**79**], J. Hansen considered a cylindrical shell which is very sensitive to imperfections under axial load. He supposed that initial deflections were the superposition of terms proportional to all the classical buckling modes. He picked out five variables connected, by simple relations, with the amplitudes of all the components of the imperfections and which had to be kept and evaluated in the asymptotic analysis.

In [**7**], J. C. Amazigo and B. Budiansky, obtained an asymptotic formula for a cylindrical shell under axial load, similar to (5.4.22). This formula characterizes the influence of axial local or random imperfections of the shell form. Similar formulae for nonsymmetric local axial imperfections and for local regular deviations of the shell thickness are obtained by S. M. Bauer [**17,18**].

5.4.6. Liapunov–Schmidt's method for initially imperfect shells. The general theory of buckling and initial post-buckling behaviour for shallow elastic shells with small imperfections of their neutral surfaces is developed by L. S. Srubshik [**202**], on the basis of Liapunov–Schmidt's method, for loads which are close to, or coincide with, the critical loads. The number of new equilibrium forms is determined and then for each of them Liapunov–Schmidt's series are constructed.

The equations of shallow shells with initial imperfections, w_0, are

$$\begin{aligned} D\Delta^2 w &= L(w_0 + w, \Phi) + \Delta_k\Phi, \\ \frac{1}{Eh}\Delta^2\Phi &= -\frac{1}{2}L(w, w) - L(w, w_0) - \Delta_k w, \end{aligned} \tag{5.4.23}$$

where w is the additional deflection, Φ is the stress function and the operators Δ, L and Δ_k are described in Section 2.6 and Subsection 5.4.3.

Let the load depend only on the parameter Λ and Λ_1 be the lowest eigenvalue which corresponds to the loading parameter of the idealized shell (for $\bar{w}_0 = 0$, where $\bar{w}_0$ is the imperfection amplitude). We distinguish the terms, in the stress function and in the deflection function, which describe the pre-buckling state (w_*, Φ_*), and represent the functions (w_*, Φ_*) in neighbourhoods of Λ_1 as a series in $\lambda = \Lambda - \Lambda_1$:

$$w_*(\Lambda) = w_*(\Lambda_1) + \sum_{j=1}^{\infty} w_j^*\lambda^j,$$

$$\Phi_*(\Lambda) = \Phi_*(\Lambda_1) + \sum_{j=1}^{\infty} \Phi_j^* \lambda^j.$$

We assume that $w = w_1 + w_*$ and $w = \Phi_1 + \Phi_*$. Then we can represent system (5.4.23) for small values of w_1, Φ_1 and λ as

$$BU = (M + M_T)U = R(U, \lambda) = M_1 U + \sum_{j=1}^{\infty} \lambda^j C_j U + \bar{w}_0 M_0 U, \tag{5.4.24}$$

where the unknown is $U = (w_1, \Phi_1)^T$. The linear operator

$$M = \begin{pmatrix} D\Delta^2 & -\Delta k \\ \Delta_k & \frac{1}{Eh}\Delta^2 \end{pmatrix}$$

describes the shell deformations. The linear differential operator M_T represents pre-buckling forces. The operator M_0 represents the initial imperfections, M_1 is a nonlinear operator and

$$C_j U = \begin{pmatrix} L(w_j^*, \Phi_1) + L(w_1, \Phi_j^*) \\ -L(w_j^*, w_1) \end{pmatrix}.$$

The asymptotic solution of system (5.4.24) is sought in the form

$$U = \sum_{j=1}^{n} U_j \xi_j + \lambda \sum_{k=1}^{n} \xi_k U_k + \sum_{j=1}^{n} \sum_{k=1}^{n} U_{jk} \xi_j \xi_k + O\left(|\xi^3|\right), \tag{5.4.25}$$

where the U_j are the buckling modes corresponding to the eigenvalue Λ_1 of multiplicity n and ξ_i are small parameters.

Substituting (5.4.25) into (5.4.24) we obtain a nonuniform system of linear equations with respect to the unknown ξ_j and U_{jk}.

In this case, the bifurcation equations can be represented in the form

$$L_j \equiv \xi_j \left(\sum_{k=1}^{n} a_{jk} \xi_k^2 + b_j \lambda \right) + w_{0j} = 0, \tag{5.4.26}$$

where $w_{0j} = \bar{w}_0(U_0, U_j)$.

In [**202**], system (5.5.26) is solved numerically for imperfect spherical and conical shells under uniform external pressure by taking into account two eigenfunctions and by simultaneously satisfying the condition

$$\det \left\| \frac{\partial L_i}{\partial \xi_k} \right\| = 0, \qquad i,\, k = 1, \dots, n,$$

for which buckling occurs.

This method allows us to compute the critical loads as a function of the n parameters w_{0j} which characterize the shell imperfections, without having to find all the branches of the solutions.

In [**204**], I. Yu. Teterin analyzes a similar case for non-shallow shells of revolution. It is noted (see also [**202**]), that, if we take into account the interactions between the eigenfunctions, the critical loads decrease as compared with the case of a one-mode buckling.

References

1. M. Abramowitz and I. A. Stegun (eds.), *Handbook of mathematical functions with formulas, graphs and mathematical tables*, Applied Mathematics Series, vol. 55, National Bureau of Standards, Washington, D.C., 1964 (1046 pp.).
2. J. D. Achenbach, A. K. Gautesen, and H. McMaken, *Ray methods for waves in elastic solids*, Pitman, Boston, London, Melbourne, 1982 (251 pp.).
3. O. K. Aksentian and I. I. Vorovich, *The state of stress in a thin plate*, Prikl. Mat. Mekh. **27** (1963), no. 6, 1057–1074; English transl., J. Appl. Math. Mech. **27** (1963), no. 6, 1621–1643.
4. N. A. Alfutov, *Certain paradoxes of the theory of thin elastic plates*, Izv. Akad. Nauk SSSR Mekh. Tverd. Tela **27** (1992), no. 3, 65–72; English transl., Mech. Solids **27** (1992), no. 3, 59–65.
5. N. A. Alumäe, *Determination of the critical load for a shell limited by a hyperboloid of one sheet*, Prikl. Mat. Mekh. **20** (1956), no. 2, 397–408. (Russian)
6. ——, *On the fundamental system of integrals of the equations of small axisymmetric stable vibrations for elastic conical shells of revolution*, Izv. Akad. Nauk Eston. SSR Ser. Tekhn. i Fiz. Nauk **10** (1960), no. 1, 3–15. (Russian)
7. J. C. Amazigo and B. Budiansky, *Asymptotic formulas for the buckling stresses of axially compressed cylinders with localized or random axisymmetric imperfections*, Trans. ASME Ser. E J. Appl. Mech. **39** (1972), no. 3, 179–84.
8. I. V. Andrianov, V. A. Lesnichaya, and L. I. Manevich, *Averaging method in ribbed shell statics and dynamics*, "Nauka", Moscow, 1985 (224 pp.). (Russian)
9. M. Ya. Antimirov, A. A. Kolyshkin, and R. Vaillancourt, *Perturbation methods in eddy current testing*, Asymptotic Methods in Mechanics (A. L. Smirnov and R. Vaillancourt, eds.), CRM Proceedings and Lecture Notes, this volume, Amer. Math. Soc., Providence, R.I., 1993, pp. 143–152.
10. A. G. Aslanyan and V. B. Lidsky, *Distribution of free frequencies of thin elastic shells*, "Nauka", Moscow, 1974 (156 pp.). (Russian)
11. A. G. Aslanyan, D. G. Vasil′ev, and V. B. Lidsky, *Frequencies of free vibrations of a thin shell interacting with a liquid*, Funktsional. Anal. i Prilozhen. **15** (1981), no. 3, 1–9; English transl., Functional Anal. Appl. **15** (1981), no. 3, 157–164.
12. V. M. Babich and V. S. Buldyrev, *Asymptotic methods in short wave diffraction problems*, "Nauka", Moscow, 1972 (456 pp.). (Russian)
13. V. M. Babich and A. P. Kiselev, *Geometric and seismologic description of a "nonray" S^* wave*, Izv Akad. Nauk SSSR Ser. Fiz. Zemli (1988), no. 10, 67–71; English transl., Izv. Acad. Sci. USSR Phys. Solid Earth **24** (1989), no. 10, 817–820.
14. ——, *Non-geometric waves—are there any? An asymptotic description in some 'non-geometrical' phenomena in seismic wave propagation*, Geophysical J. International **99** (1989), no. 2, 415–420.
15. V. M. Babich and I. A. Molotkov, *Mathematical methods in elastic wave theory*, Mekh. Deform. Tverd. Tela, vol. 10, Akad. Nauk. SSSR, Vsesoyuz. Inst. Nauchn. i Tekhn. Inform., Moscow, 1977, pp. 5–62. (Russian)
16. N. S. Bakhvalov and G. P. Panasenko, *Homogenisation: Averaging processes in periodic media*, Kluwer, Dordrecht, Boston, 1989 (366 pp.).
17. S. M. Bauer, *On the influence of the imperfections of form localized in axial direction on the stability of the shells of revolution close to cylinders in form*, Vestnik Leningrad. Univ. Mat. Mekh. Astronom. (1978), no. 13, 131–134. (Russian)

18. ———, *Stability of a cylindrical shell at variable thickness*, Vestnik Leningrad. Univ. Mat. Mekh. Astronom. (1984), no. 7, 51–55; English transl., Len. Univ. Mech. Bull. (1984), no. 2, 1–8.
19. S. M. Bauer, S. B. Filippov, A. L. Maiboroda, A. L. Smirnov and I. Yu. Teterin, *Buckling of thin cylindrical shells and shells of negative Gaussian curvature*, Asymptotic Methods in Mechanics (A. L. Smirnov and R. Vaillancourt, eds.), CRM Proceedings and Lecture Notes, this volume, Amer. Math. Soc., Providence, R.I., 1993, pp. 153–161.
20. S. M. Bauer and A. L. Smirnov, *Thermo-elastic deformations of mirrors*, Asymptotic Methods in Mechanics (A. L. Smirnov and R. Vaillancourt, eds.), CRM Proceedings and Lecture Notes, this volume, Amer. Math. Soc., Providence, R.I., 1993, pp. 163–173.
21. V. L. Berdichevskii, *Variational-asymptotic method of constructing a theory of shells*, Prikl. Mat. Mekh. **43** (1979), no. 4, 664–687; English transl., J. Appl. Math. Mech. **43** (1979), no. 4, 711–736.
22. V. L. Berdichevskii and Le Khan' Chau, *High-frequency long-wave shell vibration*, Prikl. Mat. Mekh. **44** (1980), no. 4, 737–744; English transl., J. Appl. Math. Mech. **44** (1980), no. 4, 520–525.
23. M. S. Berger, *A bifurcation theory for nonlinear elliptic partial differential equations and related systems*, Bifurcation Theory and Nonlinear Eigenvalue Problems (Joseph B. Keller and Stuart Antman, eds.), W. A. Benjamin, New York, Amsterdam, 1969, pp. 113–216.
24. R. M. Bergman, *Investigation of the free vibrations of noncircular cylindrical shells*, Prikl. Mat. Mekh. **37** (1973), no. 6, 1125–1134; English transl., J. Appl. Math. Mech. **37** (1973), no. 6, 1068–1077.
25. N. N. Bogolyubov and Yu. A. Mitropolsky, *Asymptotic methods in the theory of nonlinear vibrations*, Fizmatgiz, Moscow, 1974 (504 pp.). (Russian)
26. N. N. Bogolyubov, Yu. A.Mitropolsky, and A. M. Samoilenko, *Accelerated convergence methods in nonlinear mechanics*, "Naukova Dumka", Kiev, 1969 (247 pp.). (Russian)
27. V. V. Bolotin, *On the density of the distribution of natural frequencies of thin elastic shells*, Prikl. Mat. Mekh. **27** (1963), no. 2, 362–364; English transl., J. Appl. Math. Mech. **27** (1963), no. 2, 538–543.
28. N. G. de Bruijn, *Asymptotic methods in analysis*, 3rd ed., North-Holland & Walters Noordhoff, Amsterdam, Groningen, 1970 (200 pp.).
29. V. A. Bulygin, *On a class of shells of Gaussian curvature which changes sign*, Izv. Akad. Nauk SSSR Mekh. Tverd. Tela **12** (1977), no. 5, 97–105; English transl., Mech. Solids **12** (1977), no. 5, 82–89.
30. D. Bushnell, *Buckling of shell—Pitfall for designers*, AIAA J. **19** (1981), no. 9, 1183–1226.
31. V. F. Butuzov and T. A. Urazgil'dina, *Asymptotic solution of a quasistatic thermoelasticity problem for a slender rod*, Prikl. Mat. Mekh. **51** (1987), no. 6, 989–999; English transl., J. Appl. Math. Mech. **51** (1987), no. 6, 761–769.
32. K. F. Chernykh, *Simple boundary effect and splitting of the boundary conditions in the linear theory of thin shells*, Izv. Akad. Nauk SSSR Mekhanika (1965), no. 1, 89–98. (Russian)
33. G. N. Chernyshev, *Some properties of integrals of shell-theory dynamic equations*, Izv. Akad. Nauk SSSR Mekh. Tverd. Tela **8** (1973), no. 2, 68–76; English transl., Mech. Solids **8** (1973), no. 2, 55–62.
34. T. M. Cherry, *Uniform asymptotic formulae for functions with transition points*, Trans. Amer. Mat. Soc. **68** (1950), 224–257.
35. P. Cicala, *Asymptotic theories of elastic beams and rods*, Meccanica—J. Ital. Assoc. Theoret. Appl. Mech. **16** (1981), no. 2, 85–92.
36. A. Clebsch, *Théorie de l'élasticité des corps solides*, Dunod, Paris, 1883 (900 pp.) (French); ; Johnson Reprint Corp., New York, 1966.
37. J. D. Cole and J. Kevorkian, *Perturbation methods in applied mathematics*, Springer-Verlag, New York, 1981 (558 pp.).
38. D. A. Danielson, *Buckling and initial postbuckling behavior of spheroidal shells under pressure*, AIAA J. (1969), no. 5, 936–944.
39. R. A. DiTaranto and M. Lessen, *Coriolis acceleration effect on the vibration of a rotating thin-walled circular cylinder*, Trans. ASME Ser. E J. Appl. Mech. **31** (1964), no. 12, 700–701.
40. L. H. Donnell, *A new theory for the buckling of thin cylinders under axial compression and bending*, Trans. ASME **56** (1934), no. 12, 795–806.
41. ———, *Beams, plates and shells*, McGraw-Hill, New York, 1976 (774 pp.).

42. A. A. Dorodnitsyn, *Asymptotic laws of eigenvalue distribution for some special types of differential equations of the second order*, Uspekhi Mat. Nauk (N.S.) **7** (1952), no. 6(52), 3–96.
43. N. E. Egarmin, *Precession of the vibrational standing waves of a rotating axisymmetric shell*, Izv. Akad. Nauk SSSR Mekh. Tverd. Tela **21** (1986), no. 1, 142–148; English transl., Mech. Solids **21** (1986), no. 1, 147–154.
44. V. V. Eliseev, *Application of the asymptotic method in the curvilinear rod equilibrium problem*, Izv. Akad. Nauk SSSR Mekh. Tverd. Tela **12** (1977), no. 3, 145–150; English transl., Mech. Solids **12** (1977), no. 3, 136–142.
45. ——, *One- and three-dimensional models in elastic beam mechanics*, St. Petersburg State University, St. Petersburg, 1992 (27 pp.). (Russian)
46. A. Erdélyi, *Asymptotic Expansions*, Dover, New York, 1956 (108 pp.).
47. B. A. Ershov, Yu. A. Mochalova, and E. V. Polyakova, *A mathematical model for hydroelastic problems with a fluid memory. Part I*, Asymptotic Methods in Mechanics (A. L. Smirnov and R. Vaillancourt, eds.), CRM Proceedings and Lecture Notes, this volume, Amer. Math. Soc., Providence, R.I., 1993, pp. 175–183.
48. B. A. Ershov and I. I. Strelkovskaya, *A mathematical model for hydroelastic problems with a fluid memory. Part II*, Asymptotic Methods in Mechanics (A. L. Smirnov and R. Vaillancourt, eds.), CRM Proceedings and Lecture Notes, this volume, Amer. Math. Soc., Providence, R.I., 1993, pp. 185–192.
49. V. I. Ershov, *Low frequency vibrations of cupolas*, Prikl. Mekh., vol. 4, Leningrad University Press, Leningrad, 1979, pp. 175–182. (Russian)
50. M. A. Evgrafov, *Asymptotic estimates and entire functions*, Gordon and Breach, Moscow, 1961 (181 pp.).
51. M. V. Fedoryuk, *Saddle point methods*, "Nauka", Moscow, 1977 (468 pp.). (Russian)
52. ——, *Asymptotic methods for linear ordinary differential equations*, "Nauka", Moscow, 1983 (352 pp.). (Russian)
53. ——, *Asymptotics: Integrals and series*, "Nauka", Moscow, 1987 (544 pp.). (Russian)
54. S. F. Feshchenko, N. I. Shkil′, and L. D. Nikolenko, *Asymptotic methods in the theory of linear differential equations*, American Elsevier Pub. Co., New York, 1967 (270 pp.).
55. S. B. Filippov, *Axisymmetric vibrations of an ellipsoidal shell of revolution*, Vestnik Leningrad. Univ. Mat. Mekh. Astronom. (1981), no. 7, 90–95.
56. ——, *Low-frequency vibrations of a conical shell with spaced stiffeners*, Izv. Akad. Nauk SSSR Mekh. Tverd. Tela **24** (1989), no. 5, 138–144; English transl., Mech. Solids **24** (1989), no. 5, 143–149.
57. ——, *Low-frequency vibrations of connected conical shells*, Prikl. Mekh. (1990), no. 8, 188–199. (Russian)
58. ——, *Low-frequency vibrations of cylindrical shells.* Part I: *Shells with a slanted edge*, Asymptotic Methods in Mechanics (A. L. Smirnov and R. Vaillancourt, eds.), CRM Proceedings and Lecture Notes, this volume, Amer. Math. Soc., Providence, R.I., 1993, pp. 193–204.
59. ——, *Low-frequency vibrations of cylindrical shells.* Part II: *Connected shells*, Asymptotic Methods in Mechanics (A. L. Smirnov and R. Vaillancourt, eds.), CRM Proceedings and Lecture Notes, this volume, Amer. Math. Soc., Providence, R.I., 1993, pp. 205–215.
60. K. O. Friedrichs, *Kirchhoff's boundary conditions and the edge effect for elastic plates*, Proc. Sympos. Appl. Math., vol. 3, McGraw-Hill, New York, 1950, pp. 117–124.
61. N. Fröman and P. O. Fröman, *JWKB-approximation. Contributions to the theory*, North-Holland, Amsterdam, 1965 (138 pp.).
62. Y. C. Fung and E. E. Sechler (eds.), *Thin-shell structures; theory, experiment and design*, Prentice-Hall, Englewood Cliffs, N.J., 1974 (616 pp.).
63. V. V. Galaktionov and E. A. Tropp, *An asymptotic analytical method of calculating thermoelastic stresses in a thin rod*, Izv. Akad. Nauk SSSR Ser. Fiz. **40** (1976), no. 7, 1399–1406; English transl., Bull. Acad. Sci. USSR Phys. Ser. **40** (1976), no. 7, 62–68.
64. I. Geckeler, *Die Festigkeit achsensymmetrischer Schalen*, Ing. Arch. **1** (1930), 255–270. (German)
65. A. L. Gol′denveizer, *Asymptotic integration of partial differential equations with boundary conditions depending on a parameter*, Prikl. Mat. Mekh. **22** (1958), no. 5, 657–672. (Russian)
66. ——, *Theory of elastic thin shells* (G. Harmann, ed.), Pergamon Press, New York, Oxford, London, Paris, 1961 (658 pp.).

67. ——, *Theory of thin elastic shells*, 2nd ed., "Nauka", Moscow, 1976 (512 pp.). (Russian)

68. ——, *Mathematical rigidity of surfaces and physical rigidity of shells*, Izv. Akad. Nauk SSSR Mekh. Tverd. Tela **14** (1979), no. 6, 65–77; English transl., Mech. Solids **14** (1979), no. 6, 53–63.

69. ——, *The asymptotic method in the theory of shells*, Adv. in Mech. **5** (1982), no. 1/2, 137–182.

70. ——, *Geometrical theory of stability of shells*, Izv. Akad. Nauk SSSR Mekh. Tverd. Tela **18** (1983), no. 1, 143–154; English transl., Mech. Solids **18** (1983), no. 1, 135–145.

71. ——, *The general theory of thin elastic bodies (shells, coating, spacers)*, Izv. Akad. Nauk SSSR Mekh. Tverd. Tela **27** (1992), no. 3, 5–17; English transl., Mech. Solids **27** (1992), no. 3, 2–13.

72. A. L. Gol′denveizer and Yu. D. Kaplunov, *Dynamic boundary layer in problems of vibration of shells*, Izv. Akad. Nauk SSSR Mekh. Tverd. Tela **23** (1988), no. 4, 152–162; English transl., Mech. Solids **23** (1988), no. 4, 146–155.

73. A. L. Gol′denveizer, V. B. Lidsky and P. E. Tovstik, *Free vibrations of thin elastic shells*, "Nauka", Moscow, 1979 (383 pp.). (Russian)

74. A. L. Gol′denveizer and A. L. Radovinsky, *Asymptotic vibration and radiation analysis of shells of revolution in a liquid*, Preprint no. 275, USSR Institute of Mechanics, Moscow, 1986 (61 pp.). (Russian)

75. I. S. Gradshteyn and I. M. Ryzhik, *Tables of Integrals, Sums, Series and Products*, 4th ed., Academic Press, New York, 1980 (1160 pp.).

76. D. A. Grave, *Treaty on algebraic analysis*, vol. 1, Institute of Mathematics of AN USSR, Kiev, 1938 (210 pp.). (Russian)

77. E. I. Grigolyuk and V. V. Kabanov, *Stability of shells*, "Nauka", Moscow, 1978 (359 pp.). (Russian)

78. M. I. Gusein-Zade, *Asymptotic analysis of boundary and initial conditions in the dynamics of thin plates*, Prikl. Mat. Mekh. **42** (1978), no. 5, 899–907; English transl., J. Appl. Math. Mech. **42** (1978), no. 5, 978–987.

79. J. S. Hansen, *Influence of general imperfections in axially loaded cylindrical shells*, Intern. J. Solids and Structures **11** (1975), no. 11, 1223-1233.

80. J. Heading, *An introduction to phase-integral methods*, Methuen, Wiley, London, New York, 1962 (160 pp.).

81. J. W. Hutchinson and W. T. Koiter, *Postbuckling theory*, Applied Mechanics Review **23** (1970), no. 12, 1353-1366.

82. B. P. Ivanov, A. A. Ilyukhin, and S. V. Kozlov, *On the foundation of Kirchhoff–Clebsch's theory*, Mat. Fiz. Nelinein. Mekh. **2** (1984), no. 36, 19–24. (Russian)

83. E. Jahnke, F. Emde, and F. Lösch, *Tafeln höherer Funktionen*, 7th ed., B. G. Teubner, Stuttgart, 1966 (322 pp.). (German)

84. E. Kamke, *Differentialgleichungen*, 5th ed., Akademische Verlagsgesellschaft, Leipzig, I, 1964 (316 pp.); II, 1965 (255 pp.).

85. Yu. D. Kaplunov, *High-frequency stress-strain states of low variability in thin elastic shells*, Izv. Akad. Nauk SSSR Mekh. Tverd. Tela **25** (1990), no. 5, 147–157; English transl., Mech. Solids **25** (1990), no. 5, 150–160.

86. ——, *Integration of dynamic boundary-layer equations*, Izv. Akad. Nauk SSSR Mekh. Tverd. Tela **25** (1990), no. 1, 148–160; English transl., Mech. Solids **25** (1990), no. 1, 155–166.

87. ——, *Slowly varying high-frequency stress-strain states in immersed shells*, Prikl. Mat. Mekh. **55** (1991), no. 3, 478–485; English transl., J. Appl. Math. Mech. **55** (1991), no. 3, 390–396.

88. ——, *Vibrations of shells of revolution in response to high-frequency edge excitation*, Izv. Akad. Nauk SSSR Mekh. Tverd. Tela **26** (1991), no. 6, 151–159; English transl., Mech. Solids **26** (1991), no. 6, 147–154.

89. T. Karman and H. S. Tsien, *The buckling of spherical shells by external pressure*, J. of the Aeronautical Sciences **7** (1939), no. 2, 43–50.

90. ——, *The buckling of thin cylindrical shells under axial compression*, J. of the Aeronautical Sciences **8** (1941), no. 8, 303–312.

91. N. D. Kazarinoff, *Asymptotic theory of second order differential equations with two turning points*, Arch. Rational Mech. Anal. (1958), no. 2, 129–150.

92. K. V. Kholshevnikov, *Asymptotic methods in celestial mechanics*, Leningrad University Press, Leningrad, 1985 (206 pp.). (Russian)
93. N. A. Kil′chevsky, *Generalization of modern shell theory*, Prikl. Mat. Mekh. **2** (1939), no. 4, 427–438.
94. G. Kirchhoff, *Über das Gleichgewicht und die Bewegung eines unendlich dünnen elastischen Stabes*, J. Reine Angew. Math. **56** (1858), no. 4, 285–313. (German)
95. ——, *Mechanics*, AN USSR, Moscow, 1962 (402 pp.). (Russian)
96. W. T. Koiter, *On the nonlinear theory of thin elastic shells*, Proc. Kon. Nederl. Acad. Wetensch. **69** (1966), 1–54.
97. ——, *Current trends in the theory of buckling*, Buckling of Structures, Berlin, 1976, pp. 1–16.
98. A. V. Kolos, *Methods of refining the classical theory of bending and extension of plates*, Prikl. Mat. Mekh. **29** (1965), no. 4, 771–781; English transl., J. Appl. Math. Mech. **29** (1965), no. 4, 914 925.
99. V. M. Kornev, *On the formulation of boundary conditions for the simplified equations of shells of revolution*, Prikl. Mat. Mekh. **34** (1970), no. 1, 84–94; English transl., J. Appl. Math. Mech. **34** (1970), no. 1, 76–85.
100. M. A. Kovalevskii, *Determination of the connection between two fundamental families of solutions of a linear ordinary differential equation*, Vestnik Leningrad. Univ. Mat. Mekh. Astronom. (1981), no. 1, 32–37; English transl., Vestnik Leningrad. Univ. Math. **14** (1982), 39–46.
101. ——, *Construction of the Stokes multipliers for an equations with two singular points*, Vestnik Leningrad. Univ. Mat. Mekh. Astronom. (1981), no. 7, 49–54; English transl., Vestnik Leningrad. Univ. Math. **14** (1982), 135–141.
102. ——, *Asymptotics of solutions of a standard equation for problems with a turning point*, Sibirsk. Mat. Zh. **25** (1984), no. 4, 82–92; English transl., Siberian Math. J. **25** (1984), no. 4, 575–583.
103. ——, *On difference equations for the determination of Stokes multipliers*, Soviet Math. Dokl. **277** (1984), no. 6, 272–275.
104. N. M. Krylov and N. N. Bogolyubov, *Introduction to non-linear mechanics*, Princeton University Press, Princeton, N.J., 1947 (106 pp.).
105. B. N. Kvasnikov, *On the construction of approximate analytical methods in thin shell theory*, Prikl. Mekh., vol. 6, Leningrad University Press, Leningrad, 1984, pp. 126–138. (Russian)
106. R. E. Langer, *The asymptotic solutions of ordinary linear differential equations of the second order, with special reference to a turning point*, Trans. Amer. Math. Soc. **67** (1949), 461–490.
107. ——, *The asymptotic solutions of linear differential equations of second order with two turning points*, Trans. Amer. Math. Soc. **90** (1959), 113–142.
108. V. F. Lazutkin, *The asymptotics of the eigenfunctions of the Laplace operator, concentrated near the boundary of a region*, Zh. Vychisl. Mat. i Mat. Fiz. **7** (1967), no. 6, 1237–1249; English transl., USSR Comput. Math. and Math. Phys. **7** (1967), no. 6, 37–52.
109. ——, *The construction of asymptotic series for free functions of the bouncing ball type*, Trudy MIAN **95** (1968), 106–118. (Russian)
110. A. M. Liapunov, *On figures of equilibrium slightly different from the ellipsoids of rotating homogeneous liquid mass*, vol. 4, AN SSSR, Moscow (645 pp.). (Russian)
111. V. B. Lidsky and P. E. Tovstik, *Spectra in the theory of shells*, Adv. in Mech. (1984), no. 7, 25–54. (Russian)
112. T. V. Liiva, *On free nonaxisymmetric vibrations of shells of revolution of negative Gaussian curvature*, Transactions of the Tallinn Polytechnical Institute. Series A, vol. 5, 1970, pp. 47–60. (Russian)
113. T. V. Liiva and P. E. Tovstik, *Free nonaxisymmetric vibrations of shells of revolution of negative Gaussian curvature*, Proceedings of the 7-th All-Union Conference on the Theory of Shells and Plates, "Nauka", Moscow, 1970, pp. 365–369. (Russian)
114. ——, *On buckling in the linear approximation of shells of revolution of negative Gaussian curvature*, Problems of Solid Mechanics, Sudostroenie, Leningrad, 1970, pp. 231–238. (Russian)
115. ——, *Loss of stability of shells of revolution with negative Gaussian curvature under torsion*, Izv. Akad. Nauk SSSR Mekh. Tverd. Tela **6** (1973), no. 6, 92–98; English transl., Mech. Solids **6** (1973), no. 6, 79–85.

116. C. C. Lin and A. L. Rabenstein, *On the asymptotic solutions of a class of ordinary differential equations of the fourth order*, Trans. Amer. Math. Soc. (1960), no. 94, 24–57.
117. S. A. Lomov, *Introduction to the general theory of singular perturbations*, Amer. Math. Soc., Providence, R.I., 1992 (375 pp.).
118. R. Lorenz, *Die nicht achsensymmetrische Knickung dünnwandiger Hohlzylinder*, Physikal. Zeitschrift **12** (1911), no. 7, 241–260. (German)
119. A. E. H. Love, *A treatise on the mathematical theory of elasticity*, 4th ed., Dover Publications, New York, 1944 (643 pp.).
120. A. I. Lur′e, *General theory of thin elastic shells*, Prikl. Mat. Mekh. **4** (1940), no. 2, 7–34. (Russian)
121. ———, *Concentration of stresses in the vicinity of an aperture in the surface of a circular cylinder*, Prikl. Mat. Mekh. **10** (1946), no. 3, 397–406. (Russian)
122. I. G. Malkin, *Some problems of the theory of nonlinear oscillations*, United States Atomic Energy Commission, Oak Ridge, Tenn., 1959 (589 pp.).
123. V. M. Mal′kov, *Linear theory of a thin layer of a low-compressible material*, Dokl. Akad. Nauk SSSR **293** (1987), no. 1, 52–54. (Russian)
124. V. P. Maslov, *Complex WKB method in nonlinear equations*, "Nauka", Moscow, 1977 (384 pp.). (Russian)
125. ———, *Asymptotic methods and perturbation theory*, "Nauka", Moscow, 1988 (312 pp.). (Russian)
126. V. P. Maslov and V. E. Nazaikinskii, *Asymptotics of operator and pseudo-differential equations*, Consultant Bureau, New York, London, 1988 (313 pp.).
127. A. L. Maiboroda, *Buckling of revolution shells of a negative Gaussian curvature under asymmetric loads*, Vestnik Leningrad. Univ. Mat. Mekh. Astronom. (1989), no. 22, 61–65; English transl., Len. Univ. Mech. Bull. (1989), no. 4, 30–36.
128. ———, *Buckling of circular conical shells by forced bending*, Prikl. Mekh., vol. 8, Leningrad University Press, Leningrad, 1990, pp. 209–214. (Russian)
129. ———, *Buckling of convex shells under nonaxisymmetric loading*, Asymptotic Methods in Mechanics (A. L. Smirnov and R. Vaillancourt, eds.), CRM Proceedings and Lecture Notes, this volume, Amer. Math. Soc., Providence, R.I., 1993, pp. 217–225.
130. M. F. Mekhtiev, *Free vibrations of a hollow isotropic cylinder*, Izv. Akad. Nauk SSSR Mekh. Tverd. Tela **20** (1985), no. 5, 83–88; English transl., Mech. Solids **20** (1985), no. 5, 81–86.
131. G. I. Mikhasev, *Local loss of stability of shells of zero curvature with variable thickness and elastic modulus*, Vestnik Leningrad. Univ. Mat. Mekh. Astron. (1984), no. 7, 104–105; English transl., Len. Univ. Mech. Bull. (1984), no. 2, 39–43.
132. ———, *Local loss of stability of a thin truncated ellipsoid acted upon by a combined load*, Vestnik Leningrad Univ. Mat. Mekh. Astronom. (1984), no. 19, 85–90; English transl., Len. Univ. Mech. Bull. (1984), no. 4, 63–71.
133. G. I. Mikhasev and P. E. Tovstik, *Stability of conical shells acted upon by external pressure*, Izv. Acad. Nauk SSSR Mekh. Tverd. Tela **25** (1990), no. 4, 99-104; English transl., Mech. Solids **25** (1990), no. 4, 106–119.
134. E. F. Mishchenko and N. Kh. Rosov, *Differential equations with a small parameter and relaxation vibrations*, "Nauka", Moscow, 1975 (248 pp.). (Russian)
135. N. N. Moiseev, *Asymptotic methods in nonlinear mechanics*, "Nauka", Moscow, 1969 (480 pp.). (Russian)
136. A. I. Molchanov, *Free oscillations of noncircular shells close to shells of a zero Gaussian curvature*, Vestnik Leningrad. Univ. Mat. Mekh. Astronom. (1986), no. 19, 43–45; English transl., Len. Univ. Mech. Bull. (1986), no. 4, 63–66.
137. ———, *Asymptotic integration of a set of equations of free oscillations of shells of revolution of near-zero Gaussian curvature*, Vestnik Leningrad. Univ. Mat. Mekh. Astronom. (1987), no. 7, 106–107; English transl., Len. Univ. Mech. Bull. (1987), no. 2, 56–58.
138. W. Nachbar and N. I. Hoff, *The buckling of a free edge of an axially compressed circular cylindrical shell*, Quart. Appl. Mech. **20** (1962), no. 3, 267–277.
139. A. H. Nayfeh, *Perturbation methods*, Wiley, New York, 1973 (426 pp.).
140. ———, *Introduction to perturbation techniques*, Wiley, New Yorkx, 1981 (529 pp.).
141. S. A. Nazarov, *Introduction to asymptotic methods in elasticity theory*, Leningrad University Press, Leningrad, 1983 (117 pp.). (Russian)

142. U. K. Nigul, J. Metsaveer, N. D. Veksler, and M. Kutser, *Echo-signal from elastic objects*, Institut Kibernetiki AN ESSR, Tallinn, 1974 (345 pp.). (Russian)
143. V. V. Novozhilov, *Theory of thin shells*, Wolters–Noordhoff, Groningen, 1970 (422 pp.).
144. V. V. Novozhilov, K. F. Chernykh, and E. I. Mikhailovsky, *Linear theory of thin shells*, Polytekhnika, Leningrad, 1991 (656 pp.). (Russian)
145. V. V. Novozhilov and R. M. Finkel′shtein, *On errors of Kirchhoff's hypotheses in shell theory*, Prikl. Mat. Mekh. **7** (1943), no. 5, 331–340.
146. I. F. Obrazstov, B. V. Nerubailo, and I. V. Andrianov, *Asymptotic methods in civil engineering of thin-wall structures*, Mashinostroenie, Moscow, 1991 (416 pp.). (Russian)
147. F. W. J. Olver, *Asymptotics and special functions*, Academic Press, New York, London, 1974 (584 pp.).
148. J. Padovan, *Natural frequencies of rotating prestressed cylinders*, Journal of Sounds and Vibrations **31** (1973), no. 4, 469–482.
149. ——, *Orthogonality principle for the vibrational modes of anisotropic composite domain problems*, Trans. ASME Ser. E J. Appl. Mech. **41** (1974), no. 3, 832–834.
150. ——, *On gyroscopic problems in elasticity*, Internat. J. Engrg. Sci. **16** (1978), 1061–1073.
151. V. Panc, *Theories of elastic plates*, Noordhoff, Leyden, 1975 (716 pp.).
152. G. V. Pavilainen, *Elasto-plastic deformations of ribbed plates*, Asymptotic Methods in Mechanics (A. L. Smirnov and R. Vaillancourt, eds.), CRM Proceedings and Lecture Notes, this volume, Amer. Math. Soc., Providence, R.I., 1993, pp. 227–233.
153. D. F. Parker, *An asymptotic analysis of large deflections and rotations of elastic rods*, Internat. J. Solids and Structures **15** (1979), no. 5, 361–377.
154. G. I. Petrashen′, *On thin plate vibration theory*, Leningrad. Gos. Univ. Uchen. Zap. Ser. Mat. Nauk, vol. 24, Leningrad University Press, Leningrad, 1951, pp. 172–249. (Russian)
155. M. B. Petrov, *Study of loss of stability of shells of revolution of negative Gaussian curvature by pure bending modes*, Prikl. Mekh., vol. 4, Leningrad University Press, Leningrad, 1979, pp. 211–224. (Russian)
156. ——, *On eigen and critical sizes of shells of revolution of negative Gaussian curvature*, Prikl. Mekh., vol. 5, Leningrad University Press, Leningrad, 1981, pp. 228–238. (Russian)
157. ——, *Integral equations of super low frequency vibrations of shells of revolution of curvature which changes sign*, Prikl. Mekh., vol. 6, Leningrad University Press, Leningrad, 1984, pp. 161–168. (Russian)
158. M. B. Petrov and P. E. Tovstik, *Low-frequency vibrations of shells of revolution with a curvature which changes sign*, Proceedings of the 12-th All-Union Conference on the Theory of Shells and Plates, vol. 3, Yerevan, 1980, pp. 118–124. (Russian)
159. V. V. Pikul′, *On the construction of a physically properly conditioned theory of shells*, Izv. Akad. Nauk SSSR Mekh. Tverd. Tela **27** (1992), no. 3, 18–25; English transl., Mech. Solids **27** (1992), no. 3, 14–21.
160. A. V. Pogorelov, *Geometric theory of shell stability*, (296 pp., "Nauka", Moscow, 1966. (Russian)
161. V. V. Ponyatovsky, *An application of the asymptotic method of integration to the problem of equilibrium of a thin bar arbitrarily loaded on the side surface*, Izv. Akad. Nauk SSSR Mekh. Tverd. Tela **3** (1968), no. 5, 139–143; English transl., Mech. Solids **3** (1968), no. 5, 157–161.
162. ——, *The derivation of the equations of thin-wall beams-shells with an open profile from the equations of the theory of elasticity by the asymptotic integration method*, Issledovaniya po uprugosti i plastichnosti, vol. 13, Leningrad University Press, Leningrad, 1980, pp. 40–48. (Russian)
163. A. L. Popov and G. N. Chernyshev, *The resonance frequencies of shells oscillating in an infinite fluid*, Prikl. Mat. Mekh. **43** (1979), no. 5, 869–876; English transl., J. Appl. Math. Mech. **43** (1979), no. 5, 935–942.
164. ——, *Transition surfaces for short-wave vibrations of an ellipsoidal shell in a liquid*, Prikl. Mat. Mekh. **49** (1985), no. 5, 808–814; English transl., J. Appl. Math. Mech. **49** (1985), no. 5, 620–629.
165. G. I. Pshenichnov, *Small free vibrations of elastic shells of revolution*, Inzh. Zh. **5** (1965), no. 4, 685–690. (Russian)
166. A. L. Rabenstein, *Asymptotic solutions of* $u^{iv} + \lambda^2(zu'' + \alpha u' + \beta u) = 0$ *for large* $|\lambda|$, Arch. Rational Mech. Anal. **1** (1958), no. 5, 418–435.

167. ——, *The determination of the inverse matrix for a basic reference equation for the theory of hydrodynamic stability*, Arch. Rational Mech. Anal. **2** (1959), no. 4, 355–366.
168. A. L. Radovinsky, *Classification of free vibrations of shells containing a fluid*, Izv. Akad. Nauk SSSR Mekh. Tverd. Tela **14** (1979), no. 6, 124–135; English transl., Mech. Solids **14** (1979), no. 6, 105–114.
169. ——, *Dynamics of elastic electrically conducting shells in constant and non-stationary magnetic fields*, Prikl. Mat. Mekh. **50** (1986), no. 5, 796–803; English transl., J. Appl. Math. Mech. **50** (1986), no. 5, 614–619.
170. I. M. Rapoport, *On some asymptotic methods in the theory of differential equations*, AN USSR, Kiev, 1954 (290 pp.). (Russian)
171. M. L. Rasulov, *Methods of contour integration*, North-Holland Pub. Co., Amsterdam, 1967 (439 pp.).
172. Lord Rayleigh, *The theory of sound* I, 2nd ed., 1894 (480 pp.); II, 1899 (504 pp.); University Microfilms International, Ann Arbor, Michigan, 1985.
173. E. Reissner, *On some problems in shell theory*, Proc. First Symp. on Naval Structural Mechanics, Pergammon Press, New York, 1960, pp. 71–114.
174. E. Riekstiņč, *Asymptotic expansions of integrals* I, Zinatne, Riga, 1974 (391 pp.) (Russian); II, 1977 (370 pp.); III, 1981 (370 pp.).
175. A. Rigolot, *Théorie asymptotique des milieux curvilignes et équilibre élastique d'un cylindre infiniment élancé*, C. R. Acad. Sci. Paris Sér. I Math. **272** (1971), no. 11, 753–756. (French)
176. V. A. Rodionova, *Theory of thin anisotropic shells with transverse shifts and compression*, Leningrad University Press, Leningrad, 1983 (116 pp.). (Russian)
177. N. N. Rogacheva, *On a method of separating the state of stress in shells of negative curvature with asymptotic edges*, Prikl. Mat. Mekh. **39** (1975), no. 2, 333–341; English transl., J. Appl. Math. Mech. **39** (1975), no. 2, 311–319.
178. ——, *Precise theory of thermo-elastic shells*, Proceedings of the 10-th All-Union Conference on the Theory of Shells and Plates, Tbilisi, 1975, pp. 251–259. (Russian)
179. ——, *Classification of free piezoceramic shell vibrations*, Prikl. Mat. Mekh. **50** (1986), no. 1, 147–154; English transl., J. Appl. Math. Mech. **50** (1986), no. 1, 106–111.
180. E. W.Ross (Jr.), *Asymptotic analysis of the axisymmetric vibrations of shells*, Trans. ASME Ser. E J. Appl. Mech. **33** (1966), no. 1, 85–92.
181. ——, *On inextensional vibrations of thin shells*, Trans. ASME Ser. E J. Appl. Mech. **35** (1968), no. 3, 516–523.
182. E. W. Ross (Jr) and W. T. Matthews, *Frequencies and mode shapes for axisymmetric vibration of shells*, Trans. ASME Ser. E J. Appl. Mech. **34** (1967), no. 1, 73–80.
183. H. S. Rutten, *Theory and design of shells on the basis of asymptotic analysis*, Rutten & Kruisman, Consulting Engineers, Rijswijk, 1973 (645 pp.).
184. B. de Saint-Venant, *Mémoire sur la torsion des prismes ...*, Mémoire à l'Académie des Sciences. Mémoires des Savants étrangers, vol. 14, Paris, 1855, pp. 233–560. (French)
185. E. Sanchez-Palencia, *Non-Homogeneous Media and Vibration Theory*, Lecture Notes in Phys., vol. 127, Springer-Verlag, Berlin, New York, 1980 (398 pp.).
186. E. Schmidt, *Zur Theorie linearen und nichtlinearen Integralgleichungen*, Math. Ann. **65** (1910), no. 111, 370–399. (German)
187. A. B. Shabat, *Boundary problems with a small parameter for ordinary linear differential equations*, Uspekhi Mat. Nauk (N.S.) **17** (1962), no. 1(103), 235–241. (Russian)
188. Yu. P. Shcheviev, *Elastic wave propagation through elastic shells*, Asymptotic Methods in Mechanics (A. L. Smirnov and R. Vaillancourt, eds.), CRM Proceedings and Lecture Notes, this volume, Amer. Math. Soc., Providence, R.I., 1993, pp. 235–244.
189. E. L. Shenderov, *Wave problems of hydro-acoustics*, (352 pp., Sudostroenie, Leningrad, 1972. (Russian)
190. Shih-sen Wang and Chen Yu, *Effect of rotation on vibrations of circular cylindrical shells*, J. Acoust. Soc. Amer. **55** (1974), no. 6, 1340–1342.
191. V. P. Shirshov, *Local stability of shells*, Proceedings of the 2-nd All-Union Conference on the Theory of Shells and Plates, Kiev, 1962, pp. 314–317. (Russian)
192. Yu. B. Shneerson, *Dynamic stability and forced vibrations of a horizontal rotor with a cracked shaft*, Asymptotic Methods in Mechanics (A. L. Smirnov and R. Vaillancourt, eds.), CRM Proceedings and Lecture Notes, this volume, Amer. Math. Soc., Providence, R.I., 1993, pp. 245–253.

193. I. Ya. Shtaerman, *On the integration of the equilibrium differential equations of elastic shells*, Izvestiya Kievskogo politekhnicheskogo instituta **2** (1927), Kiev, 96–100. (Russian)
194. Y. Sibuya, *Asymptotic solutions of a linear ordinary differential equation of the n-th order at a turning point*, Funkcial. Ekvac. **4** (1962), no. 2, 83–113.
195. ——, *Simplification of a linear ordinary differential equation of the n-th order at a turning point*, Arch. Rational Mech. Anal. **13** (1963), no. 3, 206–221.
196. ——, *Uniform simplification in a full neighborhood of a transition point*, Mem. Amer. Math. Soc., vol. 149, 1974 (106 pp.).
197. E. Skudrzyk, *Simple and complex vibratory systems*, Pennsylvania State Univ. Press, University Park, Pennsylvania, 1968 (514 pp.).
198. A. L. Smirnov, *The integrals of equations of oscillations of the rotating shells of revolution*, Vestnik Leningrad. Univ. Mat. Mekh. Astronom. (1981), no. 13, 114–117.
199. ——, *Free vibrations of the rotating shells of revolution*, Trans. ASME Ser. E J. Appl. Mech. **56** (1989), no. 2, 423–429.
200. A. L. Smirnov and P. E. Tovstik, *A qualitative investigation of the dynamics of rotating shells of revolution*, Modern Problems of Mechanics and Aviation, Mashinostroenie, Moscow, 1982, pp. 280–290. (Russian)
201. R. Southwell, *On the collapse of tubes by external pressure*, Philos. Mag. Ser. 6 **25** (1913), 687–697; **26** (1913), 502–510; **29** (1915), 67–76.
202. L. S. Srubshchik, *Buckling and post-buckling shell behaviour*, Rostov University Press, Rostov, 1981 (96 pp.). (Russian)
203. R. C. Tennyson, *Buckling of circular cylindrical shells in axial compression*, AIAA J. **2** (1964), no. 7, 1351-1353.
204. I. Yu. Teterin, *On the stability of thin elastic imperfect shells of revolution*, Vestnik Leningrad. Univ. Mat. Mekh. Astronom. (1984), no. 19, 54–62; English transl., Len. Univ. Mech. Bull. (1984), no. 4, 41–52.
205. J. M. T. Thompson and G. W. Hunt, *A general theory of elastic stability*, Wiley, London, New York, Sydney, Toronto, 1973 (322 pp.).
206. A. N. Tikhonov, *Systems of differential equations containing small parameters in the derivatives*, Mat. Sb. (N.S.) **31 (73)** (1952), no. 3, 575–586. (Russian)
207. G. V. Timofeeva, *Local buckling of cylindrical shells under nonuniform axial pressure*, Prikl. Mekh., Leningrad University Press, Leningrad, 1984, pp. 184–188.
208. ——, *Loss of stability by a cylindrical shell upon bending by a force and moments*, Vestnik Leningrad. Univ. Mat. Mekh. Astronom. (1984), no. 19, 99–101; English transl., Len. Univ. Mech. Bull. (1984), no. 4, 72–76.
209. G. V. Timofeeva and P. E. Tovstik, *One representation of the solutions in problems of vibration and stability of shells*, Vestnik Leningrad. Univ. Mat. Mekh. Astronom. (1985), no. 1, 73–79; English transl. Len. Univ. Mech. Bull. (1985), no. 1, 41–50.
210. S. P. Timoshenko, *On the problem of deformation and stability of cylindrical shells*, Vestnik obshchestva tekhnologii **21** (1914), 785–792. (Russian)
211. P. E. Tovstik, *Normal degeneration of boundary problems*, Vestnik Leningrad. Univ. Mat. Mekh. Astronom. (1963), no. 19, 124–134. (Russian)
212. ——, *Vibrations of plane spiral springs*, Issledovaniya po uprugosti i plastichnosti, vol. 2, Leningrad University Press, Leningrad, 1963, pp. 105–119. (Russian)
213. ——, *Free vibrations of a thin spherical cupola*, Izv. Akad. Nauk SSSR Mekhanika (1965), no. 6, 111–113. (Russian)
214. ——, *Integrals of the equations of axisymmetric stable vibrations of shells of revolution*, Issledovaniya po uprugosti i plastichnosti, vol. 4, Leningrad University Press, Leningrad, 1965, pp. 117–122. (Russian)
215. ——, *Integrals of equations of nonaxisymmetric vibrations of shells of revolution*, Issledovaniya po uprugosti i plastichnosti, vol. 5, Leningrad University Press, Leningrad, 1966, pp. 45–56. (Russian)
216. ——, *On the frequency spectrum of vibrations of shells of revolution with a large wave number along the parallel in specific cases*, Issledovaniya po uprugosti i plastichnosti, vol. 5, Leningrad University Press, Leningrad, 1966, pp. 57–69. (Russian)
217. ——, *Asymptotic integration of the equations of free axisymmetric vibrations of a thin shell of revolution in the case of a single turning point*, Izv. Akad. Nauk SSSR Mekh. Tverd. Tela **2** (1967), no. 4, 124–132; English transl., Mech. Solids **2** (1967), no. 4, 81–86.

218. ———, *On the determination of the lowest vibration frequency of conical shells of revolution*, Issledovaniya po uprugosti i plastichnosti, vol. 6, Leningrad University Press, Leningrad, 1967, pp. 109–116. (Russian)
219. ———, *Symmetrical vibration of a shell of revolution in the case of a turning point of the second order*, Vestnik Leningrad. Univ. Mat. Mekh. Astronom. (1967), no. 1, 118–124. (Russian)
220. ———, *Buckling of shells of revolution in the linear approximation*, Raschety prostranstvennykh konstruktsii, Stroiizdat, Moscow, 1970, pp. 118–138. (Russian)
221. ———, *Integrals of linear equations of order $m + n$ with a small parameter multiplying the derivatives*, Differentsial′nye Uravneniya **6** (1970), no. 6, 989–999; English transl., Differential Equations **6** (1970), no. 6, 761–769.
222. ———, *Nonaxisymmetric vibrations of shells of revolution with small wave numbers along the parallels*, Issledovaniya po uprugosti i plastichnosti, vol. 8, Leningrad University Press, Leningrad, 1971, pp. 131–140. (Russian)
223. ———, *On the density of vibration frequencies of thin shells of revolution*, Prikl. Mat. Mekh. **36** (1972), no. 2, 291–300; English transl., J. Appl. Math. Mech. **36** (1972), no. 2, 270–278.
224. ———, *Integrals of the equations of vibrations of shells of revolution with a large wave number along the parallels and with multiple turning points*, Issledovaniya po uprugosti i plastichnosti, vol. 10, Leningrad University Press, Leningrad, 1973, pp. 103–109. (Russian)
225. ———, *Low-frequency oscillations of convex shell of revolution*, Izv. Akad. Nauk SSSR Mekh. Tverd. Tela **10** (1975), no. 6, 110–116; English transl., Mechanics of Solids **10** (1975), no. 6, 95–100.
226. ———, *On the determination of the lowest frequencies of the free vibrations of thin shells*, Asymptotic Methods in the Theory of Systems, vol. 8, Irkutsk, 1975, pp. 5–22. (Russian)
227. ———, *On the local buckling of shells*, Vestnik Leningrad. Univ. Mat. Mekh. Astronom. (1982), no. 13, 72–78. (Russian)
228. ———, *Some problems of the stability of cylindrical and conical shells*, Prikl. Mat. Mekh. **47** (1983), no. 5, 815–822; English transl., J. Appl. Math. Mech. **47** (1983), no. 5, 657–663.
229. ———, *Two-dimensional problems of stability and oscillations of shells of zero Gaussian curvature*, Dokl. Akad. Nauk SSSR **271** (1983), 69–71; English transl., Soviet Phys. Dokl. **28** (1983), no. 7, 593–594.
230. ———, *The WKB method in two-dimensional problems of buckling and vibrations of thin shells*, Proceedings of the 13-th All-Union Conference on the Theory of Shells and Plates, Tallinn, 1983, pp. 194–199. (Russian)
231. ———, *Local loss of stability by cylindrical shells under axial compression*, Vestnik Leningrad. Univ. Mat. Mekh. Astronom. (1984), no. 1, 84–89; English transl., Len. Univ. Mech. Bull. (1984), no. 1, 46–54.
232. ———, *On the vibrations and stability of shells of revolution with intervals of positive and negative Gaussian curvature*, Prikl. Mekh., vol. 6, Leningrad University Press, Leningrad, 1984, pp. 145–152. (Russian)
233. ———, *Semi-momentless buckling modes of cylindrical and conical shells*, Proceedings of the 14-th All-Union Conference on the Theory of Shells and Plates, vol. 2, Tbilisi, 1987, pp. 501–506. (Russian)
234. ———, *Influence of boundary conditions on stability of cylindrical shells*, Vestnik Leningrad. Univ. Mat. Mekh. Astron. (1989), no. 3, 66–71; English transl., Len. Univ. Mech. Bull. (1989), no. 3, 39–45.
235. ———, *Surface bending and stability of shells*, Prikl. Mekh., vol. 8, Leningrad University Press, Leningrad, 1990, pp. 137–162. (Russian)
236. ———, *Buckling of quasi-cylindrical shells*, Vestnik Leningrad. Univ. Mat. Mekh. Astronom. (1990), no. 1, 76–80; English transl., Len. Univ. Mech. Bull. (1990), no. 1, 19–23.
237. ———, *Buckling of thin shells due to weak edge restraint*, Vestnik Leningrad. Univ. Mat. Mekh. Astronom. **24** (1991), no. 3, 76–81; English transl., Len. Univ. Mech. Bull. (1991), no. 3, 16–21.
238. ———, *On the forms of local buckling of thin elastic shells*, Trans. CSME **15** (1991), no. 3, 199–211.
239. ———, *Free high-frequency vibrations of anisotropic plates of variable thickness*, Prikl. Mat. Mekh. **56** (1992), no. 3, 473–479; English transl., J. Appl. Math. Mech. **56** (1992), no. 3, 390–395.

240. ——, *Edge effect under large axisymmetric deformations of shells of revolution*, Asymptotic Methods in Mechanics (A. L. Smirnov and R. Vaillancourt, eds.), CRM Proceedings and Lecture Notes, this volume, Amer. Math. Soc., Providence, R.I., 1993, pp. 255–269.

241. ——, *Turning points and caustics in linear problems of thin shell free vibrations and buckling*, Asymptotic Methods in Mechanics (A. L. Smirnov and R. Vaillancourt, eds.), CRM Proceedings and Lecture Notes, this volume, Amer. Math. Soc., Providence, R.I., 1993, pp. 271–278.

242. V. A. Trenogin, *The development and applications of the asymptotic method of Lyusternik and Vishik*, Uspekhi Mat. Nauk (N.S.) **25** (1970), no. 4(154), 123–156; English transl., Russian Math. Surveys **25** (1970), no. 4, 119–156.

243. ——, *Functional analysis*, "Nauka", Moscow, 1980 (496 pp.). (Russian)

244. T. T. Tsiryulis, *The gradient line method for asymptotic representation of contour integrals*, Zinatne, Riga, 1973 (128 pp.). (Russian)

245. H. L. Turrittin, *Stokes multipliers for asymptotic solutions of a certain differential equation*, Trans. Amer. Math. Soc. **68** (1950), 304–329.

246. ——, *Asymptotic expansions of solutions of systems of ordinary differential equations*, Contributions to the Theory of Nonlinear Oscillations, vol. 29, Princeton University Press, Princeton, N.J., 1952, pp. 81–116.

247. B. R. Vainberg, *Asymptotic methods in equations of mathematical physics*, Moscow University Press, Moscow, 1982 (296 pp.). (Russian)

248. M. M. Vainberg and V. A. Trenogin, *Bifurcation theory of solutions of nonlinear equations*, "Nauka", Moscow, 1969 (528 pp.). (Russian)

249. M. Van Dyke, *Perturbation methods in fluid mechanics*, Applied Mathematics and Mechanics Series, vol. 8, Academic Press, New York, London, 1964 (230 pp.).

250. V. V. Vasil′ev, *The theory of thin plates*, Izv. Akad. Nauk SSSR Mekh. Tverd. Tela **27** (1992), no. 3, 26–47; English transl., Mech. Solids **27** (1992), no. 3, 22–42.

251. D. G. Vasil′ev and A. L. Gol′denveizer, *Frequency distribution of free vibrations of two- and three-dimensional elastic bodies*, Mekhanika i nauchno-tekhnicheskii progress. Mekhanika deform. tverd. tela, vol. 3, "Nauka", Moscow, 1988, pp. 223–236. (Russian)

252. D. G. Vasil′ev and V. B. Lidsky, *Vibrations of shells in contact with a liquid*, Uspekhi Mat. Nauk (N.S.) **37** (1982), no. 5(227), 224–226. (Russian)

253. A. B. Vasil′eva and V. F. Butuzov, *Asymptotic expansions of solutions of singularly perturbed equations*, "Nauka", Moscow, 1973 (272 pp.). (Russian)

254. A. B. Vasil′eva, *Singularly perturbed equations in the critical case*, University of Wisconsin, Mathematics Research Center, Madison, Wisconsin, 1980 (156 pp.).

255. N. D. Veksler, *Information problems of hydro-elasticity*, Valgus, Tallinn, 1982 (246 pp.). (Russian)

256. N. I. Vekua, *On an analytical method of prismatic shells*, Transactions of the Tbilisi Mathematical Institute, vol. 21, 1955, pp. 191–259. (Russian)

257. V. G. Vil′ke, *Inertial properties of eigenforms of an axisymmetric elastic body*, Vestnik Moskov. Univ. Ser. I Mat. Mekh. (1986), no. 2, 66–72. (Russian)

258. M. I. Vishik and L. A. Lyusternik, *Regular degeneration and boundary layer for linear differential equations with small parameter*, Uspekhi Mat. Nauk (N.S.) **12** (1957), no. 5(77), 3–122. (Russian)

259. ——, *The asymptotic behavior of solutions of linear differential equations with large or quickly changing coefficients and boundary conditions*, Uspekhi Mat. Nauk (N.S.) **15** (1960), no. 4(94), 27–95; English transl., Russian Math. Surveys **15** (1960), no. 4, 23–91.

260. V. Z. Vlasov, *General shell theory with applications to engineering*, Gostekhizdat, Moscow, 1949 (784 pp.). (Russian)

261. A. S. Vol′mir, *Buckling of deformable systems*, "Nauka", Moscow, 1967 (984 pp.). (Russian)

262. Yu. S. Vorobiev and S. I. Detistov, *On the influence of centrifugal forces on the eigenvalues of circular conical shells*, Problemy mashinostroeniya (1981), no. 14, 27-29.

263. I. I. Vorovich and V. A. Babenko, *Mixed dynamical problems of elasticity theory for nonclassical domains*, "Nauka", Moscow, 1979 (319 pp.). (Russian)

264. W. Wasow, *A study of the solutions of the differential equation* $y^{(4)} + \lambda^2(xy'' + y) = 0$ *for large values of* λ, Ann. of Math. **52** (1950), no. 2, 350–361.

265. ——, *Asymptotic solution of the differential equations of hydrodynamic stability in a domain containing a transition point*, Ann. of Math. **58** (1953), no. 2, 222-252.

266. ——, *Asymptotic expansions for ordinary differential equations*, Pure Appl. Math., vol. 14 (R. Courant, L. Bers, and J. J. Stoker, eds.), Interscience, New York, London, Sydney, 1965 (362 pp.).
267. N. Yamaki, *Elastic Stability of Circular Cylindrical Shells*, Series in Appl. Math. and Mech., vol. 27, North-Holland, Amsterdam, New York, Oxford, 1984 (558 pp.).
268. V. I. Zimin, *Small vibrations of axially loaded shells of revolution*, Samarkand State University, Samarkand, 1958 (8 pp.). (Russian)
269. P. A. Zhilin, *Mechanics of deformable directed surfaces*, Internat. J. Solids and Structures **12** (1976), no. 9/10, 635–648.
270. ——, *The plate theories of Poisson and Kirchhoff from the standpoint of contemporary plate theory*, Izv. Akad. Nauk SSSR Mekh. Tverd. Tela **27** (1992), no. 3, 48–64; English transl., Mech. Solids **27** (1992), no. 3, 43–58.
271. V. F. Zhuravliev and A. L. Popov, *Precession of the natural vibrational mode of a spherical shell upon rotation*, Izv. Akad. Nauk SSSR Mekh. Tverd. Tela **20** (1985), no. 1, 147–151; English transl., Mech. Solids **20** (1985), no. 1, 138–141.
272. A. Zohar and J. Aboudi, *The free vibrations of a thin circular finite rotating cylinder*, Int. J. Mech. Sci. **15** (1973), 269–278.

Part 2

Thirteen Papers

Centre de Recherches Mathématiques
CRM Proceedings and Lecture Notes
Volume 3, 1993

Perturbation Methods in Eddy Current Testing

Maximilian Ya. Antimirov, Andrei A. Kolyshkin, and Rémi Vaillancourt

1. Introduction

Recently, great attention has been devoted to the development of new methods for the solution of eddy current testing problems [**4, 14, 7**]. Since the mathematical analysis of the problem is not simple, numerical methods are widely used in eddy current modeling [**6, 13, 8, 10**]. In spite of the existence of powerful computers, numerical solution of three-dimensional eddy current problems is a difficult task.

On the other hand, analytical solutions to eddy current problems have been found only in very simple cases [**11, 3, 12**]. If, for example, a uniform conducting medium contains a flaw whose properties differ significantly from those of the surrounding medium, then in general one has to resort to a numerical solution. However, in some cases, analytical solutions can be found by means of the perturbation theory as, for example, in [**9**].

It is natural to apply the method of small perturbations to solve eddy current problems connected with quality control of spot welds. Spot welding produces a cast core whose electric conductivity, σ_2, and magnetic permeability, μ_2, are close to the conductivity, σ_1, and magnetic permeability, μ_1, of the surrounding medium so that $\epsilon = 1 - \sigma_2/\sigma_1$ and $\epsilon_1 = 1 - \mu_2/\mu_1$ are small parameters. This approach is used in [**5, 1, 2**] for the solution of eddy current testing problems in the case $\epsilon_1 = 0$. The formula for the change in impedance caused by a cylindrical flaw of finite length is found in [**5**]. A similar approach is used in [**1**] for an infinitely long flaw with rectangular cross-section. Finally, in [**2**] an asymmetric eddy current testing problem is solved in the case of a flaw of arbitrary shape.

If the medium is ferromagnetic then, in general, $\epsilon_1 \neq 0$. Hence the derivatives of the vector potential have a finite jump at the interface between the media. This last circumstance introduces considerable modifications in the method of perturbations.

In the present paper a formula for the change in impedance caused by a circular flaw of finite length is found in the case $\epsilon_1 \neq 0$.

2. Formulation of the problem

Let R_0 denote free space which consists of the upper half-space, $z > 0$, where the positive z-axis points upwards. The lower half-space consists of an isotropic

1991 *Mathematics Subject Classification.* Primary: 35Q60; Secondary: 65T05.

Key words and phrases. Heat source, Laplace transform.

This work was supported in part by the Natural Sciences and Engineering Research Council of Canada under grant A 7691 and the Centre de recherches mathématiques of the Université de Montréal.

This is the final form of the paper.

1065-8580/93 $1.00 + $.25 per page

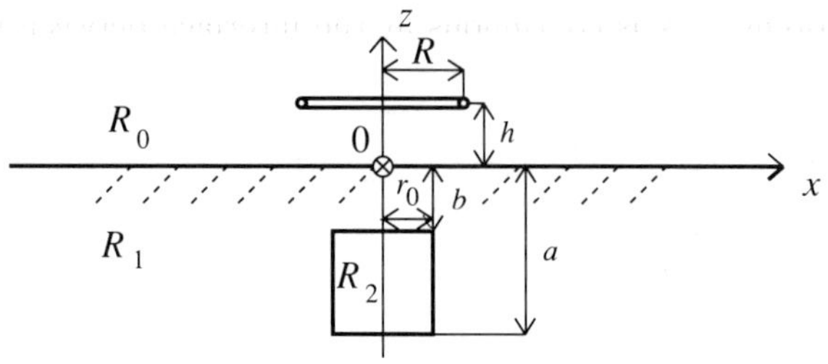

FIGURE 1. Cross-section showing circular coil in free space $R_0 : z > 0$, and conducting medium R_1 in the lower half-space $z < 0$ with cylindrical flaw R_2 of radius r_0 and height $a - b$, and at distance b below free space.

homogeneous medium in the region R_1, of electric conductivity σ_1 and magnetic permeability μ_1, containing a flaw, R_2, in the form of a circular cylinder of radius r_0 and height $a - b$. The top and bottom surfaces of the cylinder are situated on the planes $z = -b$ and $z = -a$, respectively. Hence the flaw consists of the region

$$R_2 = \{0 \le r \le r_0, \ 0 \le \varphi \le 2\pi, \ -a \le z \le -b\}.$$

The electric conductivity and magnetic permeability of the flaw are equal to σ_2 and μ_2, respectively. The geometry of the problem is shown in Fig. 1.

A circular coil of radius R, whose axis coincides with the axis of the flaw, is situated in the plane $z = h$. The coil is excited by an alternating current, $Ie^{i\omega t}$.

The problem is to determine the vector potential

$$\tilde{\mathbf{A}} = \mathbf{A}\, e^{i\omega t}$$

in each of the three regions, R_0, R_1 and R_2.

Since the region R_2 is symmetric with respect to the coil axis, the vector $\mathbf{A}$ has only one non-zero component, namely,

$$\mathbf{A} = \{0, A(r, z), 0\}, \tag{1}$$

where (r, φ, z) are cylindrical polar coordinates. In this case, Maxwell's equations reduce to the following differential equations for the function $A(r, z)$ in the three regions, R_0, R_1 and R_2:

$$LA_0 = -\mu_0 I \delta(r - R)\delta(z - h), \qquad (r, \varphi, z) \in R_0, \tag{2}$$

$$LA_1 + k_1^2 A_1 = 0, \qquad (r, \varphi, z) \in R_1 \setminus R_2, \tag{3}$$

$$LA_2 + k_2^2 A_2 = 0, \qquad (r, \varphi, z) \in R_2, \tag{4}$$

where

$$L = \frac{\partial^2}{\partial r^2} + \frac{1}{r}\frac{\partial}{\partial r} - \frac{1}{r^2} + \frac{\partial^2}{\partial z^2},$$

$$k_1^2 = -i\omega\sigma_1\mu_1, \qquad k_2^2 = -i\omega\sigma_2\mu_2,$$

A_i is the φ-component of the vector potential in the region R_i, $\delta(x)$ is the Dirac delta function and μ_0 is the primary magnetic constant.

The vector potential $\mathbf{A}$ is continuous at the interface between the regions R_0, R_1 and R_2, that is

$$A_0|_{z=0+} = A_1|_{z=0-}, \tag{5}$$

$$A_1|_{z=-b+0} = A_2|_{z=-b-0}, \qquad 0 \le r \le r_0, \tag{6}$$

$$A_1|_{z=-a-0} = A_2|_{z=-a+0}, \qquad 0 \le r \le r_0, \tag{7}$$

$$A_1|_{r=r_0+0} = A_2|_{r=r_0-0}, \qquad -a \le z \le -b. \tag{8}$$

The tangential component of the magnetic field strength, $\mathbf{H}$, is continuous at the interface between the media, so that

$$\mathbf{n} \times \frac{1}{\mu_i} \operatorname{curl} \mathbf{A}_i = \mathbf{n} \times \frac{1}{\mu_{i+1}} \operatorname{curl} \mathbf{A}_{i+1}, \qquad i = 0, 1, \tag{9}$$

where $\mathbf{n}$ is the outer unit normal to the interface. Using (1) we have

$$\operatorname{curl} \mathbf{A} = -\frac{\partial A}{\partial z}\mathbf{e}_r + \frac{1}{r}\frac{\partial}{\partial r}(rA)\mathbf{e}_z. \tag{10}$$

At the interface $z = 0$, we have $\mathbf{n} = (0, 0, 1)$; therefore from (9) and (10) we obtain

$$\left.\frac{1}{\mu_0}\frac{\partial A_0}{\partial z}\right|_{z=0+} = \left.\frac{1}{\mu_1}\frac{\partial A_1}{\partial z}\right|_{z=0-}. \tag{11}$$

Similarly,

$$\left.\frac{1}{\mu_1}\frac{\partial A_1}{\partial z}\right|_{z=-b+0} = \left.\frac{1}{\mu_2}\frac{\partial A_2}{\partial z}\right|_{z=-b-0}, \qquad 0 \le r \le r_0, \tag{12}$$

$$\left.\frac{1}{\mu_1}\frac{\partial A_1}{\partial z}\right|_{z=-a-0} = \left.\frac{1}{\mu_2}\frac{\partial A_2}{\partial z}\right|_{z=-a+0}, \qquad 0 \le r \le r_0. \tag{13}$$

On the lateral surface of the cylinder (in region R_2) we have $\mathbf{n} = (1, 0, 0)$. Hence, substituting (10) into (9) we obtain

$$\left.\frac{1}{\mu_1}\frac{1}{r}\frac{\partial}{\partial r}(rA_1)\right|_{r=r_0+0} = \left.\frac{1}{\mu_2}\frac{1}{r}\frac{\partial}{\partial r}(rA_2)\right|_{r=r_0-0}, \qquad -a < z < -b. \tag{14}$$

Thus the mathematical formulation of the problem is as follows: find the solution of system (2)–(4) which satisfies the boundary conditions (5)–(8) and (11)–(14). Because it is not possible to find the analytical solution of this problem, we consider some simplifications of the model.

3. Simplification of the model

We assume that the electric conductivity, σ_2, and the magnetic permeability, μ_2, of the flaw (region R_2) are close to those of the conducting half-space (region R_1). In this case, it is natural to introduce two small parameters:

$$\epsilon = 1 - \frac{\sigma_2}{\sigma_1}, \qquad \epsilon_1 = 1 - \frac{\mu_2}{\mu_1}, \tag{15}$$

and seek the solution to (2)–(8), (11)–(14) in the form

$$
\begin{aligned}
A_0 &= A_0^{(00)} + \epsilon A_0^{(10)} + \epsilon_1 A_0^{(01)} + \cdots, \\
A_1 &= A_1^{(00)} + \epsilon A_1^{(10)} + \epsilon_1 A_1^{(01)} + \cdots, \\
A_2 &= A_2^{(00)} + \epsilon A_2^{(10)} + \epsilon_1 A_2^{(01)} + \cdots
\end{aligned}
\tag{16}
$$

Note that (see (15))

$$\sigma_2 = \sigma_1(1-\epsilon), \qquad \mu_2 = \mu_1(1-\epsilon_1)$$

and

$$k_2^2 = -i\omega\sigma_2\mu_2 = k_1^2(1-\epsilon-\epsilon_1). \tag{17}$$

Substituting (16) and (17) into (2)–(4) and comparing the coefficients of each power of ϵ and ϵ_1 lead to a system of equations for each such power. We restrict ourselves to the determination of the first three terms in (16), namely $A_i^{(00)}$, $A_i^{(10)}$ and $A_i^{(01)}$, $i = 0, 1, 2$.

If $\epsilon = \epsilon_1 = 0$, then problem (2)–(8), (11)–(14) reduces to the following

PROBLEM I.

$$LA_0^{(00)} = -\mu_0 I\delta(r-R)\delta(z-h), \qquad z > 0, \tag{18}$$

$$LA_1^{(00)} + k_1^2 A_1^{(00)} = 0, \qquad z < 0, \tag{19}$$

with the boundary conditions

$$A_0^{(00)}\Big|_{z=0+} = A_1^{(00)}\Big|_{z=0-}, \qquad \frac{1}{\mu_0}\frac{\partial A_0^{(00)}}{\partial z}\Big|_{z=0+} = \frac{1}{\mu_1}\frac{\partial A_1^{(00)}}{\partial z}\Big|_{z=0-}. \tag{20}$$

Note that Problem I is the "unperturbed" problem. It corresponds to the case where the coil, of radius R, is situated at height h above the uniform conducting half-space (the region R_2 is absent in this case).

Next, we consider the case $\epsilon_1 = 0$, $\epsilon \neq 0$ (that is, we compare coefficients of ϵ to the first power). Problem (2)–(8), (11)–(14) reduces to the following form which will be referred to below as

PROBLEM II.

$$LA_0^{(10)} = 0, \qquad (r,\varphi,z) \in R_0, \tag{21}$$

$$LA_1^{(10)} + k_1^2 A_1^{(10)} = 0, \qquad (r,\varphi,z) \in R_1 \setminus R_2, \tag{22}$$

$$LA_2^{(10)} + k_1^2 A_2^{(10)} = k_1^2 A_2^{(00)}, \qquad (r,\varphi,z) \in R_2, \tag{23}$$

with the boundary conditions:

$$A_0^{(10)}\Big|_{z=0+} = A_1^{(10)}\Big|_{z=0-}, \qquad \frac{1}{\mu_0}\frac{\partial A_0^{(10)}}{\partial z}\Big|_{z=0+} = \frac{1}{\mu_1}\frac{\partial A_1^{(10)}}{\partial z}\Big|_{z=0-}, \tag{24}$$

$$A_1^{(10)}\Big|_S = A_2^{(10)}\Big|_S, \qquad \frac{\partial A_1^{(10)}}{\partial \mathbf{n}}\Big|_S = \frac{\partial A_2^{(10)}}{\partial \mathbf{n}}\Big|_S, \tag{25}$$

where S is the whole surface of the cylinder (the boundary of the region R_2), and $\mathbf{n}$ is the unit outer normal to S.

Finally, a third problem is obtained from (2)–(8), (11)–(14) by comparing the coefficients to the first power in ϵ_1:

PROBLEM III.

$$LA_0^{(01)} = 0, \qquad (r,\varphi,z) \in R_0, \tag{26}$$

$$LA_1^{(01)} + k_1^2 A_1^{(01)} = 0, \qquad (r,\varphi,z) \in R_1 \setminus R_2, \tag{27}$$

$$LA_2^{(01)} + k_1^2 A_2^{(01)} = k_1^2 A_2^{(00)}, \qquad (r,\varphi,z) \in R_2, \tag{28}$$

with the boundary conditions:

$$A_0^{(01)}\Big|_{z=0+} = A_1^{(01)}\Big|_{z=0-}, \qquad \frac{1}{\mu_0}\frac{\partial A_0^{(01)}}{\partial z}\Big|_{z=0+} = \frac{1}{\mu_1}\frac{\partial A_1^{(01)}}{\partial z}\Big|_{z=0-}, \tag{29}$$

$$\frac{\partial A_1^{(01)}}{\partial z}\Big|_{z=-b+0} - \frac{\partial A_1^{(00)}}{\partial z}\Big|_{z=-b+0} = \frac{\partial A_2^{(01)}}{\partial z}\Big|_{z=-b-0}, \qquad 0 \le r < r_0, \tag{30}$$

$$\frac{\partial A_1^{(01)}}{\partial z}\Big|_{z=-a-0} - \frac{\partial A_1^{(00)}}{\partial z}\Big|_{z=-a-0} = \frac{\partial A_2^{(01)}}{\partial z}\Big|_{z=-a+0}, \qquad 0 \le r < r_0, \tag{31}$$

$$\frac{\partial}{\partial r}\left[rA_1^{(01)}\right]\Big|_{r=r_0+0} - \frac{\partial}{\partial r}\left[rA_1^{(00)}\right]\Big|_{r=r_0+0} = \frac{\partial}{\partial r}\left[rA_2^{(01)}\right]\Big|_{r=r_0-0}, \qquad -a < z < -b. \tag{32}$$

4. Mathematical analysis

The solution of Problem I is found in [**11**] for the case $\mu_1 = \mu_0$ by means of the Hankel transform method. It is not difficult to obtain the solution of (18)–(20) in the case $\mu_1 \ne \mu_0$. Omitting the details, we present only the final result:

$$A_0^{(00)}(r,z) = \frac{\mu_0 IR}{2}\int_0^\infty J_1(\nu R)J_1(\nu r)e^{-\nu|z-h|}\,d\nu + \frac{\mu_0 IR}{2}\int_0^\infty \frac{\nu\mu_1 - u}{\nu\mu_1 + u}J_1(\nu R)J_1(\nu r)e^{-\nu(z+h)}\,d\nu, \tag{33}$$

$$A_1^{(00)}(r,z) = \mu_0\mu_1 IR\int_0^\infty \frac{\nu}{\nu\mu_1 + u}J_1(\nu R)J_1(\nu r)e^{uz-\nu h}\,d\nu, \tag{34}$$

where

$$u = \sqrt{\nu^2 - k_1^2}. \tag{35}$$

We now consider the solution of Problem II in greater detail. We note that the only difference between equations (22) and (23) is the non-zero right-hand side in (23). Therefore these equations can be combined into one equation with a discontinuous right-hand side in the whole region $z < 0$:

$$LA_1^{(10)} + k_1^2 A_1^{(10)} = \begin{cases} 0, & (r,\varphi,z) \in R_1 \setminus R_2, \\ k_1^2 A_1^{(00)}, & (r,\varphi,z) \in R_2, \end{cases} \tag{36}$$

where $A_1^{(00)}$ is given by (34). The subscript 1, in (36) and in the following, corresponds to values in the lower half-plane, $z < 0$. The boundary conditions (25) in this case are satisfied automatically. Hence Problem II can be formulated as follows: find the solution of system (21), (36) satisfying the boundary conditions (24).

Thus applying the Hankel transform

$$\tilde{A}_i(\lambda, z) = \int_0^\infty A_i^{(10)}(r, z) J_1(\lambda r) r\, dr, \qquad i = 0, 1,$$

to (21) and (36) we get the system of ordinary differential equations

$$\frac{d^2\tilde{A}_0}{dz^2} - \lambda^2 \tilde{A}_0 = 0, \qquad z > 0, \tag{37}$$

$$\frac{d^2\tilde{A}_1}{dz^2} - q^2 \tilde{A}_1 = \begin{cases} F(\lambda, z), & -a \le z \le -b, \\ 0, & z \notin [-a, -b], \end{cases} \tag{38}$$

with the following boundary conditions

$$\tilde{A}_0\Big|_{z=0+} = \tilde{A}_1\Big|_{z=0-}, \qquad \frac{1}{\mu_0}\frac{d\tilde{A}_0}{dz}\Big|_{z=0+} = \frac{1}{\mu_1}\frac{d\tilde{A}_1}{dz}\Big|_{z=0-}, \tag{39}$$

where

$$q = \sqrt{\lambda^2 - k_1^2} \tag{40}$$

and

$$F(\lambda, z) = k_1^2 \int_0^\infty D(\nu, \lambda)\, d\nu \tag{41}$$

is the Hankel transform of the right-hand side of (36) with

$$D(\nu, \lambda) = \mu_0\mu_1 I R r_0 \frac{\nu J_1(\nu R) e^{-\nu h}}{\nu\mu_1 + u} \frac{\nu J_1(\lambda r_0) J_0(\nu r_0) - \lambda J_0(\lambda r_0) J_1(\nu r_0)}{\lambda^2 - \nu^2}. \tag{42}$$

The solution of (37) and (38) which is bounded at infinity has the form

$$\tilde{A}_0(\lambda, z) = C_1\, e^{-\lambda z}, \qquad z > 0 \tag{43}$$

$$\tilde{A}_1(\lambda, z) = \begin{cases} C_2\, e^{qz} + C_3\, e^{-qz}, & -b \le z \le 0, \\ C_4\, e^{qz} + C_5\, e^{-qz} + f(\lambda, z), & -a \le z \le -b, \\ C_6\, e^{qz}, & z \le -a, \end{cases} \tag{44}$$

where $C_1, \dots, C_6$ are arbitrary constants and $f(\lambda, z)$ is a particular solution to (38) on the interval $[-a, -b]$. The function $f(\lambda, z)$ is sought in the form

$$f(\lambda, z) = \int_0^\infty E(\nu, \lambda)\, e^{uz}\, d\nu, \tag{45}$$

where $E(\nu, \lambda)$ is an unknown function of ν and λ. Substituting (45) into (38) we obtain

$$E(\nu, \lambda) = k_1^2 \frac{D(\nu, \lambda)}{u^2 - q^2}. \tag{46}$$

We can differentiate the integral (45) with respect to z twice because it is uniformly convergent. The values of the constants $C_1, \dots, C_6$ are determined from the boundary conditions (39) and the continuity conditions on the function $\tilde{A}_1(\lambda, z)$ and its derivatives with respect to z at $z = -b$ and $z = -a$. Since we are interested in determining only the function $\tilde{A}_0(\lambda, z)$, we present only the value of C_1:

$$C_1 = \frac{1}{\lambda\mu_1 + q}\int_0^\infty E(\nu,\lambda)(q-u)\left[e^{-(u+q)b} - e^{-(u+q)a}\right] d\nu. \tag{47}$$

Substituting (47) into (43) and taking the inverse Hankel transform we obtain the solution $A_0^{(10)}(r, z)$ in the following form:

$$\begin{aligned} A_0^{(10)}(r,z) = {} & -k_1^2\mu_0\mu_1 I R r_0 \int_0^\infty \frac{\lambda}{\lambda\mu_1 + q} J_1(\lambda r)\, e^{-\lambda z}\, d\lambda \\ & \times \int_0^\infty \frac{\nu}{\nu\mu_1 + u} J_1(\nu R)\, e^{-\nu h} \\ & \qquad \times \frac{\nu J_1(\lambda r_0) J_0(\nu r_0) - \lambda J_0(\lambda r_0) J_1(\nu r_0)}{(u+q)(\lambda^2 - \nu^2)} \\ & \qquad \times \left[e^{-(u+q)b} - e^{-(u+q)a}\right] d\nu. \end{aligned} \tag{48}$$

Note that the function $A_0^{(10)}(r, z)$ is found in [**5**] for the case $\mu_1 = \mu_0$.

As the last step of the solution, we consider Problem III. It is seen by comparing (21)–(25) and (26)–(32) that the main difference between Problem II and Problem III is in the boundary conditions at the surface S. The normal derivative of the vector potential has a finite jump on S (see (30)–(32)). As in Problem II, equations (27) and (28) can be combined into one equation (see (36)). In order to satisfy conditions (30)–(32) we add some terms to the right-hand side of (36) so that equations (27) and (28) become

$$\begin{aligned} LA_1^{(01)} + k_1^2 A_1^{(01)} = {} & \left\{ \begin{array}{ll} 0, & (r,\varphi,z) \in R_1 \setminus R_2 \\ k_1^2 A_1^{(00)}, & (r,\varphi,z) \in R_2 \end{array} \right\} \\ & + \frac{\partial A_1^{(00)}}{\partial z} \delta(z+b) 1_{[0,r_0)}(r) \\ & - \frac{\partial A_1^{(00)}}{\partial z} \delta(z+a) 1_{[0,r_0)}(r) \\ & + \frac{1}{r_0}\frac{\partial}{\partial r}(r A_1^{(00)}) \delta(r - r_0) 1_{(-a,-b)}(z), \end{aligned} \tag{49}$$

where

$$1_{(\tau_1,\tau_2)}(t) = \begin{cases} 0, & t \notin (\tau_1,\tau_2), \\ 1, & t \in (\tau_1,\tau_2). \end{cases}$$

It can easily be shown that integrating (49) with respect to z in the neighbourhoods of $z = -b$ and $z = -a$, and with respect to r in a neighbourhood of $r = r_0$, we obtain (30)–(32). Hence Problem III can be formulated as follows: find the solution of system (26) and (49) satisfying conditions (29).

Applying the Hankel transform to (26) and (49), we get

$$\frac{d^2\tilde{A}_0}{dz^2} - \lambda^2\tilde{A}_0 = 0, \qquad z > 0, \tag{50}$$

$$\frac{d^2\tilde{A}_1}{dz^2} - q^2\tilde{A}_1 = \left\{ \begin{array}{ll} F(\lambda, z), & -a \le z \le -b \\ 0, & z \notin [-a, -b] \end{array} \right\} \tag{51}$$
$$+ \int_0^\infty D(\nu, \lambda) u\, e^{uz}\, d\nu [\delta(z+b) - \delta(z+a)]$$
$$+ \mu_0\mu_1 IR\, r_0 J_1(\lambda r_0) \int_0^\infty \frac{\nu}{\nu\mu_1 + q} J_1(\lambda R)$$
$$\times e^{-\nu h} \nu\, e^{uz} J_0(\nu r_0)\, d\nu 1_{(-a,-b)}(z),$$

where $F(\lambda, z)$ and $D(\nu, \lambda)$ are given by (41) and (42), respectively. The solution of (50), (51) with boundary conditions (29) and continuity conditions on the vector potential and its derivatives with respect to z at $z = -a$ and $z = -b$, can be similarly found as it was done in Problem II.

In particular, the solution $A_0^{(01)}(r, z)$ has the form

$$A_0^{(01)}(r, z) = A_0^{(10)}(r, z) + \mu_0\mu_1 IR\, r_0 \int_0^\infty \frac{\lambda J_1(\lambda r) e^{-\lambda z}}{\lambda\mu_1 + q}\, d\lambda \tag{52}$$
$$\times \int_0^\infty \frac{\nu}{\nu\mu_1 + u} J_1(\nu R)\, e^{-\nu h} \left[e^{-(u+q)a} - e^{-(u+q)b} \right]$$
$$\times \left[u \frac{\nu J_1(\lambda r_0) J_0(\nu r_0) - \lambda J_1(\nu r_0) J_0(\lambda r_0)}{\lambda^2 - \nu^2} \right.$$
$$\left. + \frac{\nu J_1(\lambda r_0) J_0(\nu r_0)}{u+q} \right] d\nu.$$

Thus, if we take into account the linear terms with respect to the small parameters ϵ and ϵ_1, the vector potential $A_0(r, z)$ in the upper half space has the form:

$$A_0(r, z) = A_0^{(00)}(r, z) + \epsilon A_0^{(10)}(r, z) + \epsilon_1 A_0^{(01)}(r, z), \tag{53}$$

where $A_0^{(00)}(r, z)$, $A_0^{(10)}(r, z)$ and $A_0^{(01)}(r, z)$ are given by (33), (48) and (52), respectively. Note that in (33) the first term represents the vector potential of a solitary coil in unbounded free space, the second term gives the contribution from eddy currents induced in the uniform conducting lower half-space without any flaw, and the sum $\epsilon A_0^{(10)}(r, z) + \epsilon_1 A_0^{(01)}(r, z)$ in (53) represents the vector potential due to a flaw in the form of a circular cylinder of finite length.

The change in impedance, Z, due to the flaw can be found by the formula

$$Z = \frac{i\omega}{I} \oint_C \left[\epsilon A_0^{(10)}(r, z) + \epsilon_1 A_0^{(01)}(r, z) \right] dl, \tag{54}$$

where C is the contour formed by the wire of the coil.

Substituting (48) and (52) into (54) we obtain

$$Z = 2\pi\mu_0\mu_1\omega Rd\epsilon Z_1, \tag{55}$$

where

$$(56)\qquad Z_1 = \beta^2 \int_0^\infty \frac{xJ_1(\beta x)e^{-\alpha\beta x}}{x\mu_1 + \sqrt{x^2+i}}\,dx \int_0^\infty \frac{yJ_1(\beta y)e^{-\alpha\beta y}}{y\mu_1 + \sqrt{y^2+i}}$$
$$\times \left[e^{-(\sqrt{x^2+i}+\sqrt{y^2+i})\beta\delta} - e^{-(\sqrt{x^2+i}+\sqrt{y^2+i})\beta\gamma}\right]$$
$$\times \left\{ \frac{yJ_1(d\beta x)J_0(d\beta y) - xJ_0(d\beta x)J_1(d\beta y)}{x^2-y^2} \right.$$
$$\times \left[\frac{(1+\epsilon_1/\epsilon)}{\sqrt{x^2+i}+\sqrt{y^2+i}} + i\frac{\epsilon_1}{\epsilon}\sqrt{y^2+i}\right]$$
$$\left. + i\frac{\epsilon_1}{\epsilon}\frac{yJ_1(d\beta x)J_0(d\beta y)}{\sqrt{x^2+i}+\sqrt{y^2+i}} \right\} dy.$$

The parameters α, β, γ, δ and d are defined by

$$\alpha = \frac{h}{R}, \quad \beta = R\sqrt{\omega\sigma_1\mu_1}, \quad \delta = \frac{a}{R}, \quad \gamma = \frac{b}{R}, \quad d = \frac{r_0}{R}.$$

The value of Z_1 determines the change in impedance due to the flaw in the conducting half-space as a function of the parameters of the problem. This term is very important in practice because Z_1 is the value which can be measured experimentally.

5. Conclusion

An analytical formula for the change in impedance due to a flaw in the form of circular cylinder of finite length is found in this paper. If the electric conductivity and magnetic permeability of the flaw are close to those of the surrounding medium, then we can apply a perturbation method to find an analytical solution of the problem. The final result is presented in a suitable form for numerical computations.

References

1. M. Ya. Antimirov, A. A. Kolyshkin and Rémi Vaillancourt, *Eddy current nondestructive testing by a perturbation method*, J. Nondestr. Eval. **10** (1991), no. 1, 31–37.
2. ——, *A perturbation method for asymmetric problems in eddy current testing*, Can. Applied Math. Quartely **1** (Summer 1993), no. 3 (to appear).
3. C. V. Dodd and W. E. Deeds, *Analytical solutions to eddy current probe coil problems*, J. Appl. Phys. **39** (1968), no. 6, 2829–2838.
4. W. S. Dunbar, *The volume integral method of eddy current modeling*, J. Nondestr. Eval. **5** (1985), no. 1, 9–14.
5. V. S. Fastrickii, M. Ya. Antimirov and A. A. Kolyshkin, *Application of the method of perturbations for the computation of superposed transducers*, Methods and Instruments for Automatic Nondestructive Testing, Riga Polytechnical Institute, Riga, 1983, pp. 12–22. (Russian)
6. G. L. Hower and D. E. Hadlock, *An iterative calculation of eddy current responses due to three-dimensional defects*, J. Nondestr. Eval. **8** (1989), no. 4, 247–255.
7. D. McA. McKirdy, *Recent improvements to the application of the volume integral method of eddy current modeling*, J. Nondestr. Eval. **8** (1989), no. 1, 45–52.
8. T. Nakata, N. Takahashi, K. Fujiwara and Y. Shiraki, 3-D *magnetic field analysis using special elements*, IEEE Trans. on Magnetics **TM-26** (1990), no. 5, 2379–2381.
9. A. H. Nayfeh, *Introduction to perturbation techniques*, Wiley, New York, 1981 (529 pp.).
10. L. D. Sabbagh and H. A. Sabbagh, *Eddy current modeling and flaw reconstruction*, J. Nondestr. Eval. **7** (1988), no. 1/2, 95–110.

11. V. S. Sobolev, *On the theory of a method of superposed coil with eddy current testing*, Izvestia of the Siberian Branch of the Academy of Sciences of the USSR **2** (1963), 78–88. (Russian)
12. R. L. Stoll, *The analysis of eddy currents*, Clarendon Press, Oxford, 1974 (128 pp.).
13. H. Tsuboi and M. Tanaka, *Three-dimensional eddy current analysis by the boundary element method using vector potential*, IEEE Trans. on Magnetics **TM-26** (1990), no. 2, 454–457.
14. L. Udpa and S. S. Udpa, *Solution of inverse problems in eddy current nondestructive evaluation*, J. Nondestr. Eval. **7** (1988), no. 1/2, 111–120.

ABSTRACT. A perturbation method is used to solve eddy current testing problems for a flaw in the form of a finite circular cylinder situated in a conducting half-space. The axis of the cylinder coincides with the axis of the excitation coil. The solution is found in the case where the electric conductivity and the magnetic permeability of the flaw are close to those of the conducting half-space.

DEPARTMENT OF APPLIED MATHEMATICS, THE RIGA TECHNICAL UNIVERSITY, RIGA, LATVIA, LV 1010

E-mail address: podvis@bella.riga.lv

DEPARTMENT OF APPLIED MATHEMATICS, THE RIGA TECHNICAL UNIVERSITY, RIGA, LATVIA, LV 1010

E-mail address: podvis@bella.riga.lv

DEPARTMENT OF MATHEMATICS, UNIVERSITY OF OTTAWA, OTTAWA, ON, CANADA K1N 6N5

E-mail address: rxvsg@uottawa.bitnet

Centre de Recherches Mathématiques
CRM Proceedings and Lecture Notes
Volume 3, 1993

Buckling of Thin Cylindrical Shells and Shells of Negative Gaussian Curvature

Svetlana M. Bauer, Sergei B. Filippov, Aleksandr L. Maiboroda, Andrei L. Smirnov, and Igor Yu. Teterin

1. Introduction

In this report, we present the current results on the buckling of thin elastic shells of zero and negative Gaussian curvature obtained by the research group at St. Petersburg University. To analyze the shell behaviour we use the linear theory which is valid for small deformations of thin elastic shells [**3**]. The asymptotic analysis of the equations is done by means of the small parameter h of the non-dimensional shell thickness.

The position of a point, x, on the shell neutral surface is defined by its longitudinal angle φ and the arc length of the meridian (generator). The shell is limited by two parallel planes, $x = 0$ and $x = x(\varphi)$. At each point of the neutral surface we introduce a local system of Cartesian coordinates, the axes of which are the tangent to the meridian u, the tangent to the parallel v and the initial normal w. In this system, $U = (u, v, w)$ is the displacement vector. Moreover, R is the shell radius, E is Young's modulus, ν is Poisson's ratio and m is the wave number.

The non-dimensional equations describing the small free vibrations and buckling of a thin elastic shell of revolution have the following form [**4**]:

$$(L + h^2 L_h + \omega^2 + \lambda\mu L_T)\, U = 0, \qquad \Gamma U = 0. \tag{1}$$

Here L and L_h are the linear differential operators describing the elongation and the bending-twisting of the shell, L_T is a linear differential operator produced by the pre-buckling tangential stresses, λ is the critical load parameter, μ is a small parameter whose dependence on h is determined by the shell geometry and the admissible values of ω are the eigenvalues. The operator Γ provides the linear boundary conditions. All the operators in (1) are self-adjoint.

2. Stability of a thin cylindrical shell with a slanted edge under external pressure

We study the stability of a membranous initial stress-strain state of equilibrium of a thin elastic cylindrical shell, with a slanted edge, under an uniform external lateral pressure. Moreover, we find asymptotic approximations to the critical loads

1991 *Mathematics Subject Classification*. Primary: 73K15; Secondary: 73H05.
Key words and phrases. Shell buckling, asymptotic analysis.
Presented at the First European Congress of Mathematics, in Paris in July 1992.
This is the final form of the paper.

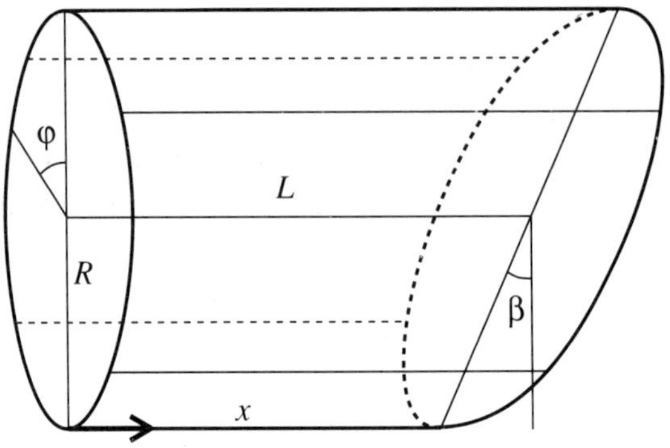

FIGURE 1. Local coordinate system on cylindrical shell.

and buckling forms. This problem is considered also in Subsection 4.4.3 of the Survey [**2**].

If, in (1), we suppose that $\omega = 0$ and $\mu = \varepsilon^6$, where $\varepsilon \sim h^{1/4}$, then the critical pressure is represented by $p = Eh\mu\lambda/R$. We assume that the shell edges, $x = 0$ and $x = l/R + \tan\beta\cos\varphi$, are clamped (see Fig. 1).

The solution, $U = (w, u, v)$, of the boundary value problem (1), according to a procedure proposed by P. E. Tovstik [**9**], can be expressed as

$$\tag{2} \begin{aligned} &\lambda = \sum_{n=0}^{\infty} \varepsilon^n \lambda_n, \qquad w = w^{(0)} + \varepsilon^2 w^{(e)}, \\ &u = \varepsilon^2 u^{(0)} + \varepsilon^4 u^{(e)}, \qquad v = \varepsilon v^{(0)} + \varepsilon^5 v^{(e)}, \\ &w^{(0)}(x,\varphi,\varepsilon) = \sum_{n=0}^{\infty} \varepsilon^n w_n(x,\varphi) \exp\left(\frac{i}{\varepsilon}\int_0^{\varphi} q(\varphi)\,d\varphi\right), \end{aligned}$$

where

$$\tag{3} \Im q_0 = 0, \quad \Im a_0 > 0, \quad q_0 = q(0), \quad a_0 = \left.\frac{dq}{d\varphi}\right|_{\varphi=0}.$$

It follows from (3) that the functions $w^{(0)}$, $u^{(0)}$ and $v^{(0)}$ are exponentially decreasing away from the line $\varphi = 0$. The edge effect functions $w^{(e)}$, $u^{(e)}$ and $v^{(e)}$ decrease exponentially away from the shell edges.

Substituting (2) into (1), one can find the unknown constants λ_n, the functions q, w_n, $w^{(e)}$, and the other functions. The zeroth-order approximation yields

$$w_0(x,\varphi) = P_0(\varphi)W_0(x), \qquad P_0 = 1 + P_{01}\varphi + P_{02}\varphi^2 + \cdots$$

The coefficients P_{0k} can be found from higher approximations. The function $W_0(x)$ is the eigenfunction for the boundary value problem

$$\tag{4} \begin{aligned} &\frac{d^4 W_0}{dx^4} - \alpha^4 W_0 = 0, \qquad \text{on } 0 < x < x_0, \\ &W_0 = \frac{dW_0}{dx} = 0, \qquad \text{at } x = 0,\ x = x_0, \end{aligned}$$

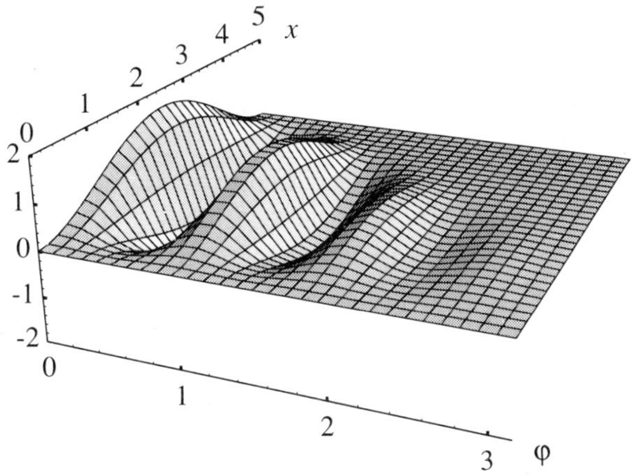

FIGURE 2. Buckling of a cylindrical shell with a slanted edge.

where $x_0 = l/R + \tan\beta$. The first eigenvalue, α_0, of (4) is the smallest positive root of the equation

$$\cosh(\alpha x_0)\cos(\alpha x_0) = 1.$$

From the higher approximation we get

$$\begin{gathered} q_0^2 = 3^{1/4}\alpha_0, \qquad \lambda_0 = \frac{4q_0^2}{3}, \qquad \lambda_1 = 2\sqrt{\lambda_0 s_0}, \\ \lambda_2 = -\frac{7}{6} + \frac{4}{3}\alpha_0^2\eta^2 + \frac{241}{54}s_0 + \frac{2\sqrt{2}\nu^2\lambda_0}{x_0} - \frac{8\alpha_0\eta(3+2\nu)}{3x_0}, \end{gathered} \tag{5}$$

where

$$s_0 = x_0^{-1}\tan\beta, \qquad \eta = \frac{\cosh z_0 - \cos z_0}{\sinh z_0 - \sin z_0}, \qquad z_0 = x_0\alpha_0.$$

We obtain two buckling forms, corresponding to the approximate value of the critical pressure p. One of them, $\Re w$, is an even function of φ and the other, $\Im w$, is an odd function of φ. This fact is caused by the existence of two eigenvalues which are so close to each other, that the proposed asymptotic method cannot distinguish between them (see Subsection 2.5.5 of the Survey [**2**]). The symmetric buckling form, $\Re w$, is shown in Fig. 2.

Similar results are obtained for other boundary conditions. For example, if the shell edges are freely supported, then $\alpha_0 = \pi/x_0$, $W_0(x) = \sin\alpha_0 x$ and

$$\lambda_2 = -\frac{7}{6} + \frac{4}{3}\alpha_0^2 + \frac{169}{54}s_0 + \frac{\sqrt{2}\lambda_0 s_0^2 x_0}{4\alpha_0^2}.$$

To evaluate q_0, λ_0 and λ_1 one can use (5).

The asymptotic method briefly described here allows one to analyze the stability and vibrations of cylindrical and conical shells when their stress-strain states are localized near some lines on the neutral surface. Some of these problems were discussed by P. E. Tovstik [**8**].

3. Stability of a rotating cylindrical shell

We now turn our attention to the buckling of the membranous stress state of a thin elastic circular cylindrical shell rotating with constant angular velocity Ω around its axis.

In this case, the term L_T in (1) is replaced by the terms introduced by the shell rotation [**7**] (see also Section 3.4 of the Survey [**2**]) to give

$$(L + h^2 L_h + \omega^2 + 2\Omega\omega L_c + \Omega^2 L_e + L_\Omega)\, U = 0, \qquad \Gamma U = 0. \tag{6}$$

Here the matrices L_c and L_e are the operators generated by the Coriolis forces and the forces of the relative motion, respectively, L_Ω is the linear differential operator describing the stresses for the initial axisymmetric state of equilibrium and the changes of the shell geometry caused by the rotation. All the operators in (6) are self-adjoint, except L_c which is antisymmetric.

Shell buckling occurs if the eigenvalue ω is equal to zero for certain values of the parameter Ω. In the linear theory, to solve this problem we assume that Ω is small and the initial deformations are bounded and of the order of Ω^2.

Taking the scalar product of (6) with U, we get the energy equation

$$\Pi_\varepsilon(U) + h^2\Pi_\varkappa(U) + \omega^2 T(U) + 2\Omega\omega T_c(U) + \Omega^2\Pi_\Omega(U) = 0.$$

The functionals $\Pi_\varepsilon(U)$ and $\Pi_\varkappa(U)$ are positive definite. Therefore, buckling occurs if the equation

$$\Pi_\varepsilon(U) + h^2\Pi_\varkappa(U) + \Omega^2\Pi_\Omega(U) = 0$$

has a real root.

In [**7**], the perturbation method is used to show that the frequencies ω and ω_0 of the rotating and non-rotating shells, respectively, are connected by the asymptotic series

$$\begin{gathered} \omega = \pm\alpha(\Omega) + \beta(\Omega), \\ \alpha(\Omega) = \sum_{i=1}^{\infty} \Omega^{2i}\alpha_{2i}, \qquad \beta(\Omega) = \sum_{i=1}^{\infty} \Omega^{2i+1}\beta_{2i+1}, \\ \beta_1 = -\frac{T_c(U_0)}{T(U_0)}. \end{gathered} \tag{7}$$

Hence shell buckling may occur for low frequency vibrations among which the so-called "pseudo-bending" modes are the most important (see Section 4.4 of the Survey [**2**]). For such vibrations, the equation

$$\Pi_\varepsilon = 0 \tag{8}$$

is precisely or approximately satisfied. From (8) it follows that $mv = w$. If the last equality holds, the functional Π_Ω is positive definite for every m different form 1. (Note that this statement is valid for the pseudo-bending of any shell of revolution.)

Therefore buckling may not occur for cylindrical shells of middle length with a strongly supported edge. On the other hand, the existence of a low frequency for $m = 1$ and, as a consequence, a lost of stability are possible either for long cylindrical shells or for shells with free or weakly supported edges. In the first case, the shell behaves like a long rotating shaft and the critical angular velocity is given by $\Omega = \omega_0$.

4. Buckling of shells of revolution of negative Gaussian curvature under nonaxisymmetric loading

We study the linear approximation to the buckling of a thin elastic shell of revolution of negative Gaussian curvature. The initial shell stress state is assumed to be nonaxisymmetric and membranous. The cases where the supports of the shell edges prevent tangential bending of the shell neutral surface are considered.

This problem has been considered in the case of initial axisymmetric stress state, for example, by T. Liiva and P. E. Tovstik [**5**] and it was found that the buckling form spreads out along the entire shell surface. The specifics of the buckling under nonaxisymmetric loading is the localization of the buckling modes in a neighbourhood of a set of asymptotic lines on the shell neutral surface as it was shown by A. L. Maiboroda [**6**]. This reduces the solution of the initial two-dimensional problem to a sequence of one-dimensional boundary value problems.

To study the buckling we use equations (1). In this case, as it was found in [**3, 5**], $\mu = h^{4/3}$. We introduce two coordinate systems, $\{x, \xi_1\}$ and $\{x, \xi_2\}$, on the neutral surface where

$$\xi_{1,2} = \varphi \pm \int_0^x l(t)\,dt, \qquad l = \sqrt{\frac{-R_2}{R_1 B^2}},$$

R_1 and R_2 are the principal radii of curvature and B is Lamé's coefficient. These coordinate systems are connected with the asymptotic lines on the shell neutral surface, since the equations

$$\xi_i(x, \varphi) = c_i = \text{const}, \qquad i = 1, 2,$$

describe two sets of asymptotic lines. When considering only one asymptotic line we shall omit the index i.

We restrict ourselves to the following types of boundary conditions:

(a) clamped-clamped shell edges,
(b) clamped-free shell edges.

The stress state consists of the main integral and the edge effect integrals. In this case, to construct the zeroth-approximation solution to the buckling mode we need only write the main integral

$$(9) \qquad \begin{gathered} U = U^*\left(x, \zeta, h^{1/6}\right) e^{ih^{-1/3}[q\xi + a(\xi-\xi_0)^2]}, \\ \zeta = h^{-1/6}(\xi - \xi_0), \qquad U^* = \sum_{n=0}^{\infty} h^{n/6} U_n^*(x, \zeta), \qquad \Im a > 0. \end{gathered}$$

Here q is real and determines the variation of the function U in the direction orthogonal to the asymptotic line. The expansion for the load parameter is

$$(10) \qquad \lambda = \lambda_0 + \lambda_2 h^{1/3} + \lambda_3 h^{1/2} + \cdots$$

Substituting (9) into the initial equations we get a system of linear equations from which we can find the unknown functions U_n^*. Moreover, for each $n > 1$, the orthogonality conditions have to be satisfied. For $n = 2$, we have

$$
(11)\qquad \begin{aligned}
&Hy_0 = \frac{d}{dx}\left[g(x)\frac{dy_0}{dx}\right] + \left[\lambda_0 q^4 f(x,\xi_0) - \frac{q^6}{g(x)}\right] y_0 = 0,\\
&y_0 = \sqrt{\frac{R_2\lambda}{B}}\, v_1, \qquad (v_0 \equiv 0),
\end{aligned}
$$

where $g(x) > 0$ is defined by the shell geometric parameters and $f(x, \xi_0)$ depends upon the initial shell stresses.

The orthogonality conditions (11) determine the dependence of the functions v_1 on the variables x. The dependence on the variable ζ is determined from the next approximations, for which the functions v_1 are polynomials in ζ.

The boundary conditions imply the conditions:

(a) $y_0(x_1) = y_0(x_2) = 0$ for the clamped-clamped edge supports,
(b) $y_0(x_1) = y_0'(x_2) = 0$ for the clamped-free edge supports.

Hence, we have the following boundary value problem for the critical load and the buckling modes

$$
\begin{aligned}
&Hy_0 = 0, \qquad y_0(x_1) = y_0(x_2) = 0, \qquad &&\text{in the case (a)},\\
&Hy_0 = 0, \qquad y_0(x_1) = \frac{dy_0}{dx}(x_2) = 0, \qquad &&\text{in the case (b)}.
\end{aligned}
$$

The zeroth approximation to the critical load is obtained as the minimum positive value of the parameter $\lambda_0(q, \xi)$ as a function of q and ξ_0.

5. Buckling of a circular cylindrical shell with initial imperfections and thickness deviations

The imperfections in the form and the local deviations in the thickness of a thin elastic shell have an important influence on the shell stability, especially for a cylindrical shell under axial compressed load. The equilibrium equation and the strain compatibility condition can be represented in the non-dimensional form

$$
(12)\qquad \begin{aligned}
&\frac{\partial^2}{\partial x^2}\left(h^3\frac{\partial^2 w}{\partial x^2}\right) + 2\frac{\partial^2}{\partial x\partial y}\left(h^3\frac{\partial^2 w}{\partial x\partial y}\right) + \frac{\partial^2}{\partial y^2}\left(h^3\frac{\partial^2 w}{\partial y^2}\right) + \nu L\left(h^3, w\right)\\
&\qquad = \Delta_k^2\Phi + L(w + w_0, \Phi),\\
&\frac{\partial^2}{\partial x^2}\left(\frac{1}{h}\frac{\partial^2 \Phi}{\partial x^2}\right) + 2\frac{\partial^2}{\partial x\partial y}\left(\frac{1}{h}\frac{\partial^2 \Phi}{\partial x\partial y}\right) + \frac{\partial^2}{\partial y^2}\left(\frac{1}{h}\frac{\partial^2 \Phi}{\partial y^2}\right) - \nu L\left(\frac{1}{h}, \Phi\right)\\
&\qquad = -\Delta_k^2 w - \frac{1}{2}L(w, w).
\end{aligned}
$$

Here Φ is the stress function and $y = R\varphi$. The operators Δ_k^2 and L are determined in Section 2.6 of the Survey [**2**]. If there is an axisymmetric local initial deflection $w_0(x)$ and the non-dimensional thickness is representable in the form $h = 1 + h_0(x)$, then the asymptotic analysis of equations (12) by means of the method described by J. C. Amazigo and B. Budiansky [**1**] and in Subsection 5.4.5 of the Survey [**2**], yields the following formula

$$
(13)\qquad 1 - \lambda = a_0\left|w_0^{(*)}(1) + \nu h_0^{(*)}(1)\right|^{2/3}
$$

for the bifurcation point in the nonaxisymmetric stress state.

Here $\lambda = T/T_{cl}$, $a_0 \approx 2.42$, w_0 is the non-dimensional initial deflection and T_{cl} is the critical load for a perfect shell. We denote by $w_0^{(*)}(u)$ the Fourier transform of $w_0(s)$:

$$w_0^{(*)}(\omega) = \frac{1}{\sqrt{2\pi}} \int_{-\infty}^{\infty} w_0(s) e^{-i\omega s}\, ds.$$

If the deviations in the shell thickness are localized in the axial direction and have the form

$$h(x, \varphi) = 1 + h_1(x)\cos(m\varphi), \qquad w_0(x, \varphi) = W_0(x)\cos(m\varphi),$$

then the ultimate (limit) point on the load-displacement diagram where the stress attains its maximum value corresponds to the critical load and the formula for the critical load in this case is

$$1 - \lambda = a_1 \left| W_0^{(*)}(0.5) + (\nu - 1) h_1^{(*)}(0.5) \right|^{8/13}, \tag{14}$$

where $a_1 \approx 0.98$.

As it was noted before, it follows from (13) and (14) that the axisymmetric local (and also regular and random) thickness deviation effect on the critical load is much smaller than the corresponding initial imperfections of the form. Nevertheless, small nonaxisymmetric thickness deviations are important, perhaps, as much as the corresponding imperfection components of the neutral surface.

6. Post-buckling behaviour of a shell

The effect of initial imperfections on the critical load and initial post-buckling behaviour of shells of revolution is studied by the method described in Subsection 5.4.6 of the Survey [**2**] and asymptotic approximations for critical loads are found.

The equations describing small finite deviations from the state of equilibrium, which is characterized by tangential stresses, have the form (see formula (5.4.24) of the Survey [**2**])

$$(M + \lambda_1 M_T)U = DU + (\lambda_1 - \lambda)U + w_0 M_0 U^0. \tag{15}$$

Here D is a nonlinear differential operator, $M_0 U^0$ is a linear operator introduced by the initial imperfections, λ is the loading parameter, and the values of the parameter λ_1 are the lowest eigenvalues corresponding to the parameter λ. The non-dimensional amplitude, w_0, of the initial imperfection and $\lambda_1 - \lambda$ are small parameters. We set $\varepsilon = \lambda_1 - \lambda$.

The asymptotic solution of system (15) corresponding to the perturbed form of equilibrium (see formula (5.4.25) in the Survey [**2**]) is represented by

$$U = \sum U_j \varkappa_j + \varepsilon \sum \varkappa_k U_k + \sum U_{jk} \varkappa_j \varkappa_k + O\left(\left|\varkappa^3\right|\right). \tag{16}$$

Here U_j are the buckling forms of tangential stresses corresponding to the eigenvalues λ_1.

Substituting (16) into (15) we get a nonhomogeneous linear equation from which the unknown "amplitudes" $\varkappa_j$ of the forms U_j and the functions U_{jk} may be found. Existence conditions on this solution yield a system of algebraic equations. It might be noted that the three plus signs in (16) provide the zeroth-order approximation to the following nonlinear algebraic system for the unknows $\varkappa_j$:

$$\Phi_j := \varkappa_j \left(\sum_k a_{jk} \varkappa_k^2 + b_j \varepsilon \right) + w_{0j} = 0. \tag{17}$$

Here w_{0j} is the parameter of the initial imperfections which is determined by

$$w_{0j} = w_0(U^0, U_j).$$

The coefficients a_j and b_j are found from integrals calculated by Laplace's asymptotic method (see Section 1.3 of the Survey [**2**]).

System (17) contains N equations which depend upon w_0 and μ in such a way that $N \sim w_0^{1/3} \mu^{-1}$. To study the effect of w_0 on ε, it is necessary to add the condition

$$\det \left\| \frac{\partial \Phi_j}{\partial \varkappa_k} \right\| = 0$$

to system (17).

The formula for ε seems to be simple if $N = 1$ or 2. Hence for a sufficiently small w_0 this formula becomes

$$\varepsilon = a w_0^{2/3}.$$

Here the coefficient a depends on the geometry and the type of the loading.

If $w_0 = 0$, i.e. for a perfect shell, ε is proportional to the load which the shell resists in the post-buckling deformation.

References

1. J. C. Amazigo and B. Budiansky, *Asymptotic formulas for the buckling stresses of axially compressed cylinders with localized or random axisymmetric imperfections*, Trans. ASME Ser. E J. Appl. Mech. **39** (1972), no. 3, 179–184.
2. S. M. Bauer, S. B. Filippov, A. L. Smirnov and P. E. Tovstik, *Asymptotic methods in mechanics with application to thin shells and plates*, Asymptotic Methods in Mechanics (A. L. Smirnov and R. Vaillancourt, eds.), CRM Proceedings and Lecture Notes, this volume, Amer. Math. Soc., Providence, R.I., 1993, pp. 3–140.
3. A. L. Gol′denveizer, V. B. Lidsky and P. E. Tovstik, *Free fibrations of thin elastic shells*, "Nauka", Moscow, 1979 (383 pp.). (Russian)
4. E. I. Grigolyuk and V. V. Kabanov, *Stability of shells*, "Nauka", Moscow, 1978 (359 pp.). (Russian)
5. T. V. Liiva and P. E. Tovstik, *On buckling in the linear approximation of shells of revolution of negative Gaussian curvature*, Problems of Solid Mechanics, Sudostroenie, Leningrad, 1970, pp. 231–238. (Russian)
6. A. L. Maiboroda, *Buckling of revolution shells of a negative Gaussian curvature under asymmetric loads*, Vestnik Leningrad. Univ. Mat. Mekh. Astronom. (1989), no. 22, 61–65; English transl., Len. Univ. Mech. Bull. (1989), no. 4, 30–36.
7. A. L. Smirnov, *Free vibrations of the rotating shells of revolution*, Trans. ASME Ser. E J. Appl. Mech. **56** (1989), no. 2, 423–429.
8. P. E. Tovstik, *Some problems of the stability of cylindrical and conical shells*, Prikl. Mat. Mekh. **47** (1983), no. 5, 815–822; English transl., J. Appl. Math. Mech. **47** (1983), no. 5, 657–663.
9. ———, *Influence of boundary conditions on stability of cylindrical shells*, Vestnik Leningrad. Univ. Mat. Mekh. Astronom. (1989), no. 15, 66–71; English transl., Len. Univ. Mech. Bull. (1989), no. 3, 39–45.

ABSTRACT. Results on the buckling of thin shells of zero and negative Gaussian curvatures, recently obtained by the research group of St. Petersburg University (Russia), are presented. The initial differential equations based on theories of Kirchhoff–Love type are studied by asymptotic methods. Buckling under external loads (axial pressure, anisotropic external pressure and combined loading) are considered by taking into account different factors (initial random imperfections, boundary conditions) as well as buckling of rotating cylinders and shells, whose forms are close to those of cylinders and cones.

INSTITUTE OF MATHEMATICS AND MECHANICS, ST. PETERSBURG STATE UNIVERSITY, 2 BIBLIOTECHNAYA SQ., STARY PETERHOF, 198904 ST. PETERSBURG, RUSSIA

E-mail address: als@hq.math.lgu.spb.su

Centre de Recherches Mathématiques
CRM Proceedings and Lecture Notes
Volume **3**, 1993

Thermo-Elastic Deformations of Mirrors

Svetlana M. Bauer and Andrei L. Smirnov

1. Introduction

It is known that very small deformations of the reflecting surfaces of the optical mirrors in a telescope affect, in an important way, the quality of the instrument. This explains why much work has been devoted to the deformation analysis and surface control of mirrors (see, for example, the large number of papers in [**8**] and [**13**] and the paper by F. B. Ray and Y. T. Chung [**15**]).

An important deforming factor is thermal loading, which inevitably acts on physical mirrors. Thermal deformations may occur not only under nonuniform heating but also under uniform heating or cooling if the mirror is manufactured from an anisotropic material and has a variable thickness or some other construction peculiarities.

The thermal properties of the different materials that are used in the construction of optical mirrors are discussed by W. J. Choyke, R. F. Farich and R. A. Hoffman [**6**]. W. P. Barnes [**1**] studies the deformation effect of specific types of construction (core configuration, lightweight mirror etc.) on the behaviour of a mirror undergoing deformation. Thermal deformations of circular multi-layer plates are analyzed by V. V. Kharitonov [**9**], C. A. Klein [**10**] and G. W. Satton [**16**].

Experimental data and results of thermal deformation analysis for some lightweight mirrors are presented by V. D. Derevenskii *et al.* [**7**]. The influence of the above mentioned factors on the thermal deformation of mirrors has also been studied by V. N. Paimushin *et al.* [**14**] and S. M. Bauer *et al.* [**4**].

It has to be noted that it is often unrealistic to try to satisfy all the various possible specifications in mirror construction.

In the present paper, we consider only one possible model of a small optical mirror and compare the analytical results obtained by the thin plate theory with the numerical results obtained by finite element methods (FEM). To this effect, we study the axisymmetric stress-strain state of a uniformly heated circular mirror. The top half-view and vertical cross-section of the mirror are shown in Fig. 1.

The maximal deflection of the mirror reflecting surface is the most important factor in our analysis.

The mirror is supported by a circular annular plate, II, of inner and outer radii $r_2 < r_0$ (see Fig. 1). Since the mirror and its support have different thermal expansion coefficients, α_1 and α_2, respectively, then, a temperature change causes a deformation of the reflecting surface of the mirror.

1991 *Mathematics Subject Classification.* Primary: 73K10; Secondary: 73V05.
Key words and phrases. Thermoelasticity, thin plate, finite elements method.
This is the final form of the paper.

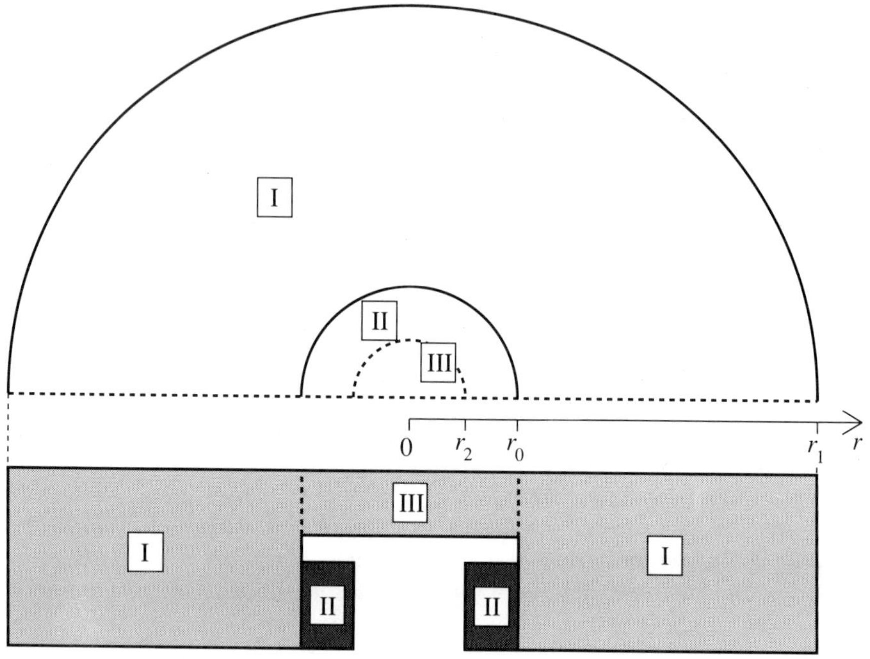

FIGURE 1. Top half-view and vertical cross-section of circular mirror

To find the equations which describe the stress-strain state of the mirror we split the mirror into two parts: a circular annular plate, I, of radii $r_0 < r_1$, and a circular plate, III, of radius r_0 (see Fig. 1), and use the theory of thin plates to analyze the stress-strain state of the three structural components, I, II and III.

2. The circular plate III

Firstly, we consider the plate III. By selecting the plate middle surface as the neutral surface, we obtain the equations, in polar coordinates, for the elastic component of the plate radial displacement, $u_3^{(e)}$, the plate deflection, w_3, the radial transverse shear force, N_3^r, and the radial bending moment, M_3^r (see W. Nowacki [**12**], A. D. Kovalenko [**11**] and Section 2.6 of the Survey [**3**]):

$$\frac{d^2u_3^{(e)}}{dr^2} + \frac{1}{r}\frac{du_3^{(e)}}{dr} - \frac{u_3^{(e)}}{r^2} = 0, \qquad \Delta^2 w_3 = 0,$$
(1)
$$N_3^r = B_3\left(\frac{du_3^{(e)}}{dr} + \frac{\nu_1}{r}u_3^{(e)}\right), \qquad M_3^r = -D_3\left(\frac{d^2w_3}{dr^2} + \frac{\nu_1}{r}\frac{dw_3}{dr}\right),$$

where the Laplacian in polar coordinates is

$$\Delta = \frac{\partial^2}{\partial r^2} + \frac{1}{r}\frac{\partial}{\partial r} + \frac{1}{r^2}\frac{\partial^2}{\partial \varphi^2},$$

the constants

$$B_3 = E_1\frac{h_3}{1-\nu_1^2}, \qquad D_3 = \frac{1}{12}\frac{h_3^3}{1-\nu_1^2},$$

are the plate stiffness coefficients, E_1 is Young's modulus, ν_1 is Poisson's ratio and h_3 is the plate thickness.

The total radial displacement, u_3, is the sum of the elastic deformation and the thermal expansion,

$$u_3 = u_3^{(e)} + \alpha_1 T r,$$

where T is the change in temperature.

For a circular plate without a central hole (i.e. a disk instead of a ring) the solution of equations (1) can be written in the form

$$u_3 = C_1 r, \tag{2}$$

$$w_3 = -\frac{C_2}{4} r^2 + C_3, \tag{3}$$

$$N_3^r = B_3 C_1 (1 + \nu_1), \tag{4}$$

$$M_3^r = \frac{C_2}{2} D_3 (1 + \nu_1). \tag{5}$$

The angle of rotation, ψ_3, of the normal, **n**, to the surface of the mirror may be found from the expression

$$\psi_3 = -\frac{dw_3}{dr} = \frac{C_2}{2} r. \tag{6}$$

Here C_1, C_2 and C_3 are constants which have to be determined from the boundary conditions or, in our case, from the joint conditions.

3. The circular annular plates I and II

Secondly, we consider the circular plates I and II. A similar problem for a circular anisotropic two-layer mirror of non-constant thickness has been considered by S. M. Bauer *et al.* [**4**], where, by taking the lower plate surface as the neutral plane, the equations for the transverse shear forces and plate bending moments have been obtained for the entire plate under the Kirchhoff hypothesis (see [**3, 11, 12**]). The system of equations obtained in [**4**], which describes an isotropic plate of constant thickness, is

$$\begin{aligned}
\frac{du_1^{(e)}}{dr} &= \left[\nu_1({L_1}^2 - D_1 B_1)\frac{u_1^{(e)}}{r} + D_1 N_1^r - L_1 M_1^r\right]\frac{1}{B_1 D_1 - {L_1}^2},\\
\frac{d\psi_1}{dr} &= \left[\nu_1({L_1}^2 - B_1 D_1)\frac{\psi_1}{r} - L_1 N_1^r + B_1 M_1^r\right]\frac{1}{B_1 D_1 - {L_1}^2},\\
\frac{dN_1^r}{dr} &= B_1\left(1-\nu_1^2\right)\frac{u_1^{(e)}}{r^2} + L_1\left(1-\nu_1^2\right)\frac{\psi_1}{r^2} - (1-\nu_1)\frac{N_1^r}{r},\\
\frac{dM_1^r}{dr} &= L_1\left(1-\nu_1^2\right)\frac{u_1^{(e)}}{r^2} + D_1\left(1-\nu_1^2\right)\frac{\psi_1}{r^2} - (1-\nu_1)\frac{M_1^r}{r},
\end{aligned} \tag{7}$$

where

$$B_1 = E_1\frac{h_1}{1-\nu_1^2}, \qquad L_1 = E_1\frac{h_1^2}{2(1-\nu_1^2)}, \qquad D_1 = E_1\frac{h_1^3}{3(1-\nu_1^2)}$$

are the stiffness coefficients for the circular plate I, h_1 is the thickness of I, $u_1^{(e)}$ is the elastic component of the radial displacement, w_1 is the deflection, ψ_1 is the angle of rotation of the normal $\mathbf{n}$, N_1^r is the radial transverse shear force and M_1^r is the bending moment of the plate. (The corresponding variables for the supporting ring II are denoted by h_2, $u_2^{(e)}$, w_2, ψ_2, N_2^r, M_2^r).

It is convenient to introduce new variables in the above system of equations by the following formulae:

$$\Psi_1 = h_1\psi_1, \qquad N_1^R = \frac{N_1^r}{B_1}, \qquad M_1^R = \frac{1}{h_1}\frac{M_1^r}{B_1}. \tag{8}$$

Then, by using

$$L_1 = h_1\frac{B_1}{2}, \qquad D_1 = \frac{h_1^2}{3}B_1,$$

equations (7) become:

$$\begin{aligned} \frac{du_1^{(e)}}{dr} &= -\nu_1\frac{u_1^{(e)}}{r} + 4N_1^R - 6M_1^R, \\ \frac{d\Psi_1}{dr} &= -\nu_1\frac{\Psi_1}{r} - 6N_1^R + 12M_1^R, \\ \frac{dN_1^R}{dr} &= \frac{1-\nu_1^2}{r^2}\left(u_1^{(e)} + \frac{\Psi}{2}\right) - (1-\nu)\frac{N_1^R}{r}, \\ \frac{dM_1^R}{dr} &= \frac{1-\nu_1^2}{r^2}\left(\frac{u_1^{(e)}}{2} + \frac{\Psi_1}{3}\right) - (1-\nu_1)\frac{M_1^R}{r}, \end{aligned} \tag{9}$$

which we rewrite in vector form,

$$\frac{d\mathbf{y}}{dr} = A(r)\mathbf{y},$$

where

$$\mathbf{y} = (u_1^{(e)}, \Psi_1, N_1^R, M_1^R)^T. \tag{10}$$

It can easily be shown that the last system has the following four linear independent solutions:

$$\begin{aligned} \mathbf{y}_1 &= \left(r, 0, 1+\nu_1, \frac{1+\nu_1}{2}\right)^T, \\ \mathbf{y}_2 &= \left(0, r, \frac{1+\nu_1}{2}, \frac{1+\nu_1}{3}\right)^T, \\ \mathbf{y}_3 &= \left(-\frac{2}{3r}, \frac{1}{r}, \frac{1-\nu_1}{6r^2}, 0\right)^T, \\ \mathbf{y}_4 &= \left(-\frac{1}{2r}, \frac{1}{r}, 0, -\frac{1-\nu_1}{12r^2}\right)^T. \end{aligned}$$

Therefore, one can find the solution of system (10) in the form

$$\mathbf{y} = G_1\mathbf{y}_1 + G_2\mathbf{y}_2 + G_3\mathbf{y}_3 + G_4\mathbf{y}_4,$$

where G_i are arbitrary constants.

Since the plate deflection, w_1, is obtained from the equation

$$-\frac{dw_1}{dr} = \psi_1,$$

then

$$w_1(r) = -G_2\frac{r^2}{2h_1} - \frac{G_3 + G_4}{h_1}\ln r + G_5,$$

with an arbitrary constant, G_5.

A system, similar to (9), can be written for the supporting ring II and its solution is

$$\mathbf{y}^k = Q_1\mathbf{y}_1^k + Q_2\mathbf{y}_2^k + Q_3\mathbf{y}_3^k + Q_4\mathbf{y}_4^k$$

for arbitrary constants Q_i.

Thus, to obtain the stress-strain state of the entire structure, one has to determine the following twelve constants:

$$C_i, \qquad i = 1, 2, 3, \qquad G_j, \qquad j = 1, \dots, 5, \qquad Q_l, \qquad l = 1, \dots, 4.$$

4. The conditions used in determining the constants

To evaluate these twelve constants, we require that the free edges satisfy four boundary conditions at $r = r_2$ and $r = r_1$ (see Fig. 1):

$$\begin{aligned} N_1^r(r_2) &= 0, \qquad N_2^r(r_1) = 0, \\ M_1^r(r_2) &= 0, \qquad M_2^r(r_1) = 0, \end{aligned} \tag{11}$$

two conditions to balance the forces and moments in the cross-section $r = r_0$:

$$\begin{aligned} N_1^r(r_0) &= N_2^r(r_0) + N_3^r, \\ M_1^r(r_0) &= M_2^r(r_0) + M_3^r + N_3^r\left(h_1 - \frac{h_3}{2}\right), \end{aligned} \tag{12}$$

and five joint conditions for "sewing" together the deflections and the rotation angles of the normal:

$$\begin{gathered} u_1^{(e)}(r_0) + \alpha_1 T r_0 = u_2^{(e)}(r_0) + \alpha_2 T r_0, \\ u_1^{(e)}(r_0) = u_3^{(e)}(r_0) - \psi_3(r_0)\left(h_1 - \frac{h_3}{2}\right), \\ \psi_1(r_0) = \psi_2(r_0) = \psi_3(r_0), \end{gathered} \tag{13}$$

and

$$w_1(r_0) = w_3(r_0). \tag{14}$$

Thus, ten constants, C_1, C_2, $G_1, \dots, G_4$, $Q_1, \dots, Q_4$, are determined from the ten linear equations (11)–(13). The constants C_3 and G_5 only affect the deflection and satisfy equation (14).

In this approach, the deflection, w, contains an arbitrary constant, the determination of which requires an extra condition. But the results obtained below are quite sufficient to find the relative deflection.

5. Determination of the constants

From the boundary condition (11) we get

$$\begin{aligned} G_3 &= -6r_2^2\frac{1+\nu_1}{1-\nu_1}\left(G_1 + \frac{G_2}{2}\right), \\ G_4 &= 12r_2^2\frac{1+\nu_1}{1-\nu_1}\left(\frac{G_1}{2} + \frac{G_2}{3}\right), \end{aligned} \tag{15}$$

and similar formulae for Q_3 and Q_4.

By (15), (8) and the joint conditions (12)–(13), we have

$$
\begin{aligned}
G_1 + \frac{G_2}{2} &= b\varkappa_2\left(Q_1 + \frac{Q_2}{2}\right) + NN,\\
\frac{G_1}{2} + \frac{G_2}{3} &= b\varkappa_2\left(\frac{Q_1}{2} + \frac{Q_2}{3}\right) + MM,\\
G_1 a_2 &= Q_1 a_1 + T r_0(\alpha_2 - \alpha_1) + UU,\\
G_2 a_2 \varkappa_2 &= Q_2 a_1 = \Psi,
\end{aligned}
\tag{16}
$$

where $\Psi = h_2\psi_3$,

$$
\begin{aligned}
NN &= N_3^r \frac{\varkappa_3}{B_1(1+\nu_1)} y,\\
UU &= u_3^{(e)} - h_1\left(1 - \frac{\varkappa_3}{2}\right)\psi_3,\\
MM &= M_3^r + h_1\left(1 - \frac{\varkappa_3}{2}\right) N_3^r \frac{\varkappa_3}{B_1 h_1(1+\nu_1)} y,
\end{aligned}
$$

and

$$
\begin{aligned}
y &= \frac{r_0^2}{r_0^2 - r_2^2}, \qquad b = \frac{E_2}{E_1}\frac{1-\nu_1}{1-\nu_2}\frac{r_0^2 - r_1^2}{r_0^2 - r_2^2},\\
\varkappa_2 &= \frac{h_2}{h_1}, \qquad \varkappa_3 = \frac{h_3}{h_1},\\
a_1 &= r_0 + \frac{r_1^2}{r_0}\frac{1+\nu_2}{1-\nu_2}, \qquad a_2 = r_0 + \frac{r_2^2}{r_0}\frac{1+\nu_1}{1-\nu_1}.
\end{aligned}
$$

It follows from equations (16) that

$$
\begin{aligned}
G_2 &= \frac{\Psi}{a_2\varkappa_2}, \qquad Q_2 = \frac{\Psi}{a_1},\\
G_1 &= \frac{UU}{a_2}, \qquad Q_1 = \frac{1}{a_1}\left[UU - T r_0(\alpha_2 - \alpha_1)\right],
\end{aligned}
\tag{17}
$$

and

$$
\begin{aligned}
NN &= c_{11}\left[u_3^{(e)} - \frac{\Psi}{\varkappa_2}(1 - \varkappa_3)\right] + c_{12}\Psi + d_1,\\
MM &= \varkappa_2 c_{12}\left[u_3^{(e)} - \frac{\Psi}{\varkappa_2}(1 - \varkappa_3)\right] + c_{22}\Psi + d_1\frac{\varkappa_2^2}{2},
\end{aligned}
\tag{18}
$$

where

$$
\begin{aligned}
c_{11} &= \frac{1}{a_2} - b\frac{\varkappa_2}{a_1}, \qquad c_{12} = \frac{1}{2}\left(\frac{\varkappa_2}{a_2} - b\frac{\varkappa_2}{a_1}\right),\\
c_{22} &= \left(\frac{1}{a_2} - b\frac{\varkappa_2}{a_1}\right)\varkappa_2, \qquad d_1 = T r_0(\alpha_2 - \alpha_1) b\frac{\varkappa_2}{a_1}.
\end{aligned}
$$

Substituting equations (2)–(5) into (18), we get two equations for the constants C_1 and C_2, the solutions of which are

$$
C_1 = d_1\frac{\varkappa_2}{2}\frac{f_{12} - f_{22}}{f_{11}f_{22} - f_{12}f_{21}}, \qquad C_2 = d_1\frac{\varkappa_2}{2}\frac{f_{21} - f_{11}}{f_{11}f_{22} - f_{12}f_{21}},
$$

where

$$
\begin{aligned}
f_{11} &= c_{11}r_0 - h_3 y,\\
f_{12} &= \frac{r_0}{2}[c_{12}\varkappa_2 - (1-\varkappa_3)c_{11}]h_1,\\
f_{21} &= \varkappa_2 c_{12} r_0 - (1-\varkappa_3)y,\\
f_{22} &= [c_{22}\varkappa_2 - (1-\varkappa_3)\varkappa_2 c_{12}]r_0 h_1 - \frac{\varkappa_3^3}{24 y h_1}.
\end{aligned}
$$

The deflection of the mirror is the most interesting factor for us. If we take into consideration formulae (3), (15) and (18), then the coefficients G_2, G_3 and G_4 are expressed through C_2. Taking $C_3 = 0$ in (3), one can find G_5 from condition (14). Thus, we finally get

$$
(19)\quad w(r) = \begin{cases} -\dfrac{C_2}{4}r^2, & \text{for} \quad r_0 > r > 0,\\[2ex] -C_2\dfrac{r_0}{2a_2}\left[\dfrac{r_0^2 - r^2}{2} + r_2^2\dfrac{1+\nu_1}{1-\nu_1}\ln\left(\dfrac{r}{r_0}\right) + \dfrac{r_0^2}{4}\right], & \text{for } r_2 > r > r_0. \end{cases}
$$

Formulae (19) describe the deflection of the surface of the mirror under the assumption that the axisymmetric deformations of the mirror satisfy the thin plate theory. These hypotheses for real optical mirrors are not always fulfilled. Nevertheless, the formulae obtained in this paper give a good approximation to the displacements for real mirrors.

6. Numerical example

As a numerical example, we consider a mirror with the following parameters:

$$
r_0 = 20\times 10^{-3}\,\text{m},\quad r_1 = 65\times 10^{-3}\,\text{m},\quad r_2 = 16\times 10^{-3}\,\text{m},\qquad h_1 = 20\times 10^{-3}\,\text{m},
$$
$$
E_1 = E_3 = 1.6\times 10^{12}\,\text{Pa},\qquad E_2 = 1.2\times 10^{12}\,\text{Pa},
$$
$$
\alpha_1 = \alpha_3 = 2.8\times 10^{-6},\qquad \alpha_2 = 2.5\times 10^{-6},\qquad \nu_1 = \nu_3 = 0.2,\qquad \nu_2 = 0.3.
$$

The parameters h_2 and h_3 vary.

We have solved the corresponding initial value problem numerically by means of the finite element method (FEM) by using package [**17**], which is a version of the well-known SAP IV program (see K. J. Bathe *et al.* [**2**] and D. S. Chehil [**5**]). To solve the axisymmetric problem, we used rectangular thin strain plane finite elements. These allowed us to compute all the components of the deformation for the entire structure. To compare the numerical results with the analytical results, we kept only the deflections and excluded all other deformations (radial and shear expansions).

We first studied the convergence of the method. A comparison of the numerical results with 15 and 24 elements, denoted by squares and plus signs, respectively, is shown in Fig. 2. Obviously, a further increase in the number of elements will not significantly change the result.

In Fig. 2 and later figures, the maximal displacements are shown in dimensionless values. Multiplying these by 10^9 meters, one gets their dimensional values.

In Figs. 3, 4 and 5, we compare the analytical and the FEM results, indicated by solid and dashed lines, and square and plus signs, respectively, for different

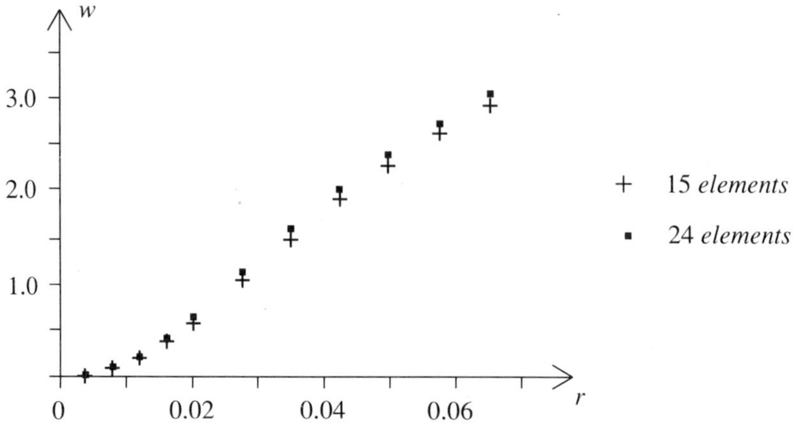

FIGURE 2. Thermo-elastic deformations of mirror computed by the finite element method

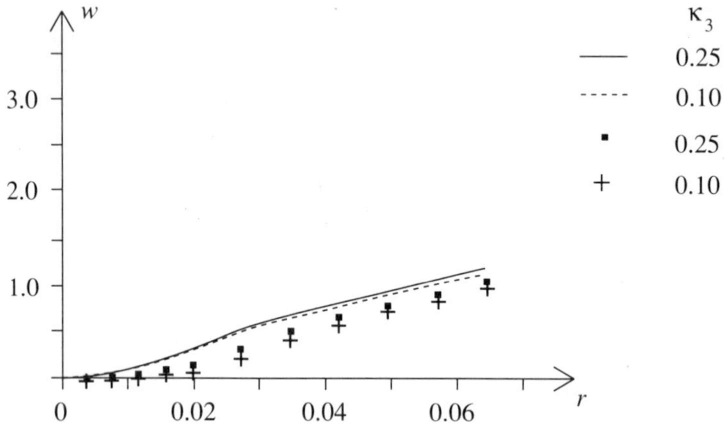

FIGURE 3. Thermo-elastic deformations of mirror, for $h_2/h_1 = 0.1$

values of the ratio $\varkappa_2 = h_2/h_1$ (solid lines and squares for[1] $\varkappa_3 = 0.25$ and dashed lines and plus signs for $\varkappa_3 = 0.1$).

For $\varkappa_2 = 0.375$, the analytical and numerical results are in good agreement. But for smaller and greater values of $\varkappa_2$, the difference between the results is not negligible, and the deformations calculated from the analytical formulae are slightly larger than the numerical ones. Nevertheless, the analytical formulae give a reliable qualitative picture of the deformations because the relative error for the maximal deformation is, in any case, less than 15%.

It is interesting to examine the effect of the parameters $\varkappa_3$ and $\varkappa_2$ on the maximal mirror deflection. In Fig. 6, one can see the effect of $\varkappa_2$ on the maximal displacements of the reflecting mirror surface, for constant values of $\varkappa_3$ such that $\varkappa_2 + \varkappa_3 \leq 1$. We see that the deflection w increases as $\varkappa_3$ increases.

[1]Note that, for the 'portability' of Figs. 3 to 7, κ_2 and κ_3 are used instead of $\varkappa_2$ and $\varkappa_3$.

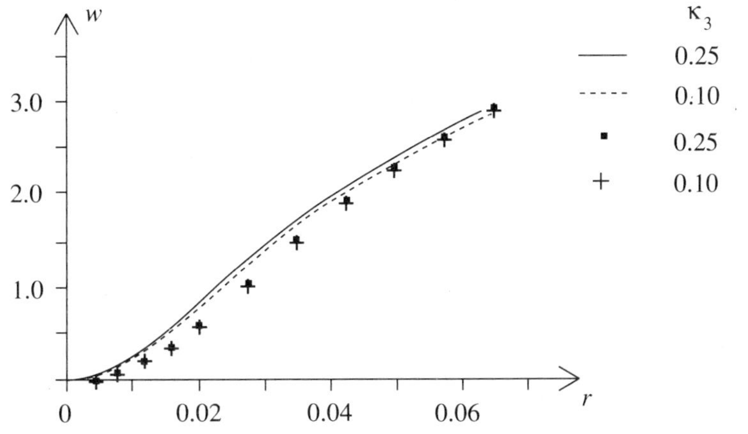

FIGURE 4. Thermo-elastic deformations of mirror, for $h_2/h_1 = 0.375$

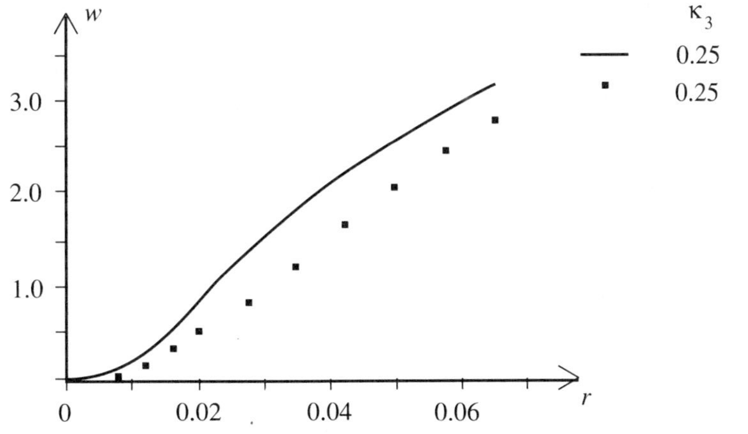

FIGURE 5. Thermo-elastic deformations of mirror, for $h_2/h_1 = 0.7$

In Fig. 7, we show the maximal mirror deflection, w, versus $\varkappa_3 = h_3/h_1$ for five different values of $\varkappa_2$:

$$\varkappa_2 = 0.1,\ 0.375,\ 0.5,\ 0.55,\ 0.7, \text{ on curves } 1,\ 2,\ 3,\ 4,\ 5, \text{ respectively.}$$

The function w is nonlinear, but it may be linearized for small constant values of $\varkappa_2$ ($\varkappa_2 \ll 1$) and small values of the variable $\varkappa_3$.

If we keep only the small quantities of the first order, we get

$$C_2 = 12T(\alpha_2 - \alpha_1)\frac{b}{h_1}\frac{a_2}{a_1}\left(1 - 2a_2 r_0 \frac{\varkappa_3}{r_2^2 - r_0^2}\right)\varkappa_2. \tag{20}$$

This formula is also valid for constant values of $\varkappa_3$ and small variable values of $\varkappa_2$.

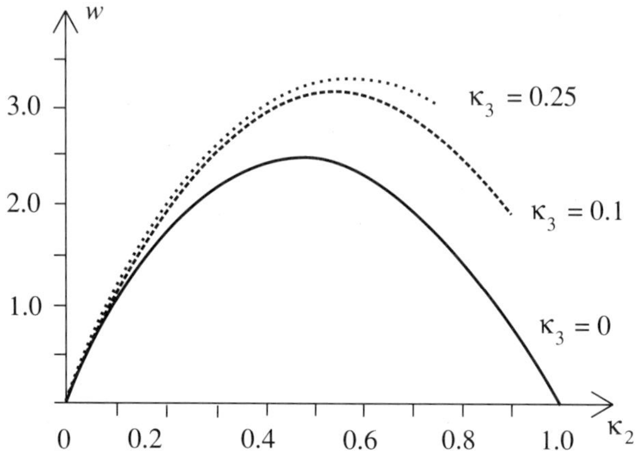

FIGURE 6. Maximal displacement of circular mirror vs. $\varkappa_2 = h_2/h_1$

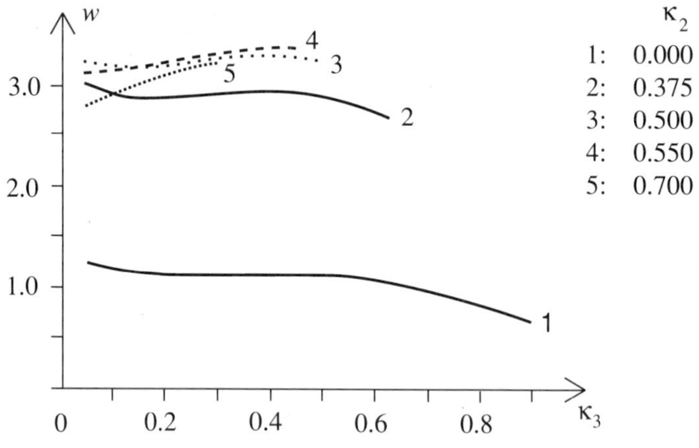

FIGURE 7. Maximal displacement of circular mirror vs. $\varkappa_3 = h_3/h_1$

For constant $\varkappa_2$ close to 1 ($\varkappa_2 = 1 - \varepsilon$ and $0 < \varepsilon \ll 1$), we find

$$C_2 = 12T(\alpha_2 - \alpha_1)\frac{b}{h_1}\frac{a_1 a_2}{(a_1 - b a_2)^2}\left[\varepsilon + \varkappa_3 a_2 \frac{r_0}{(r_2^2 - r_0^2)}\right]. \tag{21}$$

and the maximal deflection increases with $\varkappa_3$.

7. Conclusion

The application of the thin shell theory to the problem of thermo-elastic deformation of mirrors shows good agreement between the analytical results and the results obtained by the finite element method. Thus one can apply the obtained analytical formulae to the analysis of real mirror structures. One of the main problems in the construction of mirrors is to reduce the mirror deformations. These formulae allow one to minimize the mirror displacements by choosing appropriately the values of the geometric parameters of the mirror.

References

1. W. P. Barnes Jr., *Optimal design of cored mirror structures*, Applied Optics **8** (1969), no. 6, 1191–1196.
2. K. J. Bathe, E. L. Wilson, and F. E. Peterson, *SAP* IV (*A structural analysis program for static and dynamic response of linear system*), Report No. EERC 73–11, University of California, Berkeley, California, 1974 (59 pp.).
3. S. M. Bauer, S. B. Filippov, A. L. Smirnov, and P. E. Tovstik, *Asymptotic methods in mechanics with application to thin shells and plates*, Asymptotic Methods in Mechanics (A. L. Smirnov and R. Vaillancourt, eds.), CRM Proceedings and Lecture Notes, this volume, Amer. Math. Soc., Providence, R.I., 1993, pp. 3–140.
4. S. M. Bauer, A. M. Kovalev, M. B. Petrov, V. V. Tikhomirov, and P. E. Tovstik, *Numerical investigation of the temperature deformations of an optical mirror*, Optico-mekhanicheskaya promyshlennost **52** (1985), no. 9, 26–28; English transl., Sov. J. Optical Technology **52** (1985), no. 9, 529–531.
5. D. S. Chehil and R. Jategaonkar, *Determination of natural frequencies of a beam with varying section properties*, J. Sound Vibration **115** (1987), no. 3, 423–436.
6. W. J. Choyke, R. F. Farich, and R. A. Hoffman, *SiC, a new material for mirrors.* 1: *High power lasers;* 2: *VUV applications*, Applied Optics **15** (1976), no. 9, 2006–2007.
7. V. D. Derevenskii, L. Z. Dulkin, A. I. Karavashkin, V. N. Paimushin and I. Kh. Saitov, *Thermal characteristics and thermal deformations of light-weight mirrors*, Optico-mekhanicheskaya promyshlennost **50** (1983), no. 4, 18–20; English transl., Sov. J. Optical Technology **50** (1983), no. 4, 221–225.
8. A. Hewitt (ed.), *Optical and infrared telescopes for the* 1990*s*, Proceedings of the Kitt Peak National Observatory Conference, Tucson, Arizona, 1980 (506 pp.).
9. V. V. Kharitonov, *Thermo-physical mirror analysis*, MIFI, Moscow, 1986, pp. 88 (88 pp.). (Russian)
10. C. A. Klein, *Mirror figure-of-merit and material index-of-goodness for high power laser beam reflectors*, SPIE **288** (1981), 69–77.
11. A. D. Kovalenko, *Circular plates with variable thickness*, Fizmatgiz, Moscow, 1959 (296 pp.). (Russian)
12. W. Nowacki, *Thermoelasticity*, Pergamon Press, Oxford, 1962 (566 pp.).
13. F. Pachini (ed.), *Optical telescopes of the future*, ESO (European South Observatory), CERN, 1978 (432 pp.).
14. V. N. Paimushin, I. Kh. Saitov and V. D. Derevenskii, *The mathematics for a more precise calculation of the deformations of light-weight mirrors*, Optico-mekhanicheskaya promyshlennost **50** (1983), no. 1, 10–12; English transl., Sov. J. Optical Technology **50** (1983), no. 1, 11–13.
15. F. B. Ray and Y. T. Chung, *Surface analysis of an actively controlled telescope primary mirror under static loads*, Applied Optics **24** (1985), no. 4, 564–569.
16. G. W. Satton, *Model for thermal distortion of high-energy repetively pulsed laser mirrors*, J. of Energy **5** (1981), no. 6, 344–355.
17. Report, *The structural analysis of complex* 3-D *structures*, GOI, Leningrad, 1987 (346 pp.). (Russian)

Abstract. The thermal deformations of a uniformly circular multi-layer anisotropic optical mirror is studied. The mirror is split into circular and ring plates, which are analyzed by the thin plate theory. A comparison between the analytical results, obtained by this theory, and the numerical results, obtained by the finite element method, reveals good agreement between these two methods.

Institute of Mathematics and Mechanics, St. Petersburg State University, 2 Bibliotechnaya Sq., Stary Peterhof, 198904 St. Petersburg, Russia

E-mail address: als@hq.math.lgu.spb.su

Centre de Recherches Mathématiques
CRM Proceedings and Lecture Notes
Volume 3, 1993

A Mathematical Model for Hydroelastic Problems with a Fluid Memory. Part I

Boris A. Ershov, Yuliya A. Mochalova, and Ekaterina V. Polyakova

1. Introduction

In the thin airfoil theory, the solution is split into two parts. The first solution combines the unsteady motion of the airfoil and the influence of the vortex trace behind the trailing edge and completely neglects the airfoil flexibility [**3**]. In this case the unsteady hydrodynamic forces on the airfoil are calculated by Fourier analysis using the well-known results for harmonic oscillations. This analysis has been made by M. S. Home [**7**] and R. K. Amiet [**1**]. The second part of the solution takes into account the airfoil flexibility and neglects the unsteady vortex trace. The traditional method of solving hydroelastic problems includes the construction of the solution, which is usually obtained numerically (for example, see [**6**] by G. P. Guruswamy).

Our aim is to avoid these traditional limitations. In the present study, we consider the unsteady arbitrary motion of a flexible thin airfoil encountering a gust in an incompressible flow. The dynamic model of the hydroelastic problem consists of three parts, namely, a flexible airfoil, a fluid flow and a flexible airfoil interacting with a fluid flow. The model of an airfoil interacting with an ideal fluid is defined by (i) an impermeability condition and (ii) a vortex trace, which is simulated by a line of discontinuity in the velocity. The model considered here describes a wide class of hydrodynamic problems and makes it possible to investigate the hydrodynamical characteristics of a flexible airfoil, such as lifting force and moments.

One method of solution consists of a direct analytical solution of a system of integro-partial differential equations, which describe the unsteady motion of the flexible airfoil.

In this paper, an approximate analytical solution to the hydroelastic problem with a fluid memory is presented. By means of this solution, analytical expressions for the unsteady hydrodynamical forces are obtained. The analysis presented makes it possible to evaluate the influence of the inertial and elastic properties of an airfoil on the characteristics of the interaction with a fluid flow.

2. The model of an elastic airfoil

An elastic body, as a thin carrying surface, can be modeled by an infinite plane airfoil (the plane problem), subsequently referred to as a thin flexible airfoil. Let

1991 *Mathematics Subject Classification.* Primary: 76B05; Secondary: 73K10, 73D10.
Key words and phrases. Flexible airfoil, vortex trace, gust.
This is the final form of the paper.

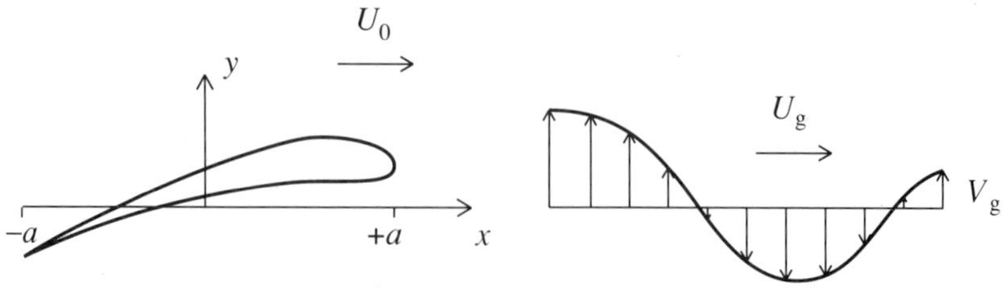

FIGURE 1. Incident gust on plate airfoil.

the (x, y)-coordinate system be attached to the plate which is moving relative to the fixed (x_1, y_1)-system so that the following relations are valid:

$$x = x_1 - U_0 t - a, \qquad y = y_1,$$

where $2a$ is the width of the plate, $U_0 = \text{const}$, and $U_0 < C$ (C is the speed of sound in undisturbed liquid).

The perturbed unsteady airfoil motion is represented by the small displacement $w(x, t)$ along the y-axis under the assumption that the derivatives

$$\frac{\partial w(x,t)}{\partial t}, \qquad \frac{\partial w(x,t)}{\partial x}$$

are small.

For the unsteady motion of a wing-shaped airfoil we can combine the motion of the entire airfoil with the deformations of the flexible wing. Using the two-dimensional theory of a thin elastic plate, we obtain the equation of the flexural vibration of the airfoil (see Section 2.6 of the Survey [**4**]):

$$\frac{\partial^2}{\partial x^2}\left[D(x)\frac{\partial^2 w(x,t)}{\partial x^2}\right] + m\frac{\partial^2 w(x,t)}{\partial t^2} = p(x,t), \tag{1}$$

where D is cylindrical rigidity, m is mass per unit wing area and $p(x, t)$ is flow pressure per unit wing area. The boundary conditions are defined by the attachment method for the wing contour (a free wing, an aileron, etc.); the initial conditions determine the displacement and the velocity of the wing middle surface at any given time.

In the most general case, a flexible airfoil encounters a gust field, the vertical and horizontal velocities of which are $V_g(x, t)$ and $U_g(x, t)$, respectively (see Fig. 1).

We suppose that the gust acts on the leading edge of the airfoil from time $t = 0$ on. The displacement of the gust is relatively slow and it is also assumed that $U_g(x, t) \ll U_0$ and $V_g(x, t) \ll U_0$. The gust vertical velocity is given by

$$V_g(x_1, t) = f(x_1 + U_g t),$$

or, in the moving coordinate system, by

$$V_g(x, t) = f(x - a + (U_g + U_0)t).$$

Neglecting the influence of the horizontal gust velocity U_g, one obtains

$$V_g(x, t) = f(x - a + U_0 t). \tag{2}$$

3. A flexible airfoil interacting with a fluid flow

The model of an airfoil interacting with an ideal fluid is defined by:

(a) A boundary condition on the profile,

$$v_n = -U_0 \frac{\partial w(x,t)}{\partial x} + \frac{\partial w(x,t)}{\partial t} + V_g(x,t). \tag{3}$$

This equation defines the normal velocity in terms of the displacement of the thin airfoil and the gust velocity.

(b) An unsteady motion of the profile producing a vortex trace behind the trailing edge. Suppose that the liquid flowing to the trailing pointed edge over the upper and lower surfaces of the airfoil has different velocities, except for several singular cases. As a result, a line of discontinuity in the velocities will be formed in the liquid behind the trailing edge. Across that line, the normal component of the liquid velocity is continuous, but the tangential component is discontinuous. The discontinuity in the velocity is defined by the integral equation

$$-\int_{-a}^{a} v_n(\xi,t)\sqrt{\frac{a-\xi}{a+\xi}}\,d\xi = \int_0^{U_0 t} u_2(\tau)\sqrt{\frac{t-\tau+t_a}{t-\tau}}\,d\tau, \qquad t_a = \frac{U_0}{2a}. \tag{4}$$

4. The model of a fluid flow

The model of a fluid flow is described as an ideal incompressible liquid at rest at infinity. An unsteady motion of the profile with a pointed trailing edge causes a perturbed unsteady parallel plane motion of the incompressible liquid. The kinematic picture of a perturbed unsteady motion of the incompressible liquid is completely defined by the velocity function:

$$\begin{aligned} \frac{dw}{dz} &= \frac{\partial \varphi}{\partial x} - i\frac{\partial \varphi}{\partial y} \\ &= \frac{1}{\pi i}\sqrt{\frac{z+a}{z-a}}\left[\int_{-a}^{a} \frac{v_n(\xi,t)}{\xi - z}\sqrt{\frac{a-\xi}{a+\xi}}\,d\xi \right. \\ &\quad \left. - \int_0^{\alpha_1} \frac{u_2(\alpha)}{\alpha_1 - \alpha + a + z}\sqrt{\frac{\alpha_1-\alpha+2a}{\alpha_1-\alpha}}\,d\alpha\right], \end{aligned}$$

where $z = x + iy$, $\alpha_1 = U_0 t$ and $\varphi(x,y,t)$ is the potential of the perturbed velocity. This equation has been derived by L. I. Sedov [**9**]. The velocity function $\frac{dw}{dz}$ is obtained in terms of the magnitude of the discontinuity in the velocity, $u_2(t)$, and the normal component, $v_n(x,t)$, of the airfoil velocity. In virtue of this velocity function the potential of the perturbed velocity is expressed as

$$\begin{aligned} \varphi(x,0,t) = -\frac{1}{\pi}&\left[\int_{-a}^{a} v_n(\xi,t)L_2(\xi,x)\,d\xi - 2\int_0^{U_0 t} u_2(U_0\tau)L_3(t-\tau,x)\,d\tau\right] \\ &- \int_0^{U_0 t} u_2(U_0\tau)\,d\tau + f(t), \end{aligned} \tag{5}$$

where $f(t)$ is an arbitrary function of time, t, and

$$L_2(\xi, x) = \ln \left| \frac{\sqrt{(a-\xi)(a+x)} + \sqrt{(a+\xi)(a-x)}}{\sqrt{(a-\xi)(a+x)} - \sqrt{(a+\xi)(a-x)}} \right|,$$

$$L_3(t, x) = \arctan \sqrt{\frac{(t+t_a)(a-x)}{t(a+x)}}.$$

We assume that

$$f(t) = \int_0^{U_0 t} u_2(u\tau)\, d\tau. \tag{6}$$

Here the function $f(t)$ is the magnitude of the potential of the perturbed velocity, calculated on the trailing edge of the airfoil. For $x = a$, equation (5) transforms into

$$\varphi(a, 0, t) = \int_0^{U_0 t} u_2(u\tau)\, d\tau + f(t). \tag{7}$$

It can be shown that

$$\varphi(a, 0, t) = 0. \tag{8}$$

Comparing (7) and (8), one obtains (6). Thus, the potential of the perturbed velocity, i.e. the unsteady motion of the liquid $\varphi(x, t)$, is given by

$$\varphi(x, 0, t) = -\frac{1}{\pi} \left[\int_{-a}^{a} v_n(\xi, t) L_2(\xi, x)\, d\xi - 2 \int_0^{U_0 t} u_2(U_0\tau) L_3(t-\tau, x)\, d\tau \right], \quad -a \le x \le a. \tag{9}$$

The model of a fluid flow is described by the potential of the perturbed velocity (9). Substitution of (9) into the linearized Lagrange integral yields the pressure jump, $p(x, t)$, across the airfoil (lower surface minus upper surface pressures) in the moving coordinate system,

$$\begin{aligned} p(x, t) = {} & \frac{2\rho}{\pi} \left[U_0 \int_{-a}^{a} v_n(\xi, t) L_1(\xi, x)\, d\xi - \int_{-a}^{a} \dot{v}_n(\xi, t) L_2(\xi, x)\, d\xi \right] \\ & - U_0 \sqrt{\frac{a-x}{a+x}} \int_0^{U_0 t} K_1(t-\tau) u_2(U_0\tau)\, d\tau, \qquad -a \le x \le a, \end{aligned} \tag{10}$$

where

$$K_1(t) = \frac{1}{\sqrt{t(t-t_a)}}, \qquad K_2(t) = \sqrt{\frac{t+t_a}{t}},$$

$$L_0(x) = \sqrt{\frac{a-x}{a+x}}, \qquad L_1(\xi, x) = \sqrt{\frac{(a-\xi)(a+x)}{(a+\xi)(a-x)}} \frac{1}{\xi - x}.$$

The airfoil pressure is expressed by the discontinuity in the velocity, $u_2(t)$, and the normal component, $v_n(x, t)$, of the airfoil velocity. The unsteady motion of the flexible airfoil is caused by a vertical gust.

5. Hydroelastic equations of motion

By means of (1), (3), (4) and (10), we can write a system of integro-partial differential equations for the deformation, $w(x,t)$, of the profile and the discontinuity in the velocity, $u_2(t)$, in the trace in the case of a gust $V_g(x,t)$:

$$
\begin{aligned}
&\frac{\partial^2}{\partial x^2}\left[D(x)\frac{\partial^2 w(x,t)}{\partial x^2}\right] + m\frac{\partial^2 w(x,t)}{\partial t^2} \\
&\qquad - \frac{2\rho}{\pi}U_0\int_{-a}^{a}\left(-U_0\frac{\partial w(\xi,t)}{\partial \xi} + \frac{\partial w(\xi,t)}{\partial t}\right)L_1(\xi,x)\,d\xi \\
&\qquad - U_0\int_{-a}^{a}\left(-U_0\frac{\partial^2 w(\xi,t)}{\partial\xi\partial t} + \frac{\partial^2 w(\xi,t)}{\partial t^2}\right)L_2(\xi,x)\,d\xi \\
&\qquad - U_0\sqrt{\frac{a-x}{a+x}}\int_0^{U_0 t}K_1(t-\tau)u_2(U_0\tau)\,d\tau = F_1(x,t), \\
&\int_{-a}^{a}\left(-U_0\frac{\partial w(\xi,t)}{\partial \xi} + \frac{\partial w(\xi,t)}{\partial t}\right)L_0(\xi)\,d\xi + \int_0^{U_0 t}K_2(t-\tau)u_2(\tau)\,d\tau = F_2(t).
\end{aligned} \tag{11}
$$

Here

$$F_1(x,t) = \frac{2\rho}{\pi}\left[U_0\int_{-a}^{a}V_g(\xi,t)L_1(\xi,x)\,d\xi - \int_{-a}^{a}\frac{\partial V_g(x,t)}{\partial x}L_2(\xi,x)\,d\xi\right],$$

$$F_2(t) = -\int_{-a}^{a}V_g(\xi,t)L_0(\xi)\,d\xi.$$

Now we introduce the initial condition

$$w(x,0) = f(x), \qquad \left.\frac{\partial w(x,t)}{\partial t}\right|_{t=0} = \psi(x). \tag{12}$$

System (11) is a set of integro-partial differential equations, which are generalizations of Fredholm and Volterra equations.

6. Setting up of the Bubnov–Galerkin proceduce and taking Laplace transforms

We assume that the deformation, $w(x,t)$, of the airfoil can be represented in the form

$$w(x,t) = \sum_{i=1}^{n} X_i(x)T_i(t), \tag{13}$$

where $\{X_i(x)\}$ is a given complete orthonormal system of functions satisfying the boundary conditions, and $\{T_i(x)\}$ are functions satisfying the initial condition, which have to be defined. We substitute (13) into system (11) and apply the Bubnov–Galerkin procedure. This yields the following system of Volterra integro-differential equations for $T(t)$ and $u_2(U_0 t)$:

$$
\begin{aligned}
&A_0\ddot{T} + A_1\dot{T} + A_2 T + B\int_0^{U_0 t}K_1(t-\tau)u_2(U_0\tau)\,d\tau = Q(t), \\
&C_1\dot{T} + C_2 T + b_{n+1}\int_0^{U_0 t}K_2(t-\tau)u_2(U_0\tau)\,d\tau = q_{n+1}(t),
\end{aligned} \tag{14}
$$

where

$$T(t) = [T_1(t), T_2(t), ..., T_n(t)]^T,$$

$$Q(t) = [q_1(t), q_2(t) \dots q_n(t)]^T,$$

$$q_i(t) = \int_{-a}^{a} X_i(x) F_1(x,t) dx, \ i = 1, ...n, \qquad q_{n+1}(t) = F_2(t), \qquad b_{n+1} = U_0,$$

and A_0, A_1, A_2, C_1, C_2 and B are matrices and vectors whose elements are beam-profile integrals given by I. M. Babakov [**2**] and B. A. Ershov *et al.* [**5**].

The kernels, $K_1(t)$ and $K_2(t)$, of system (14) have a weak singularity. We convolve this system with the function $(t-\eta)^{-1/2}$. In the integral terms, after changing the order of integration, a second convolution yields, in the inner integrals, a composition of the kernel and $(t-\eta)^{-1/2}$. Changing the sign of the variable of integration, $\eta = (t+\tau)/2 + [(t-\tau/2)] \cos 2\psi$, we can write the kernels as

$$K_1(t) = \int_0^{\pi/2} \frac{d\psi}{\sqrt{t\cos^2\psi + t_a}}, \qquad K_2(t) = \int_0^{\pi/2} \frac{\cos^2\psi}{\sqrt{t\cos^2\psi + t_a}} d\psi.$$

Now these kernels are bounded continuous functions which go to zero as $t \to \infty$. By these transformations, system (14) is converted into the equivalent system:

$$(15) \quad \begin{aligned} &A_0 \int_0^t \frac{\ddot{T}(\eta)\, d\eta}{\sqrt{(t-\eta)}} + A_1 \int_0^t \frac{\dot{T}(\eta)\, \eta}{\sqrt{(t-\eta)}} + A_2 \int_0^t \frac{T(\eta)\, d\eta}{\sqrt{(t-\eta)}} \\ &\qquad + B \int_0^t K_1(t-\tau) u_2(U_0\tau)\, d\tau = \int_0^t \frac{Q(\eta)\, d\eta}{\sqrt{(t-\eta)}}, \\ &C_1 \int_0^t \frac{\dot{T}(\eta)\, d\eta}{\sqrt{(t-\eta)}} + C_2 \int_0^t \frac{T(\eta)\, d\eta}{\sqrt{(t-\eta)}} + \pi\sqrt{t_a}\, b_{n+1} F(t) \\ &\qquad + b_{n+1} \int_0^t K_2(t-\tau) F(\tau)\, d\tau = \int_0^t \frac{q_{n+1}(\eta)\, d\eta}{\sqrt{(t-\eta)}}, \end{aligned}$$

where

$$F(t) = \int_0^t u_2(U_0\tau)\, d\tau.$$

The integral kernels $K_1(t)$ and $K_2(t)$ of system (15) are replaced by exponential approximations. Taking the Laplace transform of (15) and assuming that the initial conditions are zero, we obtain:

$$(16) \qquad P_1(p)\overline{T}(p) + P_2(p)\overline{U}_2\left(\frac{p}{U_0}\right) = \frac{1}{\sqrt{p}}\overline{Q}(p).$$

The elements of the matrices $P_1(p)$ and $P_2(p)$ are polynomials in the variable p of the Laplace transform (Laplace transforms are denoted by an overline).

System (16) establishes a relation between $\overline{T}(p)$, which defines the time variation of the airfoil deformation, and $\overline{U}_2(p)$, which determines the velocity discontinuity in the trace. The final step in the analysis of the initial system is to obtain the solution $T_i(t)$ and $u_2(t)$ by taking the inverse Laplace transforms of $\overline{T}(p)$ and $\overline{U}_2(p)$, respectively.

The solution of the system of linear non-homogeneous algebraic equations (16) for the unknown generalized coordinates $\overline{T}_i(p)$ is expressed by

$$\overline{T}_i(p) = \frac{1}{\Delta(p)} \sum_{j=1}^{n+1} \Delta^{ij}(p) Q_j(p), \qquad i = 1 \ldots n. \tag{17}$$

Here $\Delta(p)$ is the determinant of order $2(n+2)$ of system (16). Then, using the inverse Laplace transformation L^{-1}, one might get the set of unknown generalized coordinates $\overline{T}_i(p)$. In the general case, this set can be obtained by means of the convolution theorem. The elements, $\Delta^{ij}(p)$, of the determinant $\Delta(p)$ are algebraic functions, and using Heavyside's method, the transform can be carried out by taking the partial fractions.

$$L^{-1}\left\{\frac{\Delta^{ij}(p)}{\Delta(p)}\right\} = \sum_{k=1}^{2(n+2)} \frac{\Delta^{ij}(p_k)}{\Delta'(p_k)} e^{p_k \tau},$$

where the p_k are the zeros of the characteristic polynomial $\Delta(p)$ and

$$\Delta'(p_k) = \left[\frac{d\Delta(p)}{dp}\right]_{p=p_k}.$$

By the convolution theorem, the solution of system (14) is given by

$$T_i(t) = \sum_{j=1}^{n+1} \sum_{k=1}^{2(n+2)} \frac{\Delta^{ij}(p_k)}{\Delta'(p_k)} \int_0^t q_j(t-\tau) e^{p_k \tau}\, d\tau, \qquad i = 1 \ldots n. \tag{18}$$

Thus, an analytical expression for $T_i(t)$ is found. Substituting (18) into equation (13) we get the displacement $w(x,t)$,

$$w(x,t) = \sum_{i=1}^{n} \sum_{j=1}^{n+1} \sum_{k=1}^{2(n+2)} \frac{\Delta^{ij}(p_k)}{\Delta'(p_k)} X_i(x) I_j(p_k, t), \tag{19}$$

where

$$I_j(p_k, t) = \int_0^t q_j(\tau) e^{p_k(t-\tau)}\, d\tau,$$

which is the analytical solution of system (11).

It is also interesting to calculate the overall unsteady lift force on the airfoil.

NOTE. The discontinuity in the velocity $u_2(t)$ in expression (10) for the pressure jump $p(x,t)$ appears in the integrand of

$$F(u_2) = \int_0^{U_0 t} u_2(U_0 \tau) K_1(t-\tau)\, d\tau. \tag{20}$$

Hence, in defining the lift, one can neglect the discontinuity in the velocity $u_2(t)$. The pressure on the airfoil can be determined in the generalized coordinates, $T_i(t)$, and their derivatives, $\dot{T}_i(t)$, by means of the system of equations (14). Then, the lift is

$$Y(t) = \int_{-a}^{a} p(x,t)\, dx. \tag{21}$$

Let us consider the following three terms making $Y(t)$ in (21):

$$Y(t) = Y_v(t) + Y_u(t) + Y_g(t). \tag{22}$$

The first term, $Y_v(t)$, is the component of the lift force caused by the perturbed motion of the airfoil. The second term, $Y_u(t)$, is induced by the vortex trace. The third term, $Y_g(t)$, is the component caused by the gust field.

We proceed in a similar manner for the moment,

$$M(t) = M_v(t) + M_u(t) + M_g(t). \tag{23}$$

If we take into account the earlier assumptions on the motion of an airfoil in a liquid flow, the following expressions for the components of the lift force are obtained:

$$\begin{gathered} Y_v(t) = -2U_0\rho \int_{-a}^{a} \left(-U_0 \frac{\partial w(\xi,t)}{\partial \xi} + \frac{\partial w(\xi,t)}{\partial t} \right) \sqrt{\frac{a-\xi}{a+\xi}}\, d\xi \\ -2\rho \int_{-a}^{a} \left(-U_0 \frac{\partial^2 w(\xi,t)}{\partial \xi \partial t} + \frac{\partial^2 w(\xi,t)}{\partial t^2} \right) \sqrt{a^2-\xi^2}\, d\xi, \\ Y_u(t) = -2aU_0\rho \int_0^{U_0 t} u_2(U_0\tau) K_1(t-\tau)\, d\tau, \\ Y_g(t) = -2U_0\rho \int_{-a}^{a} V_g(\xi,t) \sqrt{\frac{a-\xi}{a+\xi}}\, d\xi - 2\rho \int_{-a}^{a} \frac{\partial V_g(\xi,t)}{\partial t} \sqrt{a^2-\xi^2}\, d\xi. \end{gathered} \tag{24}$$

If we substitute the expressions (13) and (3) for $w(x,t)$ and $v_n(x,t)$, respectively, into (21), we obtain the lift,

$$\begin{gathered} Y_v(t) = \sum_{i=1}^{n} \left[d_{0i}\ddot{T}_i(t) + d_{1i}\dot{T}_i(t) + d_{2i}T_i(t) \right], \\ Y_u(t) = -c_y F(u_2). \end{gathered} \tag{25}$$

The components of the moment are

$$\begin{gathered} M_v(t) = \sum_{i=1}^{n} \left[m_{0i}\ddot{T}_i(t) + m_{1i}\dot{T}_i(t) + m_{2i}T_i(t) \right], \\ M_u(t) = -c_m F(u_2), \\ M_g(t) = 2U_0\rho \int_{-a}^{a} V_g(\xi,t)\sqrt{a^2-\xi^2}\, d\xi + \rho \int_{-a}^{a} \frac{\partial V_g(\xi,t)}{\partial t} \xi \sqrt{a^2-\xi^2}\, d\xi. \end{gathered} \tag{26}$$

Here the coefficients d_{ij}, m_{ij}, c_y and c_m are evaluated by the beam profile integrals and the parameters U_0 and ρ. Thus, the expressions (25) and (26) determine, in terms of the generalized coordinates $T_i(t)$, their derivatives and the discontinuity in the velocity $u_2(t)$.

References

1. R. K. Amiet, *Gust response for flat-plate airfoil and the Kutta condition*, AIAA J. **28** (1990), 1718–1727.
2. I. M. Babakov, *Theory of vibrations*, GITTL, Moscow, 1968 (559 pp.). (Russian)
3. B. C. Basu and G. J. Hancock, *The unsteady motion of a two-dimensional aerofoil in incompressible inviscid flow*, J. Fluid Mech. **87** (1978), 159–178.

4. S. M. Bauer, S. B. Filippov, A. L. Smirnov and P. E. Tovstik, *Asymptotic methods in mechanics with application to thin shells and plates*, Asymptotic Methods in Mechanics (A. L. Smirnov and R. Vaillancourt, eds.), CRM Proceedings and Lecture Notes, this volume, Amer. Math. Soc., Providence, R.I., 1993, pp. 3–140.
5. B. A. Ershov, Yu. A. Mochalova, and E. V. Polyakova, *Unsteady motion of an elastic wing in hyposonic incompressible flow*, Vestnik Leningrad. Univ. Mat. Mekh. Astronom. (1989), no. 22, 44–50; English transl., Len. Univ. Mech. Bull. (1989), no. 4, 8–14.
6. G. P. Guruswamy, *Unsteady aerodynamic and aeroelastic calculations for wings using Euler equations*, AIAA J. **28** (1990), 461–469.
7. M. S. Home, *Contributions to the theory of sound production by vortex-airfoil interaction, with application to vortices with finite axial core velocity defect*, Proc. Roy. Soc. London Ser. A **420** (1988), 157–182.
8. A. I. Nekrasov, *Collected works*, vol. 2, GITTL, Moscow, 1962 (289 pp.). (Russian)
9. L. I. Sedov, *Plane problems of hydrodynamics and aerodynamics*, "Nauka", Moscow, 1966 (448 pp.). (Russian)

ABSTRACT. An approximate analytical method of solution of the problem for an unsteady motion of a flexible airfoil in an ideal incompressible liquid flow is proposed. The case of a flexible airfoil encountering a gust in an incompressible flow is analyzed. By taking into account the effect of a fluid memory given by the vortex trace on the hydrodynamic parameters, analytical expressions for the overall unsteady lift force and moment on the airfoil are presented.

INSTITUTE OF MATHEMATICS AND MECHANICS, ST. PETERSBURG STATE UNIVERSITY, 2 BIBLIOTECHNAYA SQ., STARY PETERHOF, 198904 ST. PETERSBURG, RUSSIA

E-mail address: als@hq.math.lgu.spb.su

Centre de Recherches Mathématiques
CRM Proceedings and Lecture Notes
Volume **3**, 1993

A Mathematical Model for Hydroelastic Problems with a Fluid Memory. Part II

Boris A. Ershov and Irina I. Strelkovskaya

1. Introduction

A plane hydroelastic problem is formulated for two flexible profiles which are moving in an ideal incompressible flow. The profiles are the median planes of a thin airfoil plate. Here we use the models considered in Section 2.6 of the Survey [**1**].

The fixed coordinates (x_1, y_1) and the moving coordinates (x, y) satisfy the following relations:

$$x = x_1 - U_0 t - a, \qquad y = y_1. \tag{1}$$

The moving coordinate system (x, y) moves with constant velocity U_0 along the x_1-axis. For the case of the undisturbed motions, the profiles are segments of length $2a$ parallel to the x-axis. They are situated such that their middles have coordinates $(0, -H)$ and $(0, H)$, respectively. It follows from relations (1) that for $t = 0$ the back edges of the profiles, for which $x = -a$ and $y = \pm H$, coincide with the y_1-axis of the motionless coordinate system.

For the unsteady motion, we take the profile displacement $w^{(i)}(x, t)$, which results from the interaction between the airfoil and the flow. We suppose that the values of $w^{(i)}$, $\partial w^{(i)}/\partial x$ and $\partial w^{(i)}/\partial t$ are small ($i = 1, 2$). The lower and upper profiles are indicated by the superscripts (1) and (2), respectively.

For the unsteady motion of a profile with pointed back and round front edges, respectively, the line of discontinuity in the velocities will be formed in the liquid behind the back edge of the profile (a vortex trace). In the case of a single profile, the velocity function is given in the monograph of L. I. Sedov [**5**]. The perturbed velocity potential $\varphi(x, y, t)$ has been obtained with the help of this function in papers by B. A. Ershov *et al.* [**2, 3**]. See also bibliography to the previous paper [**4**] in this book.

In this paper, this result is generalized to a system with two profiles. In this case, the velocity function has the following form:

$$\frac{dW}{dz} = \frac{1}{\pi i}\sqrt{\frac{z - iH + a}{z - iH - a}}\left[\int_{-a}^{a}\sqrt{\frac{a - \zeta}{a + \zeta}}\,\frac{v_n^{(2)}(\zeta, t)}{\zeta - (z - iH)}\,d\zeta\right.$$

1991 *Mathematics Subject Classification*. Primary: 76B05; Secondary: 73D10, 73K10, 76C05.
Key words and phrases. Flexible airfoil, vortex trace.
This is the final form of the paper.

(2)
$$- \int_0^{\beta_1} \sqrt{\frac{\beta_1 - \beta + 2a}{\beta_1 - \beta}} \frac{u_2^{(2)}(\beta)}{\beta_1 - \beta + a + (z - iH)} \, d\beta \Bigg]$$
$$+ \frac{1}{\pi i} \sqrt{\frac{z + iH + a}{z + iH - a}} \left[\int_{-a}^{a} \sqrt{\frac{a - \xi}{a + \xi}} \frac{v_n^{(1)}(\xi, t)}{\xi - (z + iH)} \, d\xi \right.$$
$$\left. - \int_0^{\alpha_1} \sqrt{\frac{\alpha_1 - \alpha + 2a}{\alpha_1 - \alpha}} \frac{u_2^{(1)}(\alpha)}{\alpha_1 - \alpha + a + (z + iH)} \, d\alpha \right],$$

where $u_2^{(1)}(\alpha)$ and $u_2^{(2)}(\beta)$ are the discontinuities in the velocity in the trace behind the first and second profiles, respectively.

The values of the discontinuity in the velocities behind the profiles $u_2^{(i)}$ and the normal velocities $v_n^{(i)}$ on the profiles satisfy the relation

$$\int_{-a}^{a} v_n^{(i)}(\xi, t) \sqrt{\frac{a - \xi}{a + \xi}} \, d\xi = - \int_0^{\varkappa_1} u_2^{(i)}(\varkappa) \sqrt{\frac{\varkappa_1 - \varkappa + 2a}{\varkappa_1 - \varkappa}} \, d\varkappa. \tag{3}$$

Here and below, $i = 1$, $j = 2$, $\varkappa = \alpha$, $\varkappa_1 = \alpha_1$, $\chi = \beta$, $\chi_1 = \beta_1$ for the first profile and $i = 2$, $j = 1$, $\varkappa = \beta$, $\varkappa_1 = \beta_1$, $\chi = \alpha$, $\chi_1 = \alpha_1$ for the second profile. Moreover,

$$- \Im \left(\frac{dW}{dz} \right)_{y = \pm H} = v_n^{(i)}(x, t).$$

2. The determination of the flow profile pressure

The hydrodynamical profile pressure $p(x, t)$ is determined from the linearized Lagrange integral

$$p = p_0(t) - \rho \left(\frac{\partial \varphi}{\partial t} - U_0 \frac{\partial \varphi}{\partial x} \right), \tag{4}$$

where $p_0(t)$ is a function of time and ρ is the liquid density.

Thus, the pressure on the profiles has the form

$$p_i(x, t) = -2\rho \left[U_0 \frac{\partial \varphi(x, \pm H, t)}{\partial x} - \frac{\partial \varphi(x, \pm H, t)}{\partial t} \right]. \tag{5}$$

In the case of an unsteady motion of the flexible profiles, the displacements $w^{(i)}(x, t)$ and the velocities are small and the normal velocity components of the profiles are determined by the formulae

$$v_n^{(i)}(\xi, t) = -U_0 \frac{\partial w^{(i)}(\xi, t)}{\partial \xi} + \frac{\partial w^{(i)}(\xi, t)}{\partial t}, \qquad i = 1, 2. \tag{6}$$

Using formulae (5), we finally write the profile pressure in the form

$$
\begin{aligned}
&p_i(x,t) = \\
&\quad -\frac{2\rho}{\pi}\left\{U_0\sqrt{\frac{a+x}{a-x}}\right. \\
&\qquad \times \int_{-a}^{a}\left[-U_0\frac{\partial w^{(i)}(\xi,t)}{\partial \xi}+\frac{\partial w^{(i)}(\xi,t)}{\partial t}\right]\sqrt{\frac{a-\xi}{a+\xi}}\frac{d\xi}{\xi-x} \\
&\quad -\int_{-a}^{a}\left[-U_0\frac{\partial^2 w^{(i)}(\xi,t)}{\partial \xi\partial t}+\frac{\partial^2 w^{(i)}(\xi,t)}{\partial t^2}\right]\ln|\gamma|\,d\xi \\
&\quad \left. -U_0\sqrt{\frac{a+x}{a-x}}\int_0^t u_2^{(i)}(U_0\tau)\frac{d\tau}{\sqrt{(t-\tau)(t-\tau+t_a)}}\right\} \\
&\quad -\frac{2\rho}{\pi}\left\{U_0[(x+a)^2+4H^2]^{-1/4}[(a-x)^2+4H^2]^{-3/4}\right. \\
&\qquad \times \int_{-a}^{a}\frac{\left[-U_0\frac{\partial w^{(j)}(\zeta,t)}{\partial\zeta}+\frac{\partial w^{(j)}(\zeta,t)}{\partial t}\right]}{(\zeta-x)^2+4H^2} \\
&\qquad\qquad \times\sqrt{\frac{a-\zeta}{a+\zeta}}\left[(a^2-x^2-4H^2)(\zeta-x)-8aH^2\right]d\zeta \\
&\quad -\left(c_1e^{-\lambda_1 H}+c_2e^{-\lambda_2 H}\right)\int_{-a}^{a}\left[-U_0\frac{\partial^2 w^{(j)}(\zeta,t)}{\partial\zeta\partial t}+\frac{\partial^2 w^{(j)}(\zeta,t)}{\partial t^2}\right]\ln|\gamma|\,d\zeta \\
&\quad +U_0\left[(x+a)^2+4H^2\right]^{-1/4}\left[(a-x)^2+4H^2\right]^{-3/4} \\
&\qquad \times\int_0^t\frac{u_2^{(j)}(U_0\tau)\sqrt{\frac{t-\tau+t_a}{t-\tau}}}{(\tau-t-\frac{a+x}{U_0})^2+\frac{4H^2}{U_0^2}} \\
&\qquad\qquad \times\left[\left(a^2-x^2-4H^2\right)\left(\tau-t-\frac{a+x}{U_0}\right)-\frac{8aH^2}{U_0}\right]d\tau \\
&\quad +\left(c_1e^{-\lambda_1 H}+c_2e^{-\lambda_2 H}\right)\int_0^t\frac{u_2^{(j)}(U_0\tau)}{t-\tau+\frac{a+x}{U_0}}\sqrt{\frac{(a+x)(a-x)}{(t-\tau)(t-\tau+t_a)}}\,d\tau \\
&\quad \left. -\left(c_1e^{-\lambda_1 H}+c_2e^{-\lambda_2 H}\right)\pi U_0u_2^{(j)}(U_0t)+2\pi U_0u_2^{(j)}(U_0t)+\pi U_0u_2^{(i)}(U_0t)\right\},
\end{aligned}
\tag{7}
$$

where $t_a = 2a/U_0$ and

$$\gamma = \frac{\sqrt{(a+x)(a-\zeta)}+\sqrt{(a-x)(a+\zeta)}}{\sqrt{(a+x)(a-\zeta)}-\sqrt{(a-x)(a+\zeta)}}.$$

In equation (7) the constants c_1, c_2, λ_1 and λ_2 are evaluated after one has obtained the approximations to the integrals, which contain H in the expressions for the perturbed velocity potential φ.

3. Equations of motion

In the model presented here, the airfoil equations of the disturbed motion are

$$D_i \frac{\partial^4 w^{(i)}}{\partial x^4} + m_i \frac{\partial^2 w^{(i)}}{\partial t^2} = p_i(x,t), \tag{8}$$

where D_i is the cylindrical rigidity and m_i is mass of unit area ($i = 1, 2$).

It follows from (7) that the hydrodynamical profile pressure is defined by the profile deformations and the velocity discontinuity in the trace behind the profiles. The following functions:

$$w^{(1)}(x,t), \qquad u_2^{(1)}(U_0 t), \qquad w^{(2)}(x,t), \qquad u_2^{(2)}(U_0 t)$$

satisfy the integral equations (3). If we use the differential equations (6), these integral equations become

$$\begin{aligned}\int_{-a}^{a} \left[-U_0 \frac{\partial w^{(i)}(\xi,t)}{\partial \xi} + \frac{\partial w^{(i)}(\xi,t)}{\partial t} \right] \sqrt{\frac{a-\xi}{a+\xi}}\, d\xi \\ + U_0 \int_0^t u_2^{(i)}(U_0\tau) \sqrt{\frac{t-\tau+t_a}{t-\tau}}\, d\tau = 0.\end{aligned} \tag{9}$$

If we write

$$\varkappa_1 = \int_0^t U(\tau)\, d\tau, \qquad \chi_1 = \int_0^t U(\tau)\, d\tau,$$

then for $U = U_0 = \text{const}$,

$$\varkappa_1 = U_0 t, \qquad \chi_1 = U_0 t, \qquad t_a = \frac{2a}{U_0}.$$

Hence equations (8), where p_1 and p_2 are defined by (7), and equations (9) can be written in the following system of integro-differential equations with partial derivatives of the profile deformations $w^{(i)}(x,t)$ and the velocity discontinuities $u_2^{(i)}(U_0 t)$ in the trace ($i = 1, 2$):

$$\begin{aligned} M_i\left[w(x,t), u_2(U_0 t)\right] &= D_i \frac{\partial^4 w^{(i)}}{\partial x^4} + m_i \frac{\partial^2 w^{(i)}}{\partial t^2} \\ &+ \frac{2\rho}{\pi} \Bigg\{ U_0 \sqrt{\frac{a+x}{a-x}} \int_{-a}^{a} \left[-U_0 \frac{\partial w^{(i)}(\xi,t)}{\partial \xi} + \frac{\partial w^{(i)}(\xi,t)}{\partial t} \right] \\ &\times \sqrt{\frac{a-\xi}{a+\xi}} \frac{d\xi}{\xi - x} - \int_{-a}^{a} \left[-U_0 \frac{\partial^2 w^{(i)}(\xi,t)}{\partial \xi \partial t} + \frac{\partial^2 w^{(i)}(\xi,t)}{\partial t^2} \right] \ln|\gamma|\, d\xi \\ &- U_0 \sqrt{\frac{a+x}{a-x}} \int_0^t u_2^{(i)}(U_0\tau) \frac{d\tau}{\sqrt{(t-\tau)(t-\tau+t_a)}} \Bigg\} \end{aligned}$$

(10)
$$
\begin{aligned}
&+\frac{2\rho}{\pi}\Bigg\{U_0[(x+a)^2+4H^2]^{-1/4}[(a-x)^2+4H^2]^{-3/4}\\
&\quad\times\int_{-a}^{a}\frac{\left[-U_0\frac{\partial w^{(j)}(\zeta,t)}{\partial\zeta}+\frac{\partial w^{(j)}(\zeta,t)}{\partial t}\right]}{(\zeta-x)^2+4H^2}\sqrt{\frac{a-\zeta}{a+\zeta}}\\
&\qquad\times\left[(a^2-x^2-4H^2)(\zeta-x)-8aH^2\right]d\zeta\\
&-\left(c_1e^{-\lambda_1H}+c_2\,e^{-\lambda_2H}\right)\int_{-a}^{a}\left[-U_0\frac{\partial^2 w^{(j)}(\zeta,t)}{\partial\zeta\partial t}+\frac{\partial^2 w^{(j)}(\zeta,t)}{\partial t^2}\right]\ln|\gamma|\,d\zeta\\
&+U_0\left[(x+a)^2+4H^2\right]^{-1/4}\left[(a-x)^2+4H^2\right]^{-3/4}\\
&\quad\times\int_0^t\sqrt{\frac{t-\tau+t_a}{t-\tau}}\,\frac{u_2^{(j)}(U_0\tau)}{\left(\tau-t-\frac{a+x}{U_0}\right)^2+\frac{4H^2}{U_0^2}}\\
&\qquad\times\left[(a^2-x^2-4H^2)(\tau-t-\frac{a+x}{U_0})-\frac{8aH^2}{U_0}\right]d\tau\\
&+\left(c_1e^{-\lambda_1H}+c_2e^{-\lambda_2H}\right)\int_0^t\frac{u_2^{(j)}(U_0\tau)}{t-\tau+\frac{a+x}{U_0}}sqrt\frac{(a+x)(a-x)}{(t-\tau)(t-\tau+t_a)}\,d\tau\\
&-\left(c_1e^{-\lambda_1H}+c_2e^{-\lambda_2H}\right)\pi U_0u_2^{(j)}(U_0t)+2\pi U_0u_2^{(j)}(U_0t)+\pi U_0u_2^{(i)}(U_0t)\Bigg\},
\end{aligned}
$$

and

(11)
$$
\begin{aligned}
M_k\left[w(x,t),u_2(U_0t)\right]=&\int_{-a}^{a}\left[-U_0\frac{\partial w^{(i)}(\xi,t)}{\partial\xi}+\frac{\partial w^{(i)}(\xi,t)}{\partial t}\right]\sqrt{\frac{a-\xi}{a+\xi}}\,d\xi\\
&+U_0\int_0^t u_2^{(i)}(U_0\tau)\sqrt{\frac{t-\tau+t_a}{t-\tau}}\,d\tau=0.
\end{aligned}
$$

The integro-differential equations (10)–(11) are a generalization of Fredholm and Volterra equations. This system defines the free motions of flexible profiles, and depends upon the initial conditions.

The profile deformations $w^{(1)}(x,t)$ and $w^{(2)}(x,t)$ satisfy the end conditions. We now define the initial conditions for system (10)–(11). For $t=0$, let

(12)
$$
w^{(i)}(x,0)=0,\qquad \frac{\partial w^{(i)}(x,0)}{\partial t}=\psi^{(i)}(x).
$$

System (10)–(11) does not contain the derivatives of the functions $u_2^{(1)}(U_0t)$ and $u_2^{(2)}(U_0t)$; therefore their initial values are not given.

The mathematical model of hydroelasticity for the two profiles moving in an ideal incompressible liquid is defined by the system of integro-differential equations

(10)–(11) which we rewrite in the short form:

$$
\begin{aligned}
&M_1\left(w^{(1)}(x,t), w^{(2)}(x,t), u_2^{(1)}(U_0t), u_2^{(2)}(U_0t)\right) = 0,\\
&M_2\left(w^{(1)}(x,t), w^{(2)}(x,t), u_2^{(1)}(U_0t), u_2^{(2)}(U_0t)\right) = 0,\\
&M_3\left(w^{(1)}(x,t), u_2^{(1)}(U_0t)\right) = 0,\\
&M_4\left(w^{(2)}(x,t), u_2^{(2)}(U_0t)\right) = 0.
\end{aligned}
\tag{13}
$$

This system satisfies the initial conditions (12) and end conditions (free-free, fixed-free, fixed-fixed, etc.).

4. A plan for the solution of system (13)

Since a detailed solution of system (13) by the proposed method contains many complex calculations, it is perhaps expedient only to give the plan of the solution and to present the final result.

4.1 Setting up of the Bubnov–Galerkin procedure. The profile displacements are represented in the form:

$$w^{(i)}(x,t) = X^{(i)T}(x)T^{(i)}(t),$$

where

$$X^{(i)}(x) = \left[X_1^{(i)}(x), \dots, X_n^{(i)}(x)\right]^T$$

are the orthonormalized system of the beam functions,

$$T^{(i)}(t) = \left[T_1^{(i)}(t), \dots, T_n^{(i)}(t)\right]^T$$

are generalized coordinates, and n is the number of vibration modes used. In particular, here the beam-profile integrals

$$\int_{-a}^{a}\int_{-a}^{a} X^{(i)}(x)X^{(i)T}(\xi)\ln|\gamma|\, d\xi\, dx,$$

$$\int_{-a}^{a}\int_{-a}^{a} X^{(i)}(x)X^{(i)T}(\xi)\sqrt{\frac{(a+x)(a-\xi)}{(a-x)(a+\xi)}}\,\frac{d\xi\, dx}{\xi - x}$$

are to be evaluated.

4.2 Regularizing the kernels and their approximations by sums of exponentials. For brevity, we shall not elaborate on this regularization and approximation.

4.3 Taking Laplace transforms. By taking the Laplace transform with respect to t:

$$T^{(i)^L}(q) = \int_0^\infty T^{(i)}(t)e^{-qt}\, dt,$$

$$u_2^{(i)^L}(q) = \int_0^\infty u_2^{(i)}(t)e^{-qt}\, dt,$$

we go over to the operator system:

$$
\begin{aligned}
(14)\qquad & (a_{11}q^2 + a_{12}q + a_{13})T^{(1)^L}(q) + (b_{11}q^2 + b_{12}q + b_{13})T^{(2)^L}(q) \\
& + \left[c_{12} + \frac{2c_{13}\sqrt{q}}{\sqrt{\pi t_a}} \left(\frac{N_1}{q + \frac{n_1}{t_a}} + \frac{N_2}{q + \frac{n_2}{t_a}} \right) \right. \\
& \qquad \left. + \frac{2c_{14}\sqrt{q}}{\sqrt{\pi t_a}} \left(\frac{B_1}{q + \frac{b_1}{t_a}} + \frac{B_2}{q + \frac{b_2}{t_a}} \right) \right] \frac{1}{U_0} u_2^{(2)^L}\left(\frac{q}{U_0}\right) \\
& + \frac{2c_{11}\sqrt{q}}{\sqrt{\pi t_a}} \left(\frac{D_1}{q + \frac{d_1}{t_a}} + \frac{D_2}{q + \frac{d_2}{t_a}} \right) \Bigg] \frac{1}{U_0} u_2^{(1)^L}\left(\frac{q}{U_0}\right) = \psi_1^L(q), \\
& (a_{21}q^2 + a_{22}q + a_{23})T^{(1)^L}(q) + (b_{21}q^2 + b_{22}q + b_{23})T^{(2)^L}(q) \\
& + \left[c_{22} + \frac{2c_{23}\sqrt{q}}{\sqrt{\pi t_a}} \left(\frac{N_1}{q + \frac{n_1}{t_a}} + \frac{N_2}{q + \frac{n_2}{t_a}} \right) \right. \\
& \qquad \left. + \frac{2c_{24}\sqrt{q}}{\sqrt{\pi t_a}} \left(\frac{B_1}{q + \frac{b_1}{t_a}} + \frac{B_2}{q + \frac{b_2}{t_a}} \right) \right] \frac{1}{U_0} u_2^{(1)^L}\left(\frac{q}{U_0}\right) \\
& + \frac{2c_{21}\sqrt{q}}{\sqrt{\pi t_a}} \left(\frac{D_1}{q + \frac{d_1}{t_a}} + \frac{D_2}{q + \frac{d_2}{t_a}} \right) \Bigg] \frac{1}{U_0} u_2^{(2)^L}\left(\frac{q}{U_0}\right) = \psi_2^L(q), \\
& (a_{31}q + a_{32})T^{(1)^L}(q) + \frac{1}{\sqrt{q}} \Bigg[\sqrt{\pi t_a} \\
& + \frac{1}{\sqrt{\pi t_a}} \left(\frac{P_1}{q + \frac{p_1}{t_a}} + \frac{P_2}{q + \frac{p_2}{t_a}} \right) \Bigg] \frac{1}{U_0} u_2^{(1)^L}\left(\frac{q}{U_0}\right) = \psi_3^L(q), \\
& (a_{41}q + a_{42})T^{(2)^L}(q) + \frac{1}{\sqrt{q}} \Bigg[\sqrt{\pi t_a} \\
& + \frac{1}{\sqrt{\pi t_a}} \left(\frac{P_1}{q + \frac{p_1}{t_a}} + \frac{P_2}{q + \frac{p_2}{t_a}} \right) \Bigg] \frac{1}{U_0} u_2^{(2)^L}\left(\frac{q}{U_0}\right) = \psi_4^L(q),
\end{aligned}
$$

where $(a_{\alpha\beta})$, $(b_{\alpha\beta})$ are $n\times n$ matrices and $(c_{\alpha\beta})$ is an $n\times 1$ matrices, whose elements contain the parameters of the system and the beam-profile integrals. The constants B_α, b_α, D_α, d_α, N_α, n_α, P_α, p_α are determined by the approximation of the system kernels by sums of exponential, $\left(\psi_1^{(L)}(q)\right), \left(\psi_2^{(L)}(q)\right)$ are $n\times 1$ matrices, and $\psi_3^{(L)}(q)$ and $\psi_4^{(L)}(q)$ are scalars which are determined by the initial conditions.

We write system (14) in matrix form:

$$
(15)\qquad S(q)V^L(q) = \Psi^L(q),
$$

where

$$
V^L(q) = \left[T^{(1)^L}(q), T^{(2)^L}(q), u_2^{(1)^L}\left(\frac{q}{U_0}\right), u_2^{(2)^L}\left(\frac{q}{U_0}\right) \right]^T,
$$

$$
\Psi^L(q) = \left[\psi_1^L(q), \psi_2^L(q), \psi_3^L(q), \psi_4^L(q) \right]^T,
$$

and the block matrix $S(q)$ is of the following form:

$$\begin{array}{c} \\ n \\ n \\ 1 \\ 1 \end{array} \begin{array}{c} \begin{array}{cccc} \quad n \quad & \quad n \quad & \quad 1 \quad & \quad 1 \quad \end{array} \\ \begin{pmatrix} S_{11}(q) & S_{12}(q) & S_{13}(q) & S_{14}(q) \\ S_{21}(q) & S_{22}(q) & S_{23}(q) & S_{24}(q) \\ S_{31}(q) & 0 & S_{33}(q) & 0 \\ 0 & S_{42}(q) & 0 & S_{44}(q) \end{pmatrix} \end{array},$$

where the dimensions of the blocks are as indicated.

The operator solution of equation (15) is obtained in the form:

$$V^L(q) = H^L(q)\Psi^L(q), \tag{16}$$

where $H^L(q) = S^{-1}(q)$. Thus, formula (16) determines the operator solution of the hydroelastic problem for the motion of two flexible profiles. This solution is a multiple-valued function of the complex variable q.

References

1. S. M. Bauer, S. B. Filippov, A. L. Smirnov and P. E. Tovstik, *Asymptotic methods in mechanics with application to thin shells and plates*, Asymptotic Methods in Mechanics (A. L. Smirnov and R. Vaillancourt, eds.), CRM Proceedings and Lecture Notes, this volume, Amer. Math. Soc., Providence, R.I., 1993, pp. 3–140.
2. B. A. Ershov, *The effect of a vortex trace on disturbed motion of an elastic airfoil*, Vestnik Leningrad. Univ. Mat. Mekh. Astronom. (1988), no. 15, 57–61; English transl., Len. Univ. Mech. Bull. (1988), no. 3, 18–24.
3. B. A. Ershov, Yu. A. Mochalova, and E. V. Polyakova, *Unsteady motion of an elastic wing in hyposonic incompressible flow*, Vestnik Leningrad. Univ. Mat. Mekh. Astronom. (1989), no. 22, 44–50; English transl., Len. Univ. Mech. Bull. (1989), no. 4, 8–14.
4. ——, *A mathematical model for hydroelastic problems with a fluid memory*. Part I, Asymptotic Methods in Mechanics (A. L. Smirnov and R. Vaillancourt, eds.), CRM Proceedings and Lecture Notes, this volume, Amer. Math. Soc., Providence, R.I., 1993, pp. 175–183.
5. L. I. Sedov, *Plane problems of hydrodynamics and aerodynamics*, "Nauka", Moscow, 1966 (448 pp.). (Russian)

ABSTRACT. A model for two flexible profiles moving in an ideal incompressible flow is considered. The interconnection of the profiles is provided by the flow and the vortex traces behind the back edges of the profiles. The hydroelastic problem is represented by a system of integro-differential equations which are a generalization of Fredholm and Voltera equations. The solution of the problem is given in an operator form.

INSTITUTE OF MATHEMATICS AND MECHANICS, ST. PETERSBURG STATE UNIVERSITY, 2 BIBLIOTECHNAYA SQ., STARY PETERHOF, 198904 ST. PETERSBURG, RUSSIA

E-mail address: als@hq.math.lgu.spb.su

Centre de Recherches Mathématiques
CRM Proceedings and Lecture Notes
Volume **3**, 1993

Low-Frequency Vibrations of Cylindrical Shells. Part I: Shells with a Slanted Edge

Sergei B. Filippov

1. Introduction

We take the radius, R, of the cylindrical shell as unit of measurement. Then, following W. Flügge [**6**] and A. L. Gol′denveizer [**8**], the equations of vibrations in dimensionless form can be expressed by means of the equilibrium equations:

$$\begin{aligned} T_{1,x} + S_{,\varphi} + \lambda u &= 0, \\ S_{,x} + T_{2,\varphi} + Q_2 + 2H_{,x} + \lambda v &= 0, \\ -T_2 + Q_{2,\varphi} + Q_{1,x} + \lambda w &= 0, \end{aligned} \tag{1a}$$

the elasticity relations:

$$\begin{aligned} Q_1 &= M_{1,x} + 2H_{,\varphi}, & Q_2 &= M_{2,\varphi}, \\ M_1 &= \varepsilon^8(\vartheta_{1,x} + \nu\vartheta_{2,\varphi}), & M_2 &= \varepsilon^8(\vartheta_{2,\varphi} + \nu\vartheta_{1,x}), \\ (1-\nu^2)T_1 &= u_{,x} + \nu(w + v_{,\varphi}), & (1-\nu^2)T_2 &= w + v_{,\varphi} + \nu u_{,x}, \\ H &= (1-\nu)\varepsilon^8\vartheta_{2,x}, & 2(1-\nu)S &= v_{,x} + u_{,\varphi}, \end{aligned} \tag{1b}$$

and the deformation-displacement equations:

$$\vartheta_1 = -w_{,x}, \qquad \vartheta_2 = -w_{,\varphi} + v. \tag{1c}$$

In these equations, x, φ are the coordinates on the middle surface, T_1, T_2, S, Q_1, Q_2, M_1, M_2, H are the dimensionless stress-resultant and stress-couples, ϑ_1 and ϑ_2 are the angles of rotation, u, v and w are the components of displacement, ν is Poisson's ratio, $\varepsilon^8 = h^2/[12(1-\nu^2)R^2]$ is a small parameter, h is the shell thickness, $\lambda = \rho R^2\omega^2/(\varepsilon^4 E)$ is the frequency, E is Young's modulus, ρ is the mass density, and ω is the vibration frequency.

We suppose that the straight edge, $x = x_1 = 0$, and the slanted edge, $x = x_2(\varphi)$, are clamped (see Fig. 1) and on these,

$$\begin{aligned} u = v = w = \vartheta_1 &= 0, \qquad \text{if } x = x_1, \\ u_n = v_t = w = \vartheta_n &= 0, \qquad \text{if } x = x_2, \end{aligned} \tag{2}$$

1991 *Mathematics Subject Classification.* Primary: 35B40; Secondary: 73K15.
Key words and phrases. Shell free vibrations, asymptotic methods.
This is the final form of the paper.

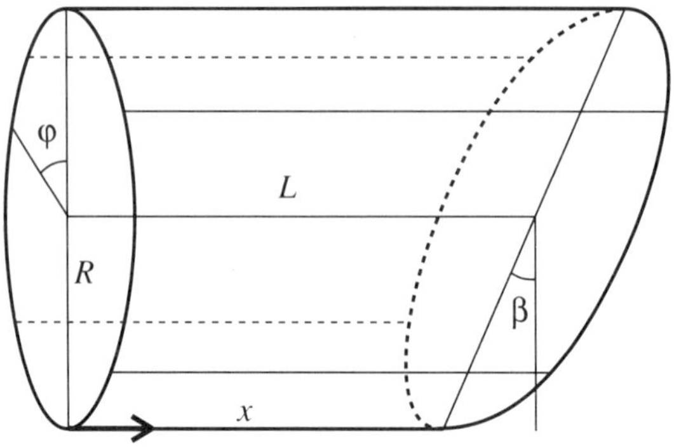

FIGURE 1. Local coordinate system on shell with slanted edge.

where

$$x_2(\varphi) = l + \tan\beta\cos\varphi, \qquad u_n = u\cos\gamma + v\sin\gamma, \qquad v_t = v\cos\gamma - u\sin\gamma,$$
$$\vartheta_n = \vartheta_1\cos\gamma + \vartheta_2\sin\gamma, \qquad \tan\gamma = \tan\beta\sin\varphi = -x_{2,\varphi}, \qquad l = L/R,$$

and L is the shell length. We shall assume that the shell is not too short, i.e.

$$l \gg \varepsilon^2. \tag{3}$$

If $\beta = 0$, then both edges of the shell are straight. In this case the vibration modes can be represented in the form

$$y(x,\varphi) = Y(x)e^{ik\varphi}, \tag{4}$$

where k is the circumferencial wave number. In this case, (1) is transformed into a system of ordinary differential equations. The vibrations of a cylindrical shell with a straight edges have been analyzed in numerous papers, for example by R. N. Arnold and G. B. Walburton [**1**], K. Forsberg [**7**] and V. L. Weingarten [**12**]. A. L. Gol′denveizer, V. B. Lidsky and P. E. Tovstik [**8**], S. B. Filippov [**5**] and others have used asymptotic methods to solve this problem.

If $\beta \neq 0$, then the solution of (1) cannot be written in the form (4) and the boundary value problem (1)–(2) becomes much more difficult. However, the vibration modes, which correspond to the lowest frequencies, rapidly decrease away from the longest generator of the cylinder. This fact makes it possible to apply the asymptotic methods of V. P. Maslov [**10**] and allowed P. E. Tovstik in [**11**] to obtain the lowest frequencies and the corresponding vibration modes. These methods produce an approximate solution to a system of partial differential equations by reducing it to some boundary value problems for ordinary differential equations.

2. Approximate solution

The approximate solution of the boundary value problem (1)–(2), which decreases away from the straight line $\varphi = 0$, can be expressed as

$$y(x,\varphi,\varepsilon) = Y^{(0)} + Y^{(b)}, \qquad \lambda = \lambda_0 + \varepsilon\lambda_1 + \varepsilon^2\lambda_2. \tag{5}$$

In this equation, y denotes one of the variables and

$$Y^{(0)} = \varepsilon^{a(y)} y^{(0)}\, e^{iq/\varepsilon}, \qquad Y^{(b)} = \varepsilon^{b(y)} y^{(b)}\, e^{iq/\varepsilon},$$

(6)
$$y^{(0)} = \sum_{k=0}^{4} \varepsilon^{k/2} y_k(x,\xi), \qquad y^{(b)} = \sum_{j=1}^{4} C_j {y_j}^{(b)} f_j(x,\varepsilon),$$

where

$$q = \sum_{k=1}^{4} \frac{\varepsilon^{k/2}\xi^k q_{k-1}}{k!}, \qquad \xi = \varepsilon^{-1/2}\varphi,$$

$$f_j = e^{xr_j/\varepsilon^2}, \quad j=1,2, \qquad f_j = e^{(x_0-x)r_j/\varepsilon^2}, \quad j=3,4,$$

$$r_{1,2} = \frac{-1\pm i}{\sqrt{2}}, \qquad r_{3,4} = \frac{1\pm i}{\sqrt{2}},$$

and

$$x_0 = x_2(0).$$

The functions $u^{(0)}$, $v^{(0)}$ and the other functions above describe the main stress-strain state of the shell. The functions describing the boundary effects, $u^{(b)}$ and $v^{(b)}$, and the other functions are very small everywhere except near the shell edges. One can determine the intensity coefficients, $a(y)$ and $b(y)$, the constants, λ_k, q_k, C_j, $y_j^{(b)}$, and the functions $y_k(x,\xi)$ by substituting (5) into (1) and (2).

If we take $a(w)=0$ and substitute $Y^{(0)}$ into (1), we obtain

(7)
$$a(v)=1, \qquad a(u)=2, \qquad a(\vartheta_1)=0, \qquad a(\vartheta_2)=-1,$$
$$u^{(0)} = -\Phi^{-2} w^{(0)}_{,x} - \varepsilon^2(2+\nu)\Phi^{-4} w^{(0)}_{,xxx}, \qquad {\vartheta_1}^{(0)} = -w^{(0)}_{,x},$$
$$v^{(0)} = i\Phi^{-1}w^{(0)} - i\nu\varepsilon^2\Phi^{-3}w^{(0)}_{,xx}, \qquad {\vartheta_2}^{(0)} = -i\Phi w^{(0)} + i\varepsilon^2\Phi^{-1}w^{(0)},$$

and

(8)
$$Gw^{(0)} - \lambda w^{(0)} = \varepsilon^2 N w^{(0)}, \qquad G = \Phi^{-4}\frac{\partial^4}{\partial x^4} + \Phi^4,$$
$$N = \Phi^2\left(4\frac{\partial^2}{\partial x^2}+2\right) + \lambda_0\Phi^{-2}\left(1-2\frac{\partial^2}{\partial x^2}\right),$$
$$\Phi = -i\varepsilon^{1/2}\frac{\partial}{\partial\xi} + \sum_{k=0}^{4}\varepsilon^{k/2}\frac{\xi^k}{k!}q_k.$$

Here Φ^{-n} is a formal asymptotic expansion of the inverse of the operator Φ^n, which satisfies the equation $\Phi^{-n}\Phi^n z = z$. For example,

$$\Phi^{-2} = \frac{1}{q_0^2} - \frac{2}{q_0^3}\varepsilon^{1/2}\left(q_1\xi - i\frac{\partial}{\partial\xi}\right) - \frac{q_2}{q_0^3}\varepsilon\, xi^2 + \frac{3}{q_0^4}\varepsilon\left(q_1\xi - i\frac{\partial}{\partial\xi}\right)^2 + \cdots$$

To satisfy the boundary conditions (2) we assume that $b(w)=2$. Then, substituting $Y^{(b)}$ into (1) we have

$$b(u_n)=4, \qquad b(\vartheta_n)=0,$$

(9)
$$w^{(b)} = \sum_{j=1}^{4} C_j f_j, \qquad u_n^{(b)} = \sum_{j=1}^{4} C_j f_j {r_j}^{-1}, \qquad \vartheta_n^{(b)} = -\sum_{j=1}^{4} C_j f_j r_j.$$

3. The boundary-value problem for the main stress-strain state

We expand the linear differential operator G, given in (8), in a power series in ε as follows:

$$G = \sum_{k=0}^{\infty} \varepsilon^{k/2} G_k, \tag{10}$$

where

$$\begin{aligned}
G_0 &= {q_0}^{-4}\frac{\partial^4}{\partial x^4} + {q_0}^4, \\
G_1 &= \frac{\partial G_0}{\partial q_0} F, \\
G_2 &= \frac{1}{2}\frac{\partial^2 G_0}{\partial {q_0}^2} F^2 + \frac{\partial G_0}{\partial q_0}\frac{q_2 \xi^2}{2}, \\
G_3 &= \frac{1}{3!}\frac{\partial^3 G_0}{\partial {q_0}^3} F^3 + \frac{1}{2}\frac{\partial^2 G_0}{\partial q_0^2} q_2(\xi^2 F - i\xi) + \frac{\partial G_0}{\partial q_0}\frac{q_3 \xi^3}{3!}, \\
G_4 &= \frac{1}{4!}\frac{\partial^4 G_0}{\partial {q_0}^4} F^4 + \frac{1}{3!}\frac{\partial^3 G_0}{\partial q_0^3} q_2 \left(\frac{\xi^2}{2}F^2 - 3i\xi F - 1\right) \\
&\quad + \frac{1}{2}\frac{\partial^2 G_0}{\partial {q_0}^2}\left(q_3\frac{\xi^3}{3}F - iq_3\frac{\xi^2}{2} + q_2^2\frac{\xi^4}{4}\right) + \frac{\partial G_0}{\partial q_0}\frac{q_4 \xi^4}{4!},
\end{aligned}$$

and

$$F = q_1 \xi - i\frac{\partial}{\partial \xi}.$$

If we substitute (10) and the expansion (6) for $w^{(0)}$ into equation (8), we obtain the following equations for λ_k and $w_k(x, \xi)$:

$$\begin{aligned}
&\sum_{k=0}^{p} \Delta_k w_{p-k} = 0, \qquad p = 0, 1, 2, 3, \qquad \sum_{k=0}^{4} \Delta_k w_{4-k} = N w_0, \\
&\Delta_k = G_k - \lambda_{k/2}, \qquad k = 0, 2, 4, \qquad \Delta_k = G_k, \qquad k = 1, 3.
\end{aligned} \tag{11}$$

The function $x_2(\varphi)$ can be expanded into the series

$$x_2(\varphi) = x_0 + \varepsilon x_0'' \frac{\xi^2}{2} + \varepsilon^2 x_0^{iv} \frac{\xi^4}{24} + \dots,$$

where

$$x_0'' = x_{2,\varphi\varphi}(0) = -\tan\beta, \qquad x_0^{iv} = x_{2,\varphi\varphi\varphi\varphi}(0) = \tan\beta.$$

Hence

$$\begin{aligned}
y_k(x_2(\varphi), \xi) =& y_k(x_0, \xi) + \frac{1}{2}\varepsilon y_{k,x}(x_0, \xi) x_0'' \xi^2 \\
&+ \frac{1}{8}\varepsilon^2 \left[y_{k,xx}(x_0, \xi)(x_0'')^2 + \frac{1}{24} y_{k,x}(x_0, \xi) x_0^{iv} \right] + \cdots
\end{aligned} \tag{12}$$

To get the boundary conditions for equations (11) we expand the variables in (2) into power series, taking into account (5), (6) and (12), and set to zero the coefficients of each power of ε. For $x = x_0$ we obtain

$$
\begin{aligned}
w_0 &= \frac{\partial w_0}{\partial x} = 0, \\
w_1 &= \frac{\partial w_1}{\partial x} = 0, \\
w_2 &= \frac{\partial w_2}{\partial x} + \frac{1}{2} x_0'' \xi^2 \frac{\partial^2 w_0}{\partial x^2} = 0, \\
w_3 &= \frac{\partial w_3}{\partial x} + \frac{1}{2} x_0'' \xi^2 \frac{\partial^2 w_1}{\partial x^2} = 0, \\
w_4 + \frac{1}{2} x_0'' \xi^2 \frac{\partial w_2}{\partial x} + \frac{1}{8}(x_0'')^2 \xi^4 \frac{\partial^2 w_0}{\partial x^2} &= \nu q_0{}^2 \frac{\partial^2 w_0}{\partial x^2}, \\
\frac{\partial w_4}{\partial x} + \frac{1}{2} x_0'' \xi^2 \frac{\partial^2 w_2}{\partial x^2} - \frac{2i}{q_0} x_0'' \xi \frac{\partial^2 w_1}{\partial x^2} &+ \frac{x_0''}{q_0}\left(2iq_1\xi - \frac{1}{q_0}\right)\frac{\partial^2 w_0}{\partial x^2} \\
+ \frac{2}{q_0{}^2} x_0'' \xi \frac{\partial^3 w_0}{\partial x^2 \partial \xi} &+ \frac{1}{8}(x_0'')^2 \xi^4 \frac{\partial^3 w_0}{\partial x^3} + \frac{1}{24} x_0^{iv} \xi^4 \frac{\partial^2 w_0}{\partial x^2} \\
&= -(2+\nu)\frac{1}{q_0{}^2}\frac{\partial^3 w_0}{\partial x^3} + q_0{}^2 u_n^{(b)}.
\end{aligned} \tag{13}
$$

The boundary conditions for $x = 0$ can be determined by assuming that $x_0'' = x_0^{iv} = 0$ in (13).

4. The boundary effect functions

Besides the boundary conditions (13), the method presented in Section 3 gives four boundary conditions for the boundary effect functions, namely,

$$
\vartheta_n^{(b)} = 0, \qquad w^{(b)} = -\nu q_0{}^{-2} w_{0,xx}, \qquad \text{for } x = 0 \quad \text{and} \quad x = x_0. \tag{14}
$$

If we substitute (9) into (14) and eliminate the terms containing the factors $f_1(x_0)$, $f_2(x_0)$, $f_3(0)$ and $f_4(0)$, which are small according to (3), then we obtain two systems of linear algebraic equations for the arbitrary constants C_j:

$$
C_j + C_{j+1} = \beta_j, \qquad r_j C_j + r_{j+1} C_{j+1} = 0, \qquad j = 1, 3,
$$

where

$$
\beta_1 = -\nu q_0{}^{-2} w_{0,xx}(0), \qquad \beta_3 = -\nu q_0{}^{-2} w_{0,xx}(x_0),
$$

and the solutions of which are

$$
C_j = \frac{r_{j+1}\beta_j}{r_{j+1} - r_j}, \qquad j = 1, 3, \qquad C_j = \frac{r_{j-1}\beta_{j-1}}{r_j - r_{j+1}}, \qquad j = 2, 4.
$$

Finally from (9) we have

$$
u_n^{(b)}(0) = -\sqrt{2}\left(\frac{\nu}{q_0}\right)^2 w_{0,xx}(0), \qquad u_n^{(b)}(x_0) = \sqrt{2}\left(\frac{\nu}{q_0}\right)^2 w_{0,xx}(x_0). \tag{15}
$$

5. Zeroth- and first-order approximations

Let us start to solve equations (11) with the boundary conditions (13). In the zeroth-order approximation ($p = 0$) the boundary value problem has the form

$$(16)\qquad (G_0 - \lambda_0)w_0 = 0, \qquad w_0 = w_{0,x} = 0, \qquad \text{for } x = 0 \quad \text{and} \quad x = x_0.$$

If we represent the solution of (16) in the form $w_0(x, \xi) = P_0(\xi)W_0(x)$, where $P(\xi)$ is a polynomial, then equation (16) becomes

$$(17)\qquad \frac{d^4W_0}{dx^4} - \alpha^4 W_0 = 0, \qquad \alpha^4 = \lambda_0 q_0^4 - q_0^8,$$

with the boundary conditions

$$W_0 = W_{0,x} = 0, \qquad \text{for } x = 0 \quad \text{and} \quad x = x_0.$$

The solution of (17) is easily seen to be

$$(18)\qquad W_0 = U(\alpha_n x) - \eta_n T(\alpha_n x),$$

where

$$T(z) = \sinh z - \sin z, \qquad U(z) = \cosh z - \cos z, \qquad \eta_n = \frac{U(\alpha_n x_0)}{T(\alpha_n x_0)},$$

and α_n is a root of the transcendential equation

$$\cosh(\alpha x_0)\cos(\alpha x_0) = 1.$$

Then

$$(19)\qquad \lambda_0^{(n)} = \left(\frac{\alpha_n}{q_0}\right)^4 + q_0^4, \qquad n = 0, 1, 2, \ldots$$

We find the value of q_0 from the compatibility condition of the first-order approximation. The boundary value problem for $p = 1$ has the form

$$(20)\qquad \begin{aligned} &(G_0 - \lambda_0)w_1 + G_1 w_0 = 0, \\ &w_1 = w_{1,x} = 0, \qquad \text{for} \quad x = 0 \quad \text{and} \quad x = x_0. \end{aligned}$$

Multiplying (20) by W_0, integrating the product from 0 to x_0 and taking into account (16), we obtain the compatibility condition

$$(21)\qquad \int_0^{x_0} \left(q_1 \xi P_0 - i\frac{\partial P_0}{\partial \xi}\right) \frac{\partial G_0}{\partial q_0} W_0 W_0 \, dx = 0.$$

Differentiating (16) with respect to q_0 gives

$$(22)\qquad \frac{\partial^k G_0}{\partial q_0^k} w_0 = \frac{\partial^k \lambda_0}{\partial q_0^k} w_0, \qquad k = 1, 2, \ldots,$$

and the compatibility condition (21) becomes

$$(23)\qquad \frac{\partial \lambda_0}{\partial q_0}\left(\frac{\partial P_0}{\partial \xi} - h_1 \xi P_0\right) I_0 = 0,$$

where

$$(24)\qquad h_1 = -iq_1, \qquad I_0 = \int_0^{x_0} W_0^2 \, dx = \frac{x_0}{4\alpha^4}\left(\frac{d^2 W_0}{dx^2}\right)^2.$$

Since the equation $P_{0,\xi} - h_1\xi P_0 = 0$ does not have a polynomial solution, then

$$\frac{\partial \lambda_0}{\partial q_0} = -\frac{4\alpha^4}{q_0^5} + 4q_0^3 = 0, \qquad q_0^2 = \alpha,$$

and from (19) we obtain

$$\lambda_0^{(n)} = 2\alpha_n^2, \qquad n = 0, 1, 2, \ldots$$

It then follows from (23) that $G_1 w_0 = 0$; hence $w_1 = P_1(\xi)W_0$.

6. Second-order approximation

The equation of the second-order approximation is

(25) $$(G_0\lambda_0)w_2 + G_1 w_1 + (G_2 - \lambda_1)w_0 = 0.$$

The compatibility condition for (25) with the appropriate boundary conditions (13), after transformations by means of (10), (22) and (24), becomes

(26) $$P_{0,\xi\xi} - 2h_1\xi P_{0,\xi} + \left[\left(h_1^2 - \frac{\Lambda_1}{\Lambda_2}\right)\xi^2 h_1 + \frac{2\lambda_1}{\Lambda_2}\right] P_0 = 0,$$

where

$$\Lambda_1 = 4\alpha^2 s_0, \qquad s_0 = -\frac{x_0''}{x_0} = \frac{\tan\beta}{l + \tan\beta},$$
$$\Lambda_2 = 32\alpha.$$

Equation (26) has polynomial solutions of degree m for

(27) $$h_1 = \sqrt{\frac{\Lambda_1}{\Lambda_2}} = \frac{1}{4}\sqrt{2\alpha s_0}, \qquad 2\lambda_1 = h_1\Lambda_2(2m+1), \qquad m = 0, 1, 2\ldots$$

Hence

(28) $$\begin{aligned}\lambda^{(n,m)} &= \lambda_0^{(n)} + \varepsilon\lambda_1^{(n,m)} + O\left(\varepsilon^2\right)\\ &= 2\alpha_n^2 + 4\varepsilon\alpha_n\sqrt{2\alpha_n s_0}\,(1+2m) + O\left(\varepsilon^2\right).\end{aligned}$$

Under the transformation of variables

(29) $$\xi = \frac{\theta}{\sqrt{h_1}},$$

equation (26), as a consequence of (27), becomes

(30) $$P_{0,\theta\theta} - 2\theta P_{0,\theta} + 2mP_0 = 0.$$

The solutions of (30), for $m = 0, 1, 2, \ldots$, are the Hermite polynomials $P_0^{(m)} = H_m(\theta)$, $(H_0 = 1, H_1 = 2\theta, \ldots)$. Hence the vibration modes are

$$w^{(n,m)}(x,\varphi) = H_m\left(\sqrt{\frac{h_1^{(n)}}{\varepsilon}}\varphi\right) W_0(x)\exp\left[\frac{1}{\varepsilon}\left(i\sqrt{\alpha_n}\varphi \frac{h_1^{(n)}}{2}\varphi^2\right)\right] + O\left(\varepsilon^{1/2}\right).$$

The functions $w^{(n,m)}$ oscillate quickly as φ varies and decrease rapidly away from the straight line $\varphi = 0$. Similar results are obtained in [**11**] for a shell with simply supported edges (see also Subsection 4.4.3 of the Survey [**3**]). Formulae (25) and (26) yield

$$\begin{aligned}(G_0-\lambda_0)w_2 &= \frac{1}{2}\Lambda_1\xi^2 P_0 W_0\\ &= \frac{1}{2}P_0\xi^2(G_0-\lambda_0)xs_0W_{0,x}.\end{aligned}$$

Hence

$$w_2 = P_2(\xi)W_0 + \frac{1}{2}P_0\xi^2 xs_0 W_{0,x}.$$

7. Third-order approximation

The equation

(31) $$(G_0-\lambda_0)w_3 + G_1w_2 + (G_2-\lambda_2)w_1 + G_3w_0 = 0$$

(see (11) for $p=3$) with the boundary conditions (13) has a nonzero solution if

(32) $$\begin{aligned}P_1'' - 2h_1\xi P_1' + 2h_1mP_1 &= iq_0{}^{-1}[(q_0q_2-3h_1{}^2)h_1\xi^3P_0\\ &\quad + (h_1{}^2-q_0q_2)(\xi P_0+\xi^2P_0') + 3h_1P_0' + 3h_1\xi P_0'' - P_0'''],\end{aligned}$$

where $P_1' = P_{1,\xi}$. The compatibility condition (32) is obtained by means of transformations similar to (23) and (26).

If we take $q_2 = 3h_1^2/q_0$, then the coefficient of the highest power of ξ on the right-hand side of (32) is equal to zero and, after the transformation of variables (29), this equation can be expressed as

(33) $$L_0P_1 = -i\frac{\sqrt{h_1}}{q_0}L_1P_0,$$

where

$$\begin{aligned}L_0P_1 &= P_1'' - 2\theta P_1' + 2mP_1, \qquad P_k' = P_{k,\theta},\\ L_1P_0 &= P_0''' - 3\theta P_0'' + (2\theta^2-3)P_0' + 2\theta P_0.\end{aligned}$$

If we transform the right-hand side of (33) with the help of the equality $P_0 = H_m$ and the following formulae for Hermite polynomials [**9**]:

(34) $$2\theta H_m = H_{m+1} + 2mH_{m-1}, \qquad H_m' = 2mH_{m-1}, \qquad m = 0,1,2,\ldots,$$

then (33) is represented in the form

$$L_0P_1 = i\frac{\sqrt{h_1}}{q_0}[-(m+1)H_{m+1} + 2m^2H_{m-1}].$$

By means of the identity

(35) $$L_0H_{m+k} = -2kH_m,$$

where k is an integer, we obtain

(36) $$P_1 = \frac{i\sqrt{h_1}}{q_0}\left[\frac{m+1}{2}H_{m+1}\left(\sqrt{h_1}\xi\right) + m^2H_{m-1}\left(\sqrt{h_1}\xi\right)\right].$$

Formulae (31) and (32) yield

$$(G_0 - \lambda_0)w_3 = \Lambda_1 \left(\frac{1}{2}\xi^2 w_1 - \frac{2i}{q_0}\xi w_0\right),$$

$$w_3 = P_3(\xi)W_0 + \left(\frac{1}{2}\xi^2 P_1 - \frac{2i}{q_0}\xi P_0\right) x s_0 W_{0,x}.$$

8. Fourth-order approximation

The compatibility condition for equation (11) for $p = 4$ with the proper boundary conditions (13) is of the following form:

(37)
$$L_0 P_2 = -\frac{i\sqrt{h_1}}{q_0} L_1 P_1 + \frac{h_1}{q_0^2} L_2 P_0,$$

where

(38)
$$L_2 P_0 = \frac{9}{4} P_0^{iv} - 9\theta P_0''' + \left(10\theta - \frac{27}{2}\right) P_0'' + (\theta^2 + 2\sigma\theta^2 + 24)\theta P_0' + \left(\sigma_m \theta^2 + \zeta\right) P_0,$$

$$\zeta_0 = \frac{2\alpha}{I_0 h_1^2 \Lambda_2}\left\{\frac{1}{\alpha}\left[u_n^{(b)} \frac{d^2 W_0}{dx^2}\right]_0^{x_0} - \frac{2(1+\nu)}{\alpha^3}\left[\frac{d^2 W_0}{dx^2}\frac{d^3 W_0}{dx^3}\right]_0^{x_0} - I_N - \lambda_2 I_0\right\},$$

where

$$\sigma_m = 4m - \frac{1}{2} + 3\sigma,$$

$$\sigma = -\frac{\alpha}{24h_1^2},$$

$$I_N = \int_0^{x_0} N W_0 W_0 \, dx,$$

$$\zeta = \zeta_0 + \frac{31}{4}.$$

If we transform the right-hand side of (37) by means of (34), we obtain

(39)
$$L_0(P_2) = \frac{h_1}{\alpha}[A(m)H_{m+2} + B(m)H_m + C(m)H_{m-2} + D(m)H_{m-4}],$$

where

(40)
$$A(m) = \frac{7 + 22m + 4m^2}{8} + \frac{3 + 2m}{4}\sigma,$$

$$B(m) = \zeta - \frac{5}{4}(1 - 10m - 10m^2) + \frac{3}{2}(1 + 2m + 2m^2)\sigma,$$

$$C(m) = m(m-1)\left[\frac{5 + 26m - 4m^2}{2} + 3(2m-1)\sigma\right],$$

$$D(m) = 2m(m-1)(m-2)(m-3)(3 + 2\sigma).$$

Equation (39) has the solution

$$P_2 = \frac{h_1}{8q_0^2}\left[-2A(m)H_{m+2} + 2C(m)H_{m-2} + D(m)H_{m-4}\right]$$

if $B(m)=0$, i.e.

$$\zeta = \frac{5}{4}\left(1-10m-10m^2\right) - \frac{3}{2}\left(1+2m+2m^2\right)\sigma. \tag{41}$$

It follows from (38), (41), (8), (15) and (18) that

$$\lambda_2^{(m,n)} = \alpha_n\left[\left(13+25m+25m^2\right)s_0 - 5 - 2m - 2m^2\right] + 8\frac{\alpha_n^2}{x_0}\left[\sqrt{2}\,\nu^2 - 2(1+\nu)\eta_n\right].$$

The lowest frequency is

(42)

$$\lambda^{(0,0)} = 2\alpha_0^2 + 4\sqrt{2s_0}\,\varepsilon\alpha_0^{3/2} + \alpha_0\varepsilon^2(-5+13s_0) + 8\frac{\alpha_0^2}{x_0}\varepsilon^2\left[\sqrt{2}\,\nu^2 - 2(1+\nu)\eta_0\right] + O\left(\varepsilon^3\right).$$

The vibration modes are represented in the form

$$w^{(n,m)}(x,\varphi) \approx \left[\left(P_0 + \varepsilon^{1/2}P_1 + \varepsilon P_2\right)W_0^{(n)} + \frac{1}{2}\varepsilon P_0 x\xi^2 s_0 W_{0,x}^{(n)}\right]e^{Z(\varphi)/\varepsilon},$$

where

$$Z(\varphi) = iq_0\varphi\left(1+\frac{h_1^2\varphi^2}{2\alpha_n}\right) - \frac{h_1\varphi^2}{2}\left[1+\left(\frac{3h_1^2}{\alpha_n}-\frac{1}{12}\right)\frac{\varphi^2}{4}\right].$$

Therefore we have two real vibration modes $\Re(w^{(n,m)})$ and $\Im(w^{(n,m)})$ corresponding to the eigenvalue $\lambda^{(n,m)}$, which is asymptotically double (see Subsection 2.5.5 of the Survey [**3**]). The vibration mode corresponding to $\lambda^{(0,0)}$ is

$$w^{(0,0)} \approx \left\{\left[1+\frac{i}{2}\sqrt{\frac{2s_0}{\alpha_n}}\,\varphi - \frac{s_0}{64}(7+6\sigma)\varphi^2\right]W_0 + \frac{1}{2}xs_0\varphi^2 W_{0,x}\right\}e^{Z(\varphi)/\varepsilon}.$$

One can use this asymptotic method for arbitrary boundary conditions. For example, if the shell edges are freely supported we have

$$\begin{aligned} T_1 &= v = w = M_1 = 0, && \text{for } x = x_1,\\ T_n &= v_t = w = M_n = 0, && \text{for } x = x_2,\\ T_n &= T_1\cos^2\gamma + 2S\sin\gamma\cos\gamma + T_2\sin^2\gamma, &&\\ M_n &= M_1\cos^2\gamma + 2H\sin\gamma\cos\gamma + M_2\sin^2\gamma. && \end{aligned}$$

Then the lowest frequency is

$$\lambda^{(0,0)} = 2\alpha_0^2 + 4\varepsilon\sqrt{2s_0}\,\alpha_0^{3/2} + \alpha_0\varepsilon^2(-5+9s_0) + \sqrt{2}\,\varepsilon^2 x_0 s_0^2 + O(\varepsilon^3), \qquad \alpha_0 = \pi/x_0.$$

9. A shell with straight edges

Approximate formulae for the lower frequencies of a cylindrical shell with straight clamped edges ($\beta = 0, x_0 = l$) are obtained in [**5**]:

$$\lambda^{(n,k)} = \frac{\alpha_n^4}{q_k^4} + q_k^4 + \varepsilon^2\left\{-4\alpha_n + 8\frac{\alpha_n^2}{l}\left[\sqrt{2}\,\nu^2 - 2(1+\nu)\eta_n\right]\right\} + O(\varepsilon^4),$$

where $q_k = \varepsilon k$ and $k \sim \varepsilon^{-1}$ is the circumferential wave number.

The smallest eigenvalue is

$$\lambda_{\min} = \min_k \lambda^{(0,k)} \approx 2\alpha_0^2 + \varepsilon^2\left\{-4\alpha_0 + 8\frac{\alpha_0^2}{l}\left[\sqrt{2}\,\nu^2 - 2(1+\nu)\eta_0\right]\right\}. \tag{43}$$

TABLE 1. The lowest frequencies
$\omega_0 = 10^2\varepsilon^2\sqrt{(1-\nu^2)\lambda^{(0,0)}}$.

$\frac{l}{R}$	Alg.	$\frac{R}{h}=200$	$\frac{R}{h}=400$	$\frac{R}{h}=600$	$\frac{R}{h}=800$
2	(43)	10.95	8.058	6.690	5.851
	(44)	11.18 (1.0%)	8.169 (0.7%)	6.763 (0.4%)	5.905 (0.7%)
	(45)	12.41 (8.9%)	8.776 (6.2%)	7.166 (5.2%)	6.205 (4.0%)
	[2]	11.30	8.230	6.789	5.948
3	(43)	7.514	5.476	4.529	3.952
	(44)	7.735 (0.6%)	5.584(0.2%)	4.600 (0.5%)	4.004 (0.1%)
	(45)	8.273 (7.0%)	5.850 (4.8%)	4.777 (4.0%)	4.136 (3.0%)
	[2]	7.692	5.570	4.577	4.001
4	(43)	5.686	4.132	3.413	2.976
	(44)	5.905 (0.7%)	4.239 (0.1%)	3.484 (0.9%)	3.028 (0.2%)
	(45)	6.205 (5.5%)	4.388 (3.5%)	3.582 (3.6%)	3.102 (2.2%)
	[2]	5.863	4.233	3.454	3.035

Expression (42) does not converge to (43) as $\beta \to 0$, but both formulae have similar structures.

The main aim of this paper was to find some high approximation to the frequencies and vibration modes of a cylindrical shell with a slanted edge. Unfortunately experimental or numerical results for the frequencies of such shells are not available to us. Therefore the influence of the correction of order ε^2 to the frequency on the error in the asymptotic formulae is studied in this paper for shell with straight edges.

Using the shallow shell equations instead of (43) we have

$$\lambda_{\min} \approx 2\alpha_0^2 + 8\varepsilon^2\frac{\alpha_0^2}{l}\left[\sqrt{2}\,\nu^2 - 2(1+\nu)\eta_0\right]. \tag{44}$$

The zeroth-order approximation yields

$$\lambda_{\min} \approx \frac{44.75}{l^2}. \tag{45}$$

In Table 1 we can see the results of numerical calculations obtained with the help of the asymptotic expressions (43)–(45) and numerical results [**4**], by means of the shallow shell theory. The relative errors in the asymptotic results compared with the numerical ones are placed inside round brackets. The relative error in (45) ranges from 2.2% to 8.9% and the relative error in (44) ranges from 0.1% to 1%. These errors decrease as the shell thickness, h, decreases and the shell length, l, increases.

Comparing the numerical results [**4**] with the asymptotic results (43) and (44), we conclude that the shallow shell theory provides values that are greater than the exact ones. The error grows with h and l.

10. Conclusion

To find third- and fourth-order approximations for the frequencies and vibration modes of cylindrical shells it was necessary to use equations (1), in contrast to [**11**] where the simpler equations of the shallow shell theory have been used. The results of this paper allow us to reduce by one the order of the relative errors of the asymptotic formulae in comparison with the formulae provided in [**11**]. One can obtain these higher terms in the asymptotic expansions for similar problems investigated in [**11**]. The analysis of the third- and fourth-order approximations for critical loads and buckling forms of thin cylindrical shells under lateral pressure is already started (see [**2**]).

References

1. R. N. Arnold and G. B. Walburton, *The flexural vibration of thin cylinders*, Proc. Inst. Mech. Engrs. **167A** (1953), 62–74.
2. S. M. Bauer, S. B. Filippov, A. L. Maiboroda, A. L. Smirnov and I. Yu. Teterin, *Buckling of thin cylindrical shells and shells of negative Gaussian curvature*, Asymptotic Methods in Mechanics (A. L. Smirnov and R. Vaillancourt, eds.), CRM Proceedings and Lecture Notes, this volume, Amer. Math. Soc., Providence, R.I., 1993, pp. 153–161.
3. S. M. Bauer, S. B. Filippov, A. L. Smirnov and P. E. Tovstik, *Asymptotic methods in mechanics with application to thin shells and plates*, Asymptotic Methods in Mechanics (A. L. Smirnov and R. Vaillancourt, eds.), CRM Proceedings and Lecture Notes, this volume, Amer. Math. Soc., Providence, R.I., 1993, pp. 3–140.
4. I. A. Birger and Y. G. Panovko, *Reference book. Strength. Stability. Vibrations*, vol. 3, Moscow, 1968 (464 pp.). (Russian)
5. S. B. Filippov, *Low-frequency vibrations of conical shells connected in parallel*, Prikl. Mekh., vol. 4, Leningrad University Press, Leningrad, 1979, pp. 145–153. (Russian)
6. W. Flügge, *Stresses in shells*, Springer-Verlag, Berlin, 1962 (499 pp.).
7. K. Forsberg, *Influence of boundary conditions of the modal characteristics of thin cylindrical shells*, AIAA J. **2** (1964), 2150–2157.
8. A. L. Gol′denveizer, V. B. Lidsky and P. E. Tovstik, *Free vibrations of thin elastic shells*, Nauka, Moscow, 1979 (383 pp.). (Russian)
9. E. Jahnke, F. Emde, and F. Lösch, *Tafeln höherer Funktionen*, 7th ed., B. G. Teubner, Stuttgart, 1966 (322 pp.). (German)
10. V. P. Maslov, *Operational methods*, Mir, Moscow, 1976 (561 pp.). (Russian)
11. P. E. Tovstik, *Some problems of the stability of cylindrical and conical shells*, Prikl. Mat. Mekh. **47** (1983), no. 5, 815–822; English transl., J. Appl. Math. Mech. **47** (1983), no. 5, 657–663.
12. V. L. Weingarten, *Free vibrations of thin cylindrical shells*, AIAA J. **2** (1964), 717–722.

ABSTRACT. The free vibrations of a thin circular cylindrical shell with a slanted edge are analyzed. The vibration modes which are localized near the longest generator of the cylinder are studied. The zeroth approximation and the next higher terms in the asymptotic expansions for the lowest frequencies and the corresponding vibration modes are found.

INSTITUTE OF MATHEMATICS AND MECHANICS, ST. PETERSBURG STATE UNIVERSITY, 2 BIBLIOTECHNAYA SQ., STARY PETERHOF, 198904 ST. PETERSBURG, RUSSIA

E-mail address: als@hq.math.lgu.spb.su

Centre de Recherches Mathématiques
CRM Proceedings and Lecture Notes
Volume **3**, 1993

Low-Frequency Vibrations of Cylindrical Shells. Part II: Connected Shells

Sergei B. Filippov

1. Introduction

Shell vibrations have been the subject of many studies (see, for example, [**4**]). Most of these studies have been limited to problems, where the variables are separable, leading to systems of ordinary differential equations. The problems of connected shells used to be considered in a similar way. In many papers, the vibrations of shells of revolution connected by parallels are analyzed (see [**1, 2, 5, 8**]). It is known that for such problems the variables may separate.

In this paper, we consider the low-frequency vibrations of a mitred pipe joint consisting of two connected circular cylindrical thin shells. In this case, to find the frequencies and vibration modes it is necessary to solve a boundary value problem for two systems of partial differential equations, in which the variables do not separate. However, the asymptotic integration method developed by P. E. Tovstik [**9, 10**] gives an approximate solution to the initial value problem by reducing it to some boundary value problems for ordinary differential equations.

We know of only a few studies on vibrations of mitred pipe joints; for example, the equilibrium state of mitred pipe joints has been studied in [**7, 11**].

2. Basic equations

We suppose that two shells, of equal radii, R, and thickness h, are connected at an angle 2β (see Fig. 1). The coordinates on the middle shell surface are $x^{(k)}$ and φ, where

$$x^{(1)} \in [x_0^{(1)}, x_*^{(1)}(\varphi)], \qquad x^{(2)} \in [x_*^{(2)}(\varphi), x_0^{(2)}],$$

$$x_*^{(k)}(\varphi) = x_0^{(k)} + (-1)^{k+1}(l^{(k)} + \tan\beta\cos\varphi),$$

$$x_0^{(1)} + l^{(1)} = x_0^{(2)} - l^{(2)}, \qquad l^{(k)} = \frac{L^{(k)}}{R}, \qquad k = 1, 2.$$

The superscript (k), $k = 1, 2$, denotes variables corresponding to the first and second shells, respectively. Later we shell omit superscripts in formulae which are valid for both shells.

1991 *Mathematics Subject Classification.* Primary: 73K15; Secondary: 35B40.
Key words and phrases. Connected shells, free vibrations, asymptotic methods.
This is the final form of the paper.

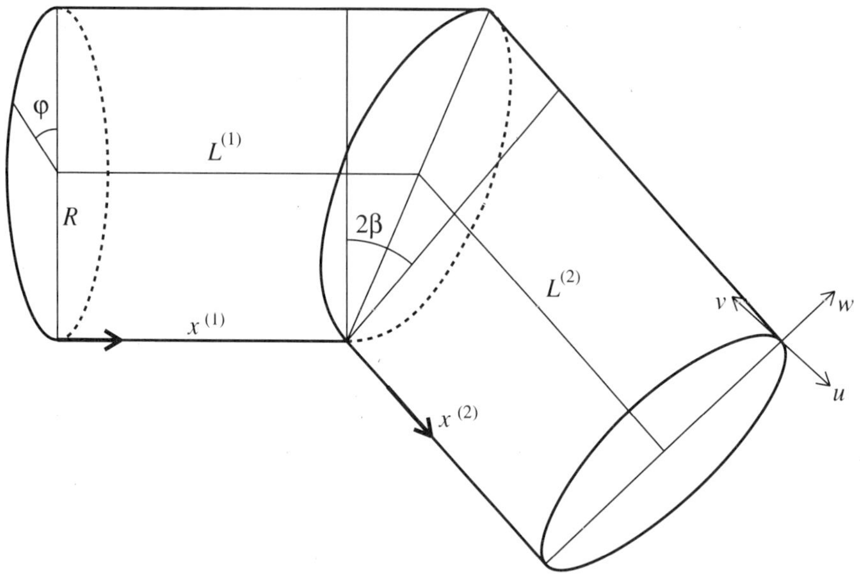

FIGURE 1. Local coordinate systems on connected shells.

The free vibrations of any of the shells are described by the following dimensionless equations (see (1) in the previous pages [**3**]), namely, the equilibrium equations:

$$\begin{aligned} T_{1,x} + S_{,\varphi} + \lambda u &= 0, \\ S_{,x} + T_{2,\varphi} + Q_2 + 2H_{,x} + \lambda v &= 0, \\ -T_2 + Q_{2,\varphi} + Q_{1,x} + \lambda w &= 0, \end{aligned} \tag{1a}$$

the elasticity relations:

$$\begin{aligned} Q_1 &= M_{1,x} + 2H_{,\varphi}, & Q_2 &= M_{2,\varphi}, \\ M_1 &= \varepsilon^8(\vartheta_{1,x} + \nu\vartheta_{2,\varphi}), & M_2 &= \varepsilon^8(\vartheta_{2,\varphi} + \nu\vartheta_{1,x}), \\ (1-\nu^2)T_1 &= u_{,x} + \nu(w + v_{,\varphi}), & (1-\nu^2)T_2 &= w + v_{,\varphi} + \nu u_{,x}, \\ H &= (1-\nu)\varepsilon^8\vartheta_{2,x}, & 2(1-\nu)S &= v_{,x} + u_{,\varphi}, \end{aligned} \tag{1b}$$

and the deformation-displacement equations:

$$\vartheta_1 = -w_{,x}, \qquad \vartheta_2 = -w_{,\varphi} + v. \tag{1c}$$

Here T_1, T_2, S, Q_1, Q_2, M_1, M_2, H are the dimensionless stress-resultants and stress-couples, ϑ_1 and ϑ_2 are the angles of rotation,

u, v and w are the components of the displacement,
ν is Poisson's ratio,
$\varepsilon^8 = h^2/[12(1-\nu^2)R^2]$ is a small parameter,
$\lambda = \rho R^2\omega^2/(\varepsilon^4 E)$ is the frequency parameter,
E is Young's modulus,
ρ is the mass density,
ω is the vibration frequency.

We suppose that the straight edges of the shells are freely supported, i.e.

$$T_1 = v = w = M_1 = 0, \qquad \text{for} \qquad x = x_0^{(k)}, \qquad k = 1, 2. \tag{2}$$

At the connection line $x^{(1)} = x_*^{(1)}$, $x^{(2)} = x_*^{(2)}$, the eight continuity conditions for the displacements, angles of rotation, stress-resultants and stress-couples are to be satisfied.

The displacement vector $\mathbf{r}^{(k)}\left(x_*^{(k)}, \varphi\right)$ of a point in the connection line can be expressed as

$$\mathbf{r}^{(k)} = u_n^{(k)}\mathbf{b}^{(k)} + u_t^{(k)}\boldsymbol{\tau} + w^{(k)}\mathbf{n}^{(k)}.$$

Here $\boldsymbol{\tau}$ is a unit vector tangent to the connection line, $\mathbf{n}^{(k)}$ is a unit vector normal to the middle surface of the shell,

$$\mathbf{b}^{(k)} = \mathbf{n}^{(k)} \times \boldsymbol{\tau},$$

and

$$\begin{aligned} u_t^{(k)} &= (-1)^k u^{(k)} \sin\gamma + v^{(k)} \cos\gamma, \\ u_n^{(k)} &= u^{(k)} \cos\gamma + (-1)^{k+1} v^{(k)} \sin\gamma, \qquad k = 1, 2, \end{aligned}$$

where

$$\tan\gamma = \tan\beta \sin\varphi.$$

Taking into account

$$\mathbf{b}^{(k)} = c_1\mathbf{b} + (-1)^{k+1}c_2\mathbf{n}, \qquad \mathbf{n}^{(k)} = (-1)^k c_2\mathbf{b} + c_1\mathbf{n},$$

where

$$c_1 = \cos\beta\cos^{-1}\gamma, \qquad c_2 = \sin\beta\cos\varphi,$$

$\mathbf{n} = \boldsymbol{\tau} \times \mathbf{b}$ and $\mathbf{b}$ is a unit vector normal to the plane separating the shells, we obtain from the equality $\mathbf{r}^{(1)} = \mathbf{r}^{(2)}$

$$\begin{aligned} u_n^{(1)} c\sin\beta + w^{(1)}\cos\beta &= -u_n^{(2)} c\sin\beta + w^{(2)}\cos\beta, \\ u_n^{(1)}\cos\beta - w^{(1)} c\sin\beta &= u_n^{(2)}\cos\beta + w^{(2)} c\sin\beta, \\ u_t^{(1)} &= u_t^{(2)}, \end{aligned} \tag{3}$$

where

$$c = \cos\gamma\cos\varphi.$$

Three more connection conditions follow from the continuity of stress-resultants and are given as

$$\begin{aligned} T_n^{(1)} c\sin\beta + Q^{(1)}\cos\beta &= -T_n^{(2)} c\sin\beta + Q^{(2)}\cos\beta, \\ T_n^{(1)}\cos\beta - Q^{(1)} c\sin\beta &= T_n^{(2)}\cos\beta + Q^{(2)} c\sin\beta, \\ S_{nt}^{(1)} &= S_{nt}^{(2)}. \end{aligned} \tag{4}$$

Here

$$\begin{aligned} T_n^{(k)} &= T_1^{(k)}\cos^2\gamma + 2(-1)^{k+1}S^{(k)}\sin\gamma\cos\gamma + T_2^{(k)}\sin^2\gamma, \\ S_{nt}^{(k)} &= (-1)^{k+1}(T_2^{(k)} - T_1^{(k)})\sin\gamma\cos\gamma + S^{(k)}(\cos^2\gamma - \sin^2\gamma), \qquad k = 1, 2, \\ Q_n^{(k)} &= Q_1^{(k)}\cos\gamma + (-1)^{k+1}Q_2^{(k)}\sin\gamma. \end{aligned}$$

Finally from the continuity of the angle of rotation and the stress-couple we have

$$\vartheta_n^{(1)} = \vartheta_n^{(2)}, \qquad M_n^{(1)} = M_n^{(2)}, \tag{5}$$

where

$$M_n^{(k)} = M_1^{(k)} \cos^2\gamma + 2(-1)^{k+1} H^{(k)} \sin\gamma \cos\gamma + M_2^{(k)} \sin^2\gamma,$$
$$\vartheta_n^{(k)} = \vartheta_1^{(k)} \cos\gamma + (-1)^{k+1} \vartheta_2^{(k)} \sin\gamma, \qquad k = 1, 2.$$

3. Approximate solution

To solve the boundary value problem (1)–(5) we use the asymptotic method of P. E. Tovstik [**9, 10**]. An eigenvalue for this problem can be expanded into a power series in a small parameter, ε, as follows:

$$\lambda = \lambda_0 + \varepsilon\lambda_1 + \varepsilon^2\lambda_2 + \cdots \tag{6}$$

The approximate solution of equations (1), which decreases away from the straight line $\varphi = 0$, has the form

$$y = y_a + y_b, \tag{7}$$

where y denotes one of the variables. The first term, y_a, in (7) describes the main stress-strain state of the shell. The second term, y_b, describes the boundary effects near the shell edges $x^{(k)} = x_0^{(k)}$ and the connection line $x^{(k)} = x_*^{(k)}$.

The approximate solution, y_a, can be expressed as

$$y_a = \varepsilon^{a(y)} \tilde{y} e^{iq/\varepsilon}, \tag{8}$$

where

$$\tilde{y} = \sum_{n=0}^{\infty} \varepsilon^{n/2} y_n(x, \xi), \qquad q = \sum_{n=1}^{\infty} \frac{\varepsilon^{n/2} \xi^n q_{n-1}}{n!}, \qquad \xi = \varepsilon^{-1/2}\varphi. \tag{9}$$

One can determine the intensity coefficient, $a(y)$, the constants, λ_n and q_n, and the functions $y_n(x, \xi)$ by substituting (6) and (8) into (1). Assuming that $a(w) = 0$, we obtain $a(v) = 1$, $a(u) = 2$, $a(T_1) = 2$, and $a(S) = 3$. The function $\tilde{w}$ is the solution of the equation

$$G\tilde{w} - \lambda\tilde{w} = \varepsilon^2 N\tilde{w}. \tag{10}$$

Here the operators G and N are given by formulae (8) of [**3**].

If $\tilde{w}$ is obtained, then one can find the other unknown functions for the main stress-strain state,

$$\tilde{u} = -\Phi^{-2}\tilde{w}_{,x} - \varepsilon^2(2+\nu)\Phi^{-4}\tilde{w}_{,xxx} + \cdots,$$
$$\tilde{T}_1 = -\Phi^{-2}\tilde{w}_{,xx} + 2\varepsilon^2(q_0^4 - \lambda_0)\tilde{w} + \cdots,$$
$$\tilde{v} = i\Phi^{-1}\tilde{w} - i\nu\varepsilon^2\Phi^{-3}\tilde{w}_{,xx} + \cdots,$$
$$\tilde{S} = -i\Phi^{-3}\tilde{w}_{,xxx} + 2i\varepsilon^2(q_0^4 - \lambda_0)\Phi^{-1}\tilde{w}_{,x} + \cdots$$

We expand the linear differential operator G, given in (10), in a power series in ε as follows:

$$G = \sum_{n=0}^{\infty} \varepsilon^{n/2} G_n. \tag{11}$$

The operators G_n are determined by (10) in [**3**]. Substituting (9) and (11) into equation (10), we get the following equations for λ_n and $w_n(x,\xi)$:

$$(12)\qquad \sum_{n=0}^{s}\Delta_n w_{s-n}=0,\qquad s=0,1,2,3,\qquad \sum_{n=0}^{4}\Delta_n w_{4-n}=Nw_0,$$
$$\Delta_n=G_n-\lambda_{n/2},\qquad n=0,2,4,\qquad L_n=G_n,\qquad n=1,3.$$

The boundary effect functions y_b can be defined by the formulae

$$(13)\qquad y_b^{(k)}=\varepsilon^{b(y)}\hat{y}^{(k)}e^{iq/\varepsilon},\qquad \hat{y}^{(k)}=\sum_{j-1}^{4}C_j{}^{(k)}\hat{y}_j f_j{}^{(k)}(x,\varepsilon),\qquad k=1,2,$$

where $\hat{y}_j$ and $C_j{}^{(k)}$ are unknown constants,

$$f_j{}^{(k)}=\exp\left[(-1)^k(x_0^{(k)}-x^{(k)})r_j/\varepsilon^2\right],\qquad r_j=-\left[1+(-1)^j i\right]/\sqrt{2}\quad j=1,2,$$
$$f_j{}^{(k)}=\exp\left[(-1)^k(x_*^{(k)}(0)-x^{(k)})r_j/\varepsilon^2\right],\quad r_3=-r_2,\quad r_4=-r_1,\qquad j=3,4.$$

One can determine the constants $\hat{y}_j$ and the relation for the intensity coefficients, $b(y)$, by substituting (13) into (1). For example,

$$\hat{w}_j=1,\qquad \hat{\vartheta}_{nj}=-r_j,\qquad \widehat{M}_{nj}=-r_j^2,$$
$$\widehat{Q}_{nj}=-r_j^3,\qquad \widehat{S}_{ntj}=\frac{iq_0}{r_j},\qquad \widehat{T}_{nj}=-\frac{q_0}{r_j^2}+\frac{\tan\beta}{r_j}.$$

We obtain definite values for the intensity coefficients:

$$b(w)=2,\quad b(\vartheta_n)=0,\quad b(M_n)=6,\quad b(Q_n)=4,\quad b(S_{nt})=3,\quad b(T_n)=4,$$

and the boundary conditions for equations (12) by substituting (7) into (2)–(5).

4. The boundary-value problem of the zeroth- and first-order approximations

We start by solving the boundary value problem for equations (12). In the zeroth-order approximation ($s=0$) the boundary value problem for the connected shells splits into two separate problems:

$$(14)\qquad \begin{aligned}&(G_0-\lambda_0)w_0^{(k)}=0,\\ &w_0^{(k)}=w_{0,xx}^{(k)}=0,\qquad \text{for } x^{(k)}=x_0^{(k)}\qquad \text{and } x^{(k)}=x_*^{(k)}.\end{aligned}$$

If we represent the solution of (14) in the form

$$w_0^{(k)}(x,\xi)=P_0^{(k)}(\xi)W_0^{(k)}(x),$$

where $P_0^{(k)}(\xi)$ is a polynomial, then equation (14) becomes

$$(15)\qquad \frac{d^4W_0^{(k)}}{dx^4}-\alpha^4W_0^{(k)}=0,\qquad \alpha^4=\lambda_0q_0^4-q_0^8,$$

with the boundary conditions

$$(16)\qquad W_0^{(k)}=W_{0,xx}^{(k)}=0,\qquad \text{for } x^{(k)}=x_0^{(k)}\qquad \text{and } x^{(k)}=x_*^{(k)}(0).$$

The eigenvalues for the boundary value problem (15)–(16) can be split into two countable sets, $\alpha = \alpha_n^{(1)}$, $n = 1, 2, \dots$, and $\alpha = \alpha_p^{(2)}$ $p = 1, 2, \dots$, where

$$\alpha_n^{(1)} = \frac{\pi n}{l_0^{(1)}}, \qquad \alpha_p^{(2)} = \frac{\pi p}{l_0^{(2)}}, \qquad l_0^{(k)} = l^{(k)} + t, \qquad t = \tan\beta. \tag{17}$$

The eigenfunctions for problem (15)–(16) are

$$W_0^{(1)} = D_1 \sin\left[\alpha_n^{(1)}(x^{(1)} - x_0^{(1)})\right], \qquad W_0^{(2)} = D_2 \sin\left[\alpha_p^{(2)}(x_0^{(2)} - x^{(2)})\right].$$

There arc three cases:

(1) $D_1 \neq 0,\ D_2 = 0,$ if $\alpha = \alpha_n^{(1)}$ and $\alpha \neq \alpha_p^{(2)}$ for all p,
(2) $D_1 = 0,\ D_2 \neq 0,$ if $\alpha = \alpha_p^{(2)}$ and $\alpha \neq \alpha_n^{(1)}$ for all n,
(3) $D_1 \neq 0,\ D_2 \neq 0,$ if $\alpha = \alpha_n^{(1)} = \alpha_p^{(2)}$.

In Cases 1 and 2, the vibrations of one of the shells are stronger than those of the other. In Case 3, the ratio D_1/D_2 depends on the parameters of the shells and may be found from the higher approximations. Case 3 takes place for the lowest frequency ($n = p = 1$) only if $l^{(1)} = l^{(2)}$.

We begin with Case 1. It follows from (15) and (17) that

$$\lambda_0^{(1)}(n) = \left(\frac{\alpha_n^{(1)}}{q_0}\right)^4 + q_0^4, \qquad n = 1, 2, \dots \tag{18}$$

We find the value of q_0 from the compatibility condition for the first-order approximation. The boundary value problem for $s = 1$ has the form

$$\begin{gathered}\left(G_0 - \lambda_0^{(1)}\right) w_1^{(k)} + G_1 w_0^{(k)} = 0, \\ w_1^{(k)} = w_{1,xx}^{(k)} = 0, \qquad \text{for} \qquad x^{(k)} = x_0^{(k)} \qquad \text{and} \qquad x^{(k)} = x_*^{(k)}(0).\end{gathered} \tag{19}$$

For the boundary value problem

$$\begin{gathered}(G_0 - \lambda_0)\, z + f(x) = 0, \\ z + g_1 = z_{,xx} + g_2 = 0, \qquad \text{for} \qquad x = x_0 \qquad \text{and} \qquad x = x_*,\end{gathered}$$

the compatibility conditions are represented by the following equation

$$\int_{x_0}^{x_*} z f(x)\, dx + \frac{1}{q^4}\left[g_1 w_{0,xxx} + g_2 w_{0,x}\right]\Big|_{x_0}^{x_*} = 0, \tag{20}$$

where w_0 is the solution of the boundary value problem (14). Condition (20) for problem (19) may be written as

$$\begin{gathered}\frac{\partial \lambda_0^{(1)}}{\partial q_0}\left(\frac{\partial P_0^{(1)}}{\partial \xi} - h_1^{(1)} \xi P_0^{(1)}\right) I_0^{(1)} = 0, \\ h_1^{(1)} = -i q_1^{(1)}, \qquad I_0^{(1)} = \int_{x_0^{(1)}}^{x_*^{(1)}} \left(W_0^{(1)}\right)^2 dx.\end{gathered} \tag{21}$$

Since the equation

$$P_{0,\xi} - h_1 \xi P_0 = 0$$

does not have a polynomial solution, then

$$\frac{\partial \lambda_0^{(1)}}{\partial q_0} = -\frac{4}{q_0^5}\left(\alpha_n^{(1)}\right)^4 + 4q_0^3 = 0, \qquad q_0^2 = \alpha_n^{(1)},$$

and from (18) we obtain

$$\lambda_0^{(1)}(n) = 2\left(\alpha_n^{(1)}\right)^2, \qquad n = 1, 2, \dots$$

It then follows from (19) and (21) that

$$w_1^{(1)} = P_1^{(1)}(\xi) W_0^{(1)}, \qquad w_1^{(2)} \equiv 0.$$

5. Boundary value problems for higher-order approximations

The boundary value problem for the second-order approximation is

$$\left(G_0 - \lambda_0^{(1)}\right) w_2^{(k)} + G_1 w_1^{(k)} + \left(G_2 - \lambda_1^{(1)}\right) w_0^{(k)} = 0, \tag{22}$$

$$\begin{aligned} &w_2^{(k)} + \frac{1}{2}(-1)^k t\xi^2 w_{0,x}^{(k)} = 0, \\ &w_{2,xx}^{(k)} + \frac{1}{2}(-1)^k t\xi^2 w_{0,xxx}^{(k)} = 0, \qquad \text{for } x^{(k)} = x_*^{(k)}(0), \qquad k = 1, 2 \\ &w_2^{(k)} = w_{2,xx}^{(k)} = 0, \qquad \text{for } x^{(k)} = x_0^{(k)}. \end{aligned} \tag{23}$$

Condition (20) for this problem, after transformations, yields

$$P_{0,\xi\xi}^{(1)} - 2h_1^{(1)}\xi P_{0,\xi}^{(1)} + \left\{\left[\left(h_1^{(1)}\right)^2 - \frac{\Lambda_1^{(1)}}{\Lambda_2^{(1)}}\right]\xi^2 - h_1^{(1)} + \frac{2\lambda_1^{(1)}}{\Lambda_2^{(1)}}\right\} P_0^{(1)} = 0, \tag{24}$$

where

$$\Lambda_1^{(1)} = 4\left(\alpha_n^{(1)}\right)^2 s_0^{(1)}, \qquad \Lambda_2^{(1)} = 32\alpha_n^{(1)}, \qquad s_0^{(1)} = \frac{t}{l_0^{(1)}}.$$

Equation (24) has polynomial solutions of degree m for

$$\begin{aligned} &h_1^{(1)} = \sqrt{\frac{\Lambda_1^{(1)}}{\Lambda_2^{(1)}}} = \frac{1}{4}\sqrt{2\alpha_n^{(1)} s_0^{(1)}}, \\ &2\lambda_1^{(1)} = h_1^{(1)}\Lambda_2^{(1)}(2m+1), \qquad m = 0, 1, 2\dots \end{aligned} \tag{25}$$

Hence

$$\lambda^{(1)}(n, m) = 2\left(\alpha_n^{(1)}\right)^2 + 4\varepsilon\alpha_n^{(1)}\sqrt{2\alpha_n^{(1)} s_0^{(1)}}(1 + 2m) + O\left(\varepsilon^2\right). \tag{26}$$

Since formula (26) agrees with formula (28) in [**3**], the latter can be used for the first shell with freely supported edges, if $\alpha_n = \pi n / l_0^{(1)}$. Then both formulae give the same approximate value of λ. Hence, in Case 1, the difference between the frequencies of connected shells and the frequencies of the first shell with freely supported edges is of the order $O\left(\varepsilon^2\right)$.

After the transformation of variables

$$\xi = \frac{\theta}{\sqrt{h_1^{(1)}}},$$

equation (24) gives, by (25),

$$(27)\quad L_0P_0 = 0, \qquad P_0(\theta) = P_0^{(1)}(\xi), \qquad L_0P = P'' - 2\theta P' + 2mP, \qquad P' = P_{,\theta}.$$

The solutions of (27), for $m = 0, 1, 2, \ldots$, are Hermite's polynomials $P_0(\theta) = H_m(\theta)$, $(H_0 = 1, H_1 = 2\theta, \ldots)$. From formulae (22)–(24) we find

$$w_2^{(1)} = P_2^{(1)} W_0^{(1)} + \frac{1}{2} P_0^{(1)} \xi^2 \left(x^{(1)} - x^{(0)}\right) s_0^{(1)} W_{0,x}^{(1)}, \qquad w_2^{(2)} \equiv 0.$$

Since the construction of the third-order approximations for connected shells is similar to that for a shell with a slanted edge (see Section 7 of [**3**]), we obtain

$$P_1^{(1)} = \frac{i\sqrt{h_1^{(1)}}}{q_0}\left[\frac{m+1}{2} H_{m+1}\left(\sqrt{h_1^{(1)}}\,\xi\right) + m^2 H_{m-1}\left(\sqrt{h_1^{(1)}}\,\xi\right)\right],$$
$$w_3^{(1)} = P_3^{(1)} W_0^{(1)} + \left(\frac{1}{2}\xi^2 P_1^{(1)} - \frac{2i}{q_0}\xi P_0^{(1)}\right)\left(x^{(1)} - x_0^{(1)}\right) s_0^{(1)} W_{0,x}^{(1)} + w_{30}^{(1)},$$
$$w_3^{(2)} \equiv 0,$$

where

$$w_{30}^{(1)} = \frac{2i\xi D_1 P_0^{(1)} \alpha_n^{(1)} t \sinh\left[\alpha_n^{(1)}\left(x^{(1)} - x_0^{(1)}\right)\right] \cos \pi n}{q_0 \sinh \pi n}.$$

To construct the fourth-order approximations we have to determine the constants $C_j^{(k)}$ in the boundary effect functions (13). If we substitute (7) into (2)–(5), we get

$$(28)\qquad \hat{w}^{(k)} = \widehat{M}_n^{(k)} = 0, \qquad \text{for } x^{(k)} = x_0^{(k)},$$

and

$$(29)\qquad \begin{aligned} &\hat{w}^{(1)} = \hat{w}^{(2)}, \qquad \hat{\vartheta}^{(1)} - \hat{\vartheta}^{(2)} = -\alpha_n^{(1)}, \\ &\widehat{M}^{(1)} = \widehat{M}^{(2)}, \qquad \widehat{S}_{nt}^{(1)} - \widehat{S}_{nt}^{(2)} = \frac{i\left(\alpha_n^{(1)}\right)^3}{q_0^3}, \qquad \text{for } x^{(k)} = x_*^{(k)}(0). \end{aligned}$$

From (28), we have $C_1^{(k)} = 0$ and $C_2^{(k)} = 0$. Therefore at the freely supported straight edge there are no boundary effects. Equations (29), which are a system of linear algebraic equations for the unknown constants $C_3^{(k)}$ and $C_4^{(k)}$, have the solutions

$$C_3^{(1)} = C_4^{(2)} = -C_3^{(2)} = -C_4^{(1)} = -\frac{i\alpha_n^{(1)}}{2\sqrt{2}}.$$

We find the constant λ_2 and the polynomial P_2 from the compatibility condition

(20) for the fourth-order approximation

$$
\begin{aligned}
\lambda_2^{(1)}(m,n) &= \frac{\alpha_n^{(1)}}{l_0^{(1)}}\Bigg[(9+25m+25m^2)t-(5+2m+2m^2)l_0^{(1)} \\
&\qquad -\frac{4}{\sin 2\beta}+\sqrt{2}\alpha_n^{(1)}\Bigg], \\
P_2(\theta) &= \frac{h_1^{(1)}}{8q_0^2}\left[-2A(m)H_{m+2}(\theta)+2C(m)H_{m-2}(\theta)\right. \\
&\qquad \left.+D(m)H_{m-4}(\theta)\right],
\end{aligned}
\tag{30}
$$

where $A(m)$, $C(m)$, $D(m)$ are defined by formulae (40) in [**3**], if we assume that

$$
\sigma = -\frac{\alpha_n^{(1)}}{24\left(h_1^{(1)}\right)^2}.
$$

It then follows from (6) and (26) that

$$
\lambda^{(1)}(m,n) = 2\left(\alpha_n^{(1)}\right)^2 + 4\varepsilon\alpha_n^{(1)}\sqrt{2\alpha_n^{(1)}s_0^{(1)}}(1+2m)+\varepsilon^2\lambda_2^{(1)}(m,n)+O\left(\varepsilon^3\right). \tag{31}
$$

The lowest eigenvalue is

$$
\lambda^{(1)}(0,1) = \frac{\pi}{\left(l_0^{(1)}\right)^2}\left[2\pi+4\varepsilon\sqrt{2\pi t}+\varepsilon^2\left(\frac{5-9\cos 2\beta}{\sin 2\beta}-5l_0^{(1)}+\frac{\sqrt{2}\pi}{l_0^{(1)}}\right)\right]+O\left(\varepsilon^3\right).
$$

The eigenfunctions have the form

$$
\begin{aligned}
w^{(1)} &\approx \Bigg[\left(P_0^{(1)}+\varepsilon^{1/2}P_1^{(1)}+\varepsilon P_2^{(1)}\right)W_0^{(1)} \\
&\qquad +\frac{1}{2}\varepsilon P_0^{(1)}\left(x^{(1)}-x_0^{(1)}\right)\xi^2 s_0^{(1)}W_{0,x}^{(1)}\Bigg]e^{Z(\varphi)/\varepsilon}, \\
w^{(2)} &= O\left(\varepsilon^2\right),
\end{aligned}
\tag{32}
$$

where

$$
Z(\varphi) = iq_0\varphi\left(1-\frac{\varphi^2}{48\sigma}\right)-\frac{h_1^{(1)}\varphi^2}{2}\left[1-\left(\frac{1}{2\sigma}+\frac{1}{3}\right)\frac{\varphi^2}{16}\right].
$$

The eigenfunction corresponding to $\lambda^{(1)}(0,1)$ is

$$
\begin{aligned}
w^{(1)} &\approx \left\{\left[1+\frac{i}{2}\sqrt{\frac{2s_0^{(1)}}{\alpha_n^{(1)}}}\varphi-\frac{s_0^{(1)}}{64}(7+6\sigma)\varphi^2\right]W_0^{(1)}\right. \\
&\qquad \left.+\frac{1}{2}(x^{(1)}-x_0^{(1)})s_0^{(1)}\varphi^2W_{0,x}^{(1)}\right\}e^{Z(\varphi)/\varepsilon} \\
w^{(2)} &= O\left(\varepsilon^2\right).
\end{aligned}
$$

Therefore as in [**3**], we have two real vibrations modes, $\Re(w)$ and $\Im(w)$, corresponding to one approximate value, (31), of the eigenvalue. This fact is due to the existence of two eigenvalues, the difference between which is of the order $O(\varepsilon^N)$ for

TABLE 1. The minimum frequencies (in Hz) for different $L^{(2)}$ (in cm).

$L^{(2)}$	40	48	56	64	72	80
f_1	282	282	282	268	242	222
f_2	382	336	298	282	282	282
f_3	383	382	382	362	328	300

arbitrary N. These eigenvalues cannot be distinguished by the proposed asymptotic method.

The functions $w^{(1)}$ oscillate quickly as φ varies and their amplitudes decrease rapidly away from the longest generator of the first shell ($\varphi = 0$).

Case 2 is similar to Case 1. One can obtain the approximate formulae for the eigenvalue $\lambda^{(2)}(m, n)$ and the corresponding eigenfunction in Case 2 by permuting the superscripts 1 and 2 in formulae (31) and (32). Case 3, which is special, is not analyzed in this paper.

6. Numerical results

To evaluate the lowest frequencies for a mitred pipe joint, we use formulae (30) and (31). We consider shells with the following values of the parameters: $R = 20$ cm, $L^{(1)} = 60$ cm, $h = 0.2$ cm, $\beta = \pi/4$, $\nu = 0.3$, $E = 2.07 \cdot 10^{11}$ N/m^2, $\rho = 7860$ kg/m^3, where $L^{(1)}$ is the mean length of the first shell (see Fig. 1). Then $l^{(1)} = L^{(1)}/R = 4$, $l_0(1) = 5$.

The three lowest frequencies $f_p = \omega_p/(2\pi)$, $p = 1, 2, 3$, for different mean lengths $L^{(2)}$ of the second shell are shown in Table 1.

The frequencies 282 Hz and 382 Hz correspond to the eigenvalue $\lambda^{(1)}(0, 1)$ and $\lambda^{(1)}(1, 1)$. They do not depend on $L^{(2)}$. The frequencies corresponding to $\lambda^{(2)}(0, 1)$ and $\lambda^{(2)}(1, 1)$ decrease as $L^{(2)}$ increases. If $L^{(2)} > L^{(1)}$, then f_1 and f_3 correspond to $\lambda^{(2)}(0, 1)$ and $\lambda^{(2)}(1, 1)$.

7. Conclusion

The main conclusions from these results are as follows.

(a) The vibrations modes corresponding to the lowest frequencies of a mitred pipe joint with freely supported edges are localized near the longest generators of the cylindrical shells.

(b) The lowest frequencies are separated into two sets and they differ little from the frequencies of the first and the second shells, respectively, with freely supported edges.

(c) If we consider the vibrations with frequencies from the first set, which differ from each of the frequencies from the second set (Case 1), then the vibrations of the first shell are stronger than the vibrations of the second shell. Case 2 is similar to Case 1.

(d) Simple approximate formulae for the lowest eigenvalues and the corresponding eigenfunctions are obtained. With the help of these formulae, in Cases 1 and 2, one may examine the influence of different shell parameters on the shell frequencies and modes.

Splitting the frequencies into two sets is also typical for connected shells of revolution (see [**1**, **2**]).

If two frequencies from different sets are equal (Case 3), then formula (30) is invalid and one can use only formula (26). A more detailed investigation shows that formula (30) is also invalid if the difference between the frequencies from the two different sets is of the order $O(\varepsilon^2)$. Therefore, Conclusion (c) is valid if the difference between frequencies is much larger than ε^2.

Similar results can be obtained for other boundary conditions at the shell straight edges. If, for example, the straight edges are clamped, then in Case 1 the lowest frequencies of connected shells are slightly different from the frequencies of the first shell with a clamped straight edge $x^{(1)} = x_0^{(1)}$ and a freely supported slanted edge $x^{(1)} = x_*^{(1)}(\varphi)$. Besides, since $C_j^{(k)} \neq 0$ for $j = 1, 2$, the boundary effects occur at the clamped edges.

The proposed asymptotic method also allows one to study the buckling of a mitred pipe joint under lateral pressure.

References

1. S. B. Filippov, *Low-frequency vibrations of conical shells connected in parallel*, Prikl. Mekh., vol. 4, Leningrad University Press, Leningrad, 1979, pp. 145–153. (Russian)
2. ——, *Free axisymmetric vibrations of connected shells of revolution. Comparison of asymptotic and numerical results*, Prikl. Mekh., vol. 5, Leningrad University Press, Leningrad, 1981, pp. 164–171. (Russian)
3. ——, *Low-frequency vibrations of cylindrical shells.* Part I: *Shells with a slanted edge*, Asymptotic Methods in Mechanics (A. L. Smirnov and R. Vaillancourt, eds.), CRM Proceedings and Lecture Notes, this volume, Amer. Math. Soc., Providence, R.I., 1993, pp. 193–204.
4. A. L. Gol′denveizer, V. B. Lidsky and P. E. Tovstik, *Free vibrations of thin elastic shells*, "Nauka", Moscow, 1979 (383 pp.). (Russian)
5. W. Hu and J. Raney, *Experimenlal and analytical study of vibrations of joined shells*, AIAA J. **5** (1967), no. 5, 976–980.
6. E. Jahnke, F. Emde, and F. Lösch, *Tafeln höherer Funktionen*, 7th ed., B. G. Teubner, Stuttgart, 1966 (322 pp.). (German)
7. V. M. Malkov, *The bending from the plane of symmetry and twisting of crossing cylindrical shells*, Raschet prostranstvennykh konstruktsii **12** (1969), 112–118. (Russian)
8. V. I. Myachenkov and I. V. Grigorjev, *Reference book. The design of block shell structures*, Mashinostroenie, Moscow, 1981 (212 pp.). (Russian)
9. P. E. Tovstik, *Some problems of the stability of cylindrical and conical shells*, Prikl. Mat. Mekh. **47** (1983), no. 5, 815–822; English transl., J. Appl. Math. Mech. **47** (1983), no. 5, 657–663.
10. ——, *On the forms of local buckling of thin elastic shells*, Trans. CSME **15** (1991), no. 3, 199–211.
11. M. V. Vedavyasa Murthy, *Stresses in mitred pipe joint under uniform internal pressure*, Int. J. of Mech. Sci. **6** (1964), no. 5, 361–370.

ABSTRACT. The free vibrations of a mitred pipe joint consisting of two connected circular cylindrical thin shells are analyzed. The asymptotic integration method for equations with a small parameter developed by P. E. Tovstik is used. Approximate formulae for the lowest frequencies and the corresponding vibration modes, the amplitudes of which decrease away from the longest generators of the cylinders, are found.

INSTITUTE OF MATHEMATICS AND MECHANICS, ST. PETERSBURG STATE UNIVERSITY, 2 BIBLIOTECHNAYA SQ., STARY PETERHOF, 198904 ST. PETERSBURG, RUSSIA

E-mail address: als@hq.math.lgu.spb.su

Centre de Recherches Mathématiques
CRM Proceedings and Lecture Notes
Volume 3, 1993

Buckling of Convex Shells under Nonaxisymmetric Loading

Aleksandr L. Maiboroda

1. Introduction

The buckling of a convex shell is studied by means of its linear approximation. This approach was discussed by many authors. The paper by D. Bushnell [**2**] could be emphasized, where the shell buckling in linear and nonlinear approximations has been analyzed by numerical methods. W. Koiter [**5**] developed the general theory of nonlinear elastic buckling for shell structures. The influence of initial deformations on the buckling load was studied in particular by L. I. Shkutin [**8**]. We consider the cases where this influence is small and suppose that initial stresses are membranous. Special attention is payed to shells with nonaxisymmetric initial stresses. These stresses can be caused, for example, by a nonaxisymmetric loading. The buckling of a shell with nonaxisymmetric initial stresses was studied by means of numerical methods by J. Wang and D. Billington [**12**]. The approach proposed in this paper allows one to study the buckling of arbitrary convex shells.

The asymptotic methods were successfully applied by A. L. Gol'denveizer, V. B. Lidsky and P. E. Tovstik [**3**] to shell vibrations and shell bucklings described by ordinary differential equations. The asymptotic approach to this problem has some advantages. For example, it gives simple formulae for the critical load and the buckling forms. In this paper, these methods are used to solve two-dimensional boundary value problems. This study continues research initiated by P. E. Tovstik [**10**].

We suppose that the shell edges are clamped. The boundary conditions essentially affect the buckling load. This question was studied before by many authors including E. I. Grigolyuk and V. V. Kabanov [**4**], W. Nashbar and N. J. Hoff [**7**]. Specifically, if the edges are weakly supported, the asymptotic order of the critical load could be reduced, as it has been shown by P. E. Tovstik [**11**].

2. The equations of shell buckling

We consider a shell of revolution (see Fig. 1).

The position of a point on the shell neutral surface is defined by the longtitudinal angle, φ, and the length, s, of the arc of a meridian. To describe the displacements in each point α of a shell we introduce a local orthogonal coordinate system $\{u, v, w\}$. To define the critical load and the buckling mode we use the

1991 *Mathematics Subject Classification.* Primary: 35B40; Secondary: 73H05, 73K15.
Key words and phrases. Shell buckling, asymptotic integration method.
This is the final form of the paper.

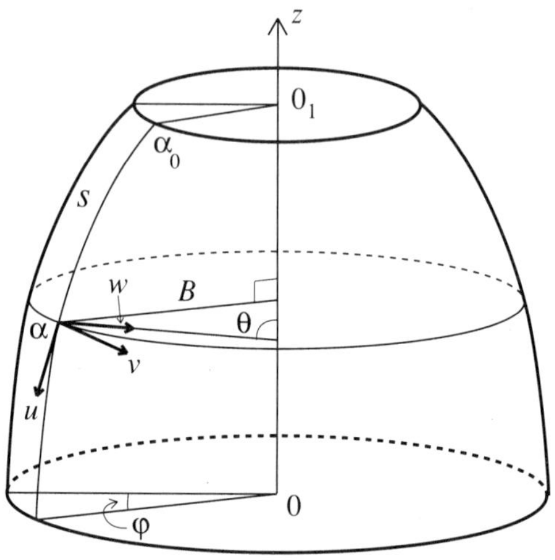

FIGURE 1. Shell of revolution with local coordinate system $\{u, v, w\}$

following linear system of partial differential equations in dimensionless form [**6**]:

$$(1)\quad \begin{gathered} \frac{\partial U}{\partial \varphi} - \frac{W}{R_1} = T_1 - \sigma T_2, \qquad \frac{1}{B}\frac{\partial V}{\partial \varphi} + \frac{B'}{B}U - \frac{W}{R_2} = T_2 - \sigma T_1, \\ \frac{1}{B}\frac{\partial U}{\partial \varphi} + B\frac{\partial}{\partial s}\left(\frac{V}{B}\right) = 2(1+\sigma)T_{12}, \qquad \frac{\partial}{\partial s}(BT_1) - B'T_2 + \frac{\partial T_{12}}{\partial \varphi} = 0, \\ \frac{T_1}{R_1} + \frac{T_2}{R_2} - \mu^4\Delta^2 W + \Lambda\mu^2\Delta_T W = 0, \qquad \frac{\partial T_2}{\partial \varphi} + \frac{\partial}{\partial s}(BT_{12}) + B'T_{12} = 0, \\ \Delta_T W = B^{-1}\left[\frac{\partial}{\partial s}\left(Bt_1\frac{\partial W}{\partial s} + t_{12}\frac{\partial W}{\partial \varphi}\right) + \frac{\partial}{\partial \varphi}\left(\frac{t_2}{B}\frac{\partial W}{\partial \varphi} + t_{12}\frac{\partial W}{\partial s}\right)\right]. \end{gathered}$$

Here U, V, W are the projetions of the displacements on the axes of the local orthogonal coordinate system,

R_1, R_2 are the principal radii of curvature,

B is the distance between a point on the shell and the axis of symmetry (Lamé's coefficient),

R is the characteristic shell radius,

T_1, T_2, T_{12} are membranous stresses.

The functions $t_1(s,\varphi)$, $t_2(s,\varphi)$ and $t_{12}(s,\varphi)$ are connected with the initial stresses t_i^* and t_{12}^* by the formula

$$(2)\qquad [t_i^*, t_{12}^*] = \Lambda\mu^2 Eh[t_i, t_{12}].$$

Here E is Young's modulus, h is the shell thickness, Λ is the dimensionless loading, $\mu^4 = h^2/\left[12(1-\sigma^2)R^2\right]$ is a small parameter and σ is Poisson's ratio. The functions R_1, R_2, B, t_i, t_2 and t_{12} are supposed to be slowly varying and to be sufficiently smooth.

3. Zeroth approximation to the critical load

We fix a point, α_*, on the shell middle surface and assume that the functions R_i, t_i, t_{12} and B are equal to $R_i(\alpha_*)$, $t_i(\alpha_*)$, $t_{12}(\alpha_*)$ and $B(\alpha_*)$, respectively. Therefore system (1) has constant coefficients. The solution of this system is sought in the form (see Subsection 4.4.1 of the Survey [**1**])

$$
\begin{aligned}
&[W, U, V, T_1, T_2, T_{12}] = [W^*, \mu U^*, \mu V^*, T_1^*, T_2^*, T_{12}^*]\, e^{iz}, \\
&z = \frac{1}{\mu}\,(ps + q\varphi),
\end{aligned} \tag{3}
$$

where W^*, U^*, V^*, T_1^*, T_2^* and T_{12}^* are unknown constants. Substituting (3) into (1) we get the following equation:

$$
\begin{aligned}
H(p, q, \alpha^*, \Lambda) = \left[p^2 + \left(\frac{q}{B}\right)^2\right]^2 + \left[k_2 p^2 + k_1\left(\frac{q}{B}\right)^2\right]^2 \left[p^2 + \left(\frac{q}{B}\right)^2\right]^{-2} \\
- \Lambda\left[t_1 p^2 + 2t_{12} p\frac{q}{B} + t_2\left(\frac{q}{B}\right)^2\right] = 0,
\end{aligned} \tag{4}
$$

which is a necessary and sufficient condition for the solution to exist. Shell buckling occurs when p and q are real. P. E. Tovstik has shown [**9**] that this hypothesis was valid for clamped shell edges.

We rewrite formula (4) in the form

$$\Lambda_0 = f(p, q, \alpha_*)$$

and find the lowest positive value of Λ_0, for which equation (4) has a real root:

$$\Lambda_0 = \min_{p,q,\alpha_*}{}^{+} f(p, q, \alpha_*) = f(p_0, q_0, \alpha_0).$$

There are four cases:

(a) f has a minimum at

$$p = 0, \qquad q = \sqrt{k_1}\, B, \qquad \Lambda_0 = \frac{2k_1}{t_2}, \tag{5a}$$

if

$$t_{12}(\alpha_*) = 0, \qquad [k_2 t_2 - k_1 t_1](\alpha_*) > 0,$$

(b) f has a minimum at

$$p = q = \sqrt{k_2}, \qquad q = 0, \qquad \Lambda_0 = \frac{2k_2}{t_1}, \tag{5b}$$

if

$$t_{12}(\alpha_*) = 0, \qquad [k_1 t_1 - k_2 t_2](\alpha_*) > 0,$$

(c) f has a minimum at

$$p = r\cos\beta, \qquad q = r\sin\beta, \qquad \Lambda_0 = \frac{2k_2}{t_1}, \tag{5c}$$

where β is an arbitrary constant and

$$r = \sqrt{k_2\cos^2\beta + k_1\sin^2\beta},$$

if

$$t_{12}(\alpha_*) = 0, \qquad [k_1 t_1 - k_2 t_2](\alpha_*) = 0,$$

(d) f has a minimum at

$$p = r\cos\beta, \qquad q = r\sin\beta, \qquad r = \sqrt{k_2\cos^2\beta + k_1\sin^2\beta}, \tag{5d}$$

where

$$\tan\beta = \frac{1}{2t_{12}k_1}\left[-t_1k_1 + t_2k_2 + \sqrt{(k_1t_1 - k_2t_2)^2 + 4k_1k_2t_{12}^2}\right]$$

and

$$\Lambda_0 = 4k_1k_2\left[t_1k_1 + t_2k_2 + \sqrt{(k_1t_1 - k_2t_2)^2 + 4k_1k_2t_{12}}\right]^{-1},$$

if $t_{12}(\alpha_*) \neq 0$

Case (c) is special and will not be considered in this paper. The point α_0, where the function f attains its positive minimum, is called the weakest point. P. E. Tovstik [**10**] studied the case where this point was far from the edges of the shell, and found the buckling modes near the weakest point (see Subsection 4.4.1 of the Survey [**1**]). These modes decrease as one moves away from the point α_0. The case where the point α_0 lies on the shell edge is considered below. Initial shell stresses could not be assumed to be membranous near the shell edges but this fact affects the critical load and reduces it by values of the order $O(\mu)$ in the Cases (a) and (d). Therefore this simplified approach to this problem seems to be correct.

To find the asymptotic solution of the boundary value problem one has to construct the integrals of equations (1) and to satisfy the boundary conditions on the shell edge. However, one is not required to satisfy boundary conditions on the opposite edge of the shell as the solution decreases exponentially as one gets away from the weakest point. The solution is the sum of the main integral and the integrals of the edge effect. The integrals of the edge effect correspond to roots $p = p_k$, $k = 1, 2, 3$, of the equation

$$H(p_k, q_0, \alpha_*, \Lambda_0) = 0, \qquad \Im p_k > 0. \tag{6}$$

In this case the integrals of the edge effect decrease as the distance from the edge $s = 0$ increases. It must be underlined that, in the Case (b), equation (4) degenerates into a fourth-order equation and has no roots with positive imaginary part.

4. The main integral structure

We consider the main integral structure in detail ($p = p_0$, $q = q_0$). We shall evaluate all functions containing H at $\alpha = \alpha_0$, $p = p_0$, $q = q_0$, $\Lambda = \Lambda_0$ and denote them by h. Partial derivatives of functions will be denoted by the corresponding subscript, for example $h_p = \partial h/\partial p$. It follows from (4) that

$$h = h_p = h_q = h_\varphi = 0, \qquad h_s \geq 0. \tag{7}$$

If $h_s > 0$ and the matrix

$$A = \begin{bmatrix} h_{pp} & h_{pq} & h_{p\varphi} \\ h_{qp} & h_{qq} & h_{q\varphi} \\ h_{\varphi p} & h_{\varphi q} & h_{\varphi\varphi} \end{bmatrix}$$

is positive definite, we find the main integral in the form:

$$(8)\qquad \begin{aligned}[W,U,V,T_1,T_2,T_{12}](s,\varphi,\mu) &= [w,\mu u,\mu v,\tau_1,\tau_2,\tau_{12}](s,\varphi,\mu)e^{iz},\\ z &= \mu^{-1}(ps+q\varphi).\end{aligned}$$

An analysis of system (1) reveals that the buckling mode is localized in an area of size:

$$(9)\qquad s = O(\mu^{2/3}), \qquad \varphi = O(\mu^{1/2}).$$

We introduce the following fast variables:

$$(10)\qquad \nu = s\mu^{-2/3}, \qquad \xi = \varphi\mu^{-1/2},$$

and seek an asymptotic expansion for the load, Λ, in the form of a power series in $\mu^{1/6}$, as follows:

$$(11)\qquad \Lambda = \Lambda_0 + \mu^{2/3}\Lambda_1 + \mu\Lambda_3 + \mu^{7/6}\Lambda_4 + \cdots$$

If we solve system (1), we obtain the following equation for the unknown function w:

$$(12)\qquad \begin{aligned} &w(h_s\nu - \Lambda_1 h_T) - \frac{1}{2}h_{pp}\frac{\partial^2 w}{\partial\nu^2} - \mu^{1/6}\left(h_{pq}\frac{\partial^2 w}{\partial\nu\partial\xi} + ih_{p\varphi}\frac{\partial w}{\partial\nu}\xi\right)\\ &+\mu^{1/3}\left\{\frac{i}{6}h_{ppp}\frac{\partial^3 w}{\partial\nu^3} - ih_{ps}\frac{\partial w}{\partial\nu}\nu - \frac{1}{2}h_{qq}\frac{\partial^2 w}{\partial\xi^2} - i\xi h_{q\varphi}\frac{\partial w}{\partial\xi}\right.\\ &\left.+\left[\frac{1}{2}h_{\varphi\varphi}\xi^2 - \Lambda_3 h_T - \frac{i}{2}(h_{ps}+h_{q\varphi})\right]w\right\} + O\left(\mu^{1/2}\right) = 0,\end{aligned}$$

where

$$h_T = t_1(\alpha_0)p_0^2 + 2t_{12}(\alpha_0)p_0\frac{q_0}{B} + t_2(\alpha_0)\frac{q_0}{B}^2.$$

If we consider the zeroth approximation to equation (12) and apply a linear transformation to the independent variable we obtain Airy's equation. (see Sections 1.6 and 1.7 of the Survey [**1**]). Therefore the solution of equation (12) is represented in the form:

$$(13)\qquad w = b(\nu,\xi,\mu)U(\eta) + c(\nu,\xi,\mu)\eta' U'(\eta),$$

$$(14)\qquad b = \sum_{k=0}^{\infty}\mu^{k/6}b_k, \qquad c = \sum_{k=0}^{\infty}\mu^{k/6}c_k,$$

where $U(\eta)$ is Airy's function satisfying the equation

$$\frac{d^2U}{d\eta^2} = \eta U,$$

with U decreasing exponentially as $\eta \to \infty$. Here $\eta = K\eta + \gamma$, where K and γ are constants to be defined. Substituting (13) into equation (12) and equating the coefficients of U and U' to zero we obtain the sequence of systems:

$$L_1^0(b_0, c_0) = 0, \qquad L_2^0(b_0, c_0) = 0,$$

(15) $$L_1^0(b_k, c_k) + \sum_{i=1}^{k} L_1^i(b_{k-i}, c_{k-i}) = 0,$$

(16) $$L_2^0(b_k, c_k) + \sum_{i=1}^{k} L_2^i(b_{k-i}, c_{k-i}) = 0, \qquad k = 1, 2, \dots$$

System (15) has a non-zero polynomial solution in ν, if the following condition holds:

$$K = \left(\frac{2h_s}{h_{pp}}\right)^{1/3}, \qquad \Lambda_1 = -\frac{h_{pp}\gamma K^2}{2h_T},$$

where the constant γ may be obtained from the boundary conditions. The equation for the function c_0 is obtained from (15) in the form

(17) $$\frac{d^3 c_0}{d\nu^3} - 4K^2\eta\frac{dc_0}{d\nu} - 2K^3 c_0 = 0.$$

The only polynomial solution of this equation is the zero polynomial

$$c_0 = 0.$$

It follows that $b_0 = b_0(\xi)$. It is clear from the foregoing that the polynomials c_k are uniquely determined by (16). The polynomials b_k are also determined by equation (16) but they contain an arbitrary function. Thus, it follows from (15) and (16) that

(18) $$c_1 = 0, \qquad b_1 = \left(-h_{pq}\frac{\partial b_0}{\partial \xi} - ih_{p\varphi}\xi b_0\right)\frac{\nu}{h_{pp}},$$

(19) $$c_2 = d_1\frac{\partial^2 b_0}{\partial \xi^2} + id_2\xi\frac{\partial b_0}{\partial \xi} + (d_3\xi^2 + d_4)b_0,$$

where

$$d_1 = \frac{1}{2}\frac{h_{pq}^2 - h_{pp}h_{qq}}{h_{pp}h_s},$$

$$d_2 = \frac{h_{pq}h_{p\varphi} - h_{q\varphi}h_{pp}}{h_{pp}h_s},$$

$$d_3 = \frac{h_{\varphi\varphi}h_{pp} - h_{p\varphi}{}^2}{2h_{pp}h_s},$$

$$d_4 = \frac{1}{2h_s}\left[i\frac{h_{p\varphi}h_{pq} - h_{q\varphi}h_{pp}}{h_{pp}} + \frac{i}{6}\frac{\partial^3 h}{\partial p^3}K^3 - 2\Lambda_3 h_T\right].$$

5. The integrals of the edge effect

Now we consider the integrals of the edge effect. These integrals have the following structure (see Subsection 3.1.2 of the Survey [**1**]):

(20) $$\begin{gathered}[W^k, U^k, V^k, T_1^k, T_2^k, T_{12}^k] = [w^k, u^k\mu, v^k\mu, \tau_1^k, \tau_2^k, \tau_{12}^k](\nu^*, \xi, \mu)\, e^{iz_k}, \\ z_k = \mu^{-1}(p_k s + q_0\varphi), \qquad \nu^* = s\mu^{-1}, \qquad k = 1, 2, 3.\end{gathered}$$

The asymptotic expansion is valid for all the unknown functions x^k:

$$x^k(\nu^*, \xi, \mu) = \sum_{k=1}^{\infty} \mu^{k/6} x_i^k(\nu^*, \xi). \tag{21}$$

The polynomials, x_i^k in ν^*, contain the additional arbitrary function $x_i^k(\xi)$. Specifically $w_0^k = w_0^k(\xi)$.

6. The asymptotic formulae for the critical load and the buckling modes

Now we try to construct an iterative procedure which satisfies the boundary condition. Substituting a linear combination of the main integral and the edge effect integrals into the boundary condition we obtain a system of four linear equations for the four arbitrary functions $b_0^0(\xi)$, $w_0^k(\xi)$. This system has a solution if

$$U(\gamma) = 0. \tag{22}$$

Equation (22) defines the set of Airy's function roots. Moreover, the lowest value Λ_1 corresponds to $\gamma = -2.338\ldots$. Proceeding from this, we find the asymptotic solution of the boundary value problem in the form

$$W = W^0 + \mu^{1/3} \sum_{k=1}^{3} W^k, \tag{23}$$

where W^0 is the main integral and W^k are the integrals of the edge effect. Substituting (23) into the boundary condition and eliminating the functions w_0^k, $k = 1, 2, 3$, from the above system, we get the following equation for b_0:

$$c_2 + Q\, b_0 = 0, \tag{24}$$

where Q is a constant. Equation (24) has an exponentially decreasing solution, as $\xi \to \infty$, of the following form

$$b_{0,m} = P_{0,m}(\xi)\, e^{a\xi^2/2}, \tag{25}$$

where

$$a = -i\frac{h_{q\varphi}h_{pp} - h_{pq}h_{p\varphi}}{h_{qq}h_{pp} - h_{pq}h_{pq}} - \sqrt{\frac{h_{pp}\det(A)}{(h_{qq}h_{pp} - h_{pq}h_{pq})^2}}.$$

Here $P_{0,m}$ are polynomials of degree m such that:

$$G_0(P_{0,m}) = g_1\frac{d^2P_{0,m}}{d\xi^2} + g_2\xi\frac{dP_{0,m}}{d\xi} + g_3, \qquad P_{0,m} = 0, \tag{26}$$

where

$$g_1 = d_1, \qquad g_2 = \frac{1}{h_s}\sqrt{\frac{\det(A)}{h_{pp}}},$$

and

$$g_3 = \frac{1}{2h_s}\left[\frac{i}{6}\frac{\partial^3 h}{\partial p^3}K^3 - 2\Lambda_3 h_T + \sqrt{\frac{\det(A)}{h_{pp}}}\right] + Q. \tag{27}$$

Equation (26) has a solution, if

$$\Lambda_3 = \frac{1}{h_T}\left[\left(m+\frac{1}{2}\right)\sqrt{\frac{\det(A)}{h_{pp}}} + Q^*\right], \tag{28}$$

where

$$Q^* = Qh_s - \frac{i}{12}\frac{\partial^3 h}{\partial p^3}K^3$$

is a real constant. The case $m = 0$, $P_0 = 1$ is the most interesting one, because it corresponds to the lowest value of Λ_3. The functions $b_{i,m}$ are given by the formulae:

$$b_{i,m}(\xi) = e^{a\xi^2/2}P_{i,m}, \tag{29}$$

$$G_0(P_{i,m}) = f_i(\xi, \Lambda_{3+i}), \qquad i = 1, 2, \dots, \tag{30}$$

where the polynomials f_i were determined by the previous steps of the iterative process. The corrections to the critical load, Λ_i, $i = 4, 5, \dots$, are determined from the orthogonality conditions. Therefore shell buckling occurs if Λ satisfies

$$\Lambda = \Lambda_0 + \mu^{2/3}\Lambda_1 + \mu\Lambda_3 + O\left(\mu^{7/6}\right), \tag{31}$$

where Λ_0 is defined by (5),

$$\Lambda_1 = -\frac{h_{pp}\gamma_0 K^2}{2h_T}, \qquad \gamma_0 = -2.338\dots, \qquad K = \left(\frac{2h_s}{h_{pp}}\right)^{1/3},$$

$$\Lambda_3 = \frac{1}{h_T}\left[\left(m+\frac{1}{2}\right)\sqrt{\frac{\det(A)}{h_{pp}}} + Q^*\right], \qquad m = 0, 1, 2\dots$$

The eigenvalue Λ is asymptotically twofold.

7. Conclusion

The buckling of a convex shell under nonaxisymetrical loading is studied in the linear approximation. We demonstrate that an uneven loading leads to localization of the buckling form near the weakest point. The buckling form decreases as one moves away from this point.

References

1. S. M. Bauer, S. B. Filippov, A. L. Smirnov and P. E. Tovstik, *Asymptotic methods in mechanics with application to thin shells and plates*, Asymptotic Methods in Mechanics (A. L. Smirnov and R. Vaillancourt, eds.), CRM Proceedings and Lecture Notes, this volume, Amer. Math. Soc., Providence, R.I., 1993, pp. 3–140.
2. D. Bushnell, *Shell buckling – Pitfall for designer*, AIAA J. **19** (1981), 1183–1226.
3. A. L. Gol′denveizer, V. B. Lidsky and P. E. Tovstik, *Free vibrations of thin elastic shells*, "Nauka", Moscow, 1979 (383 pp.). (Russian)
4. E. I. Grigolyuk and V. V. Kabanov, *Stability of shells*, "Nauka", Moscow, 1978 (359 pp.). (Russian)
5. W. T. Koiter, *Elastic stability and postbuckling behavior*, Proceedings of the Symposium on Nonlinear Problems, University of Wisconsin Press, Madison, WI, 1962, pp. 257-275.
6. A. L. Maiboroda, *Construction of principal integral in the problem of loss of stability of convex shells near the edge*, Vestnik Leningrad. Univ. Mat. Mekh. Astronom. (1986), no. 1, 123–126; English transl., Len. Univ. Mech. Bull. (1986), no. 1, 57–61.

7. W. Nachbar and N. I. Hoff, *The buckling of a free edge of an axially compressed circular cylindrical shell*, Quart. Appl. Mech. **20** (1962), no. 3, 267–277.
8. L. I. Shkutin, *The effect of precritical strains on the stability of a longitudinally compressed cylindrical shell*, Prikl. Mekh. **3** (1967), no. 1, 127–130; English transl., Soviet Appl. Mech. **3** (1967), no. 1, 74–75.
9. P. E. Tovstik, *Buckling of shells of revolution in the linear approximation*, Raschety prostranstvennykh konstruktsii, vol. 13, Stroiizdat, Moscow, 1970, pp. 118–138. (Russian)
10. ——, *On the local buckling of shells*, Vestnik Leningrad. Univ. Mat. Mekh. Astronom. (1982), no. 13, 72–78. (Russian)
11. ——, *Surface bending and stability of shells*, Prikl. Mekh., vol. 8, Leningrad University Press, Leningrad, 1990, pp. 137–162. (Russian)
12. J. Wang and D. Billington, *Buckling of cylindrical shells under wind pressure*, Proceedings of the ASCE, Division of Mechanical Engineering, vol. 100, pp. 1005–1022.

ABSTRACT. The buckling of a convex shell with clamped edges is studied by means of its linear approximation. The initial stresses are supposed to be nonaxisymmetric and membranous. The problem may be reduced to a two-dimensional boundary value problem. The asymptotic method is used, under the assumption that the shell thickness parameter is small. It is shown that the buckling modes are localized near the weakest point. The case where the weakest point lies on the shell edge is considered in detail. Asymptotic formulae for the critical load and the buckling modes are obtained.

INSTITUTE OF MATHEMATICS AND MECHANICS, ST. PETERSBURG STATE UNIVERSITY, 2 BIBLIOTECHNAYA SQ., STARY PETERHOF, 198904 ST. PETERSBURG, RUSSIA

E-mail address: als@hq.math.lgu.spb.su

Centre de Recherches Mathématiques
CRM Proceedings and Lecture Notes
Volume **3**, 1993

Elasto-Plastic Deformations of Ribbed Plates

Galina V. Pavilainen

1. Introduction

We consider the elasto-plastic deformation of a structure consisting of a thin rectangular plate reinforced by ribs. The structure is placed in a uniform thermal field. It is assumed that the lower ends of the ribs are clamped in the foundation which is not deformed by the thermal field.

As usual, this structure is investigated either by computational or experimental methods (see, for example, the papers by G. R. Heppler and J. S. Hansen [**3**], K. P. Soldatos [**14**] and others [**4, 11, 15**] and FEM analyses by P. Lardeur and J. L. Batoz [**8**], Shi Li and Liangxin Sun [**9**], V. B. Petrov and A. S. Saimanin [**11**], and Chandra Ram *et al.* [**12**]). We have only a small number of analytical solutions [**13, 16**] and this in insufficient. In this paper we shall find the analytical solution with the help of the classical theory of beam and plate bending (see Chapter 4 of the Survey [**1**]), S. Timoshenko's theory [**2, 3, 18, 19**], A. D. Kovalenko's theory of thermoelasticity [**7, 17**] and the elasto-plastic theory by R. Hill and A. A. Ilyshin [**5, 6**].

We study the uniform heating-cooling of the plate in order to estimate the effect of the thermal field, modeled by a load P, on the geometrical parameters and the physical properties of the plate. If the plate has no ribs and expands uniformly in the thermal field, bending is possible only under its own weight [**9**]. The ribs resist the uniform plate expansion and cause additional plate bending. If P is larger than some specified value, the ribs are in the elasto-plastic state. This means that it is necessary to determine the influence of the load on the vertical displacement of the plate and the horizontal displacement of the ribs caused by bending.

As in [**10**], we consider, to first approximation, the structure as a system of horizontal and vertical beams and split this system into elements (see the shaded part in Fig. 1) consisting of bar I and beam II.

The elastic bar I is stretching and bending, and the elasto-plastic beam II is only bending under the load (see Fig. 2).

In considering this problem, we neglect the transverse shear deformation of beam II.

Bar I and beam II are connected at right angle and have rectangular normal cross-sections. The lower end of beam II is clamped. We introduce the local coordinate system XOY for each element, where $v(x)$ is the deflection of bar I and $w(y)$ is the deflection of beam II. The positive direction of the deflections is shown in Fig. 2.

1991 *Mathematics Subject Classification.* Primary: 73K05; Secondary: 73E99.
Key words and phrases. Elasto-plastic deformations, reinforced plate, thermal load.
This is the final form of the paper.

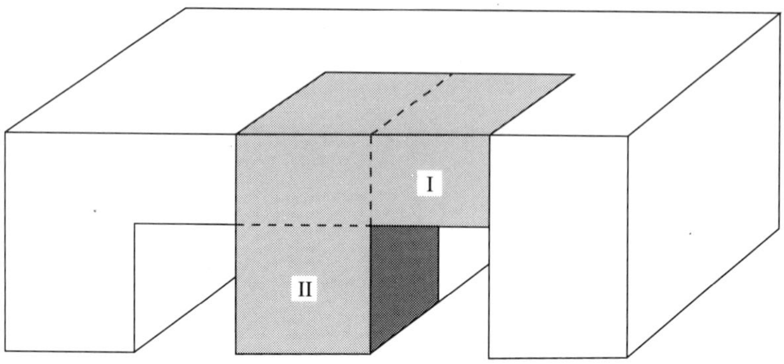

FIGURE 1. Beam structure.

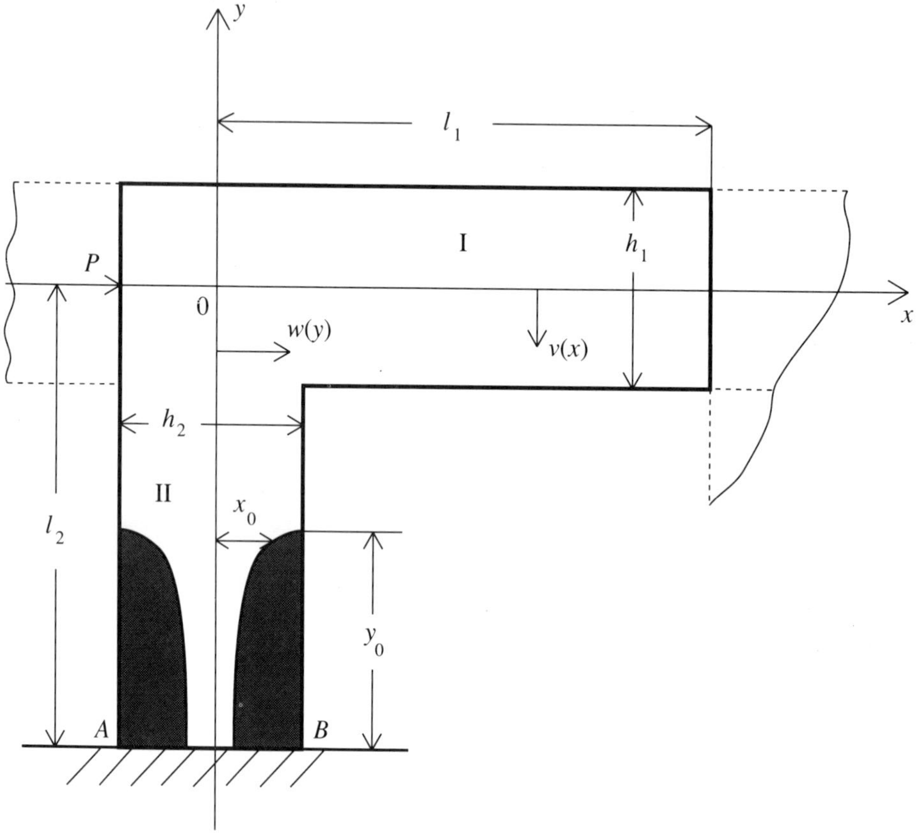

FIGURE 2. Element of beam structure with plastic zone.

2. The solution of the elastic problem

We start with the elastic problem. The expression describing the bending of an elastic beam is the solution of a fourth-order differential equation (see Section 2.6 in [**1**] and [**10, 18**]). The bending of the element (see Fig. 2) is represented by the solutions of the following system of differential equations:

$$EJ_1\frac{d^4v}{dx^4} = 0, \qquad EJ_2\frac{d^4w}{dy^4} = 0, \tag{1}$$

where E is Young's modulus and J_1 and J_2 are the moments of inertia of the cross-sections of the beams. This system satisfies the following boundary conditions:

$$w(-l_2) = \frac{dw}{dy}(-l_2) = 0, \tag{2}$$

$$v(0) = v(l_2) = 0, \tag{3}$$

$$EJ_2\frac{d^3w}{dy^3}(0) = P, \tag{4}$$

$$\frac{dv}{dx}(0) = \frac{dw}{dy}(0), \tag{5}$$

$$M_0 = EJ_2\frac{d^2w}{dy^2}(0) = EJ_1\left[\frac{d^2v}{dx^2}(-0) + \frac{d^2v}{dx^2}(+0)\right], \tag{6}$$

where M_0 is the bending moment of the rigid connection of bar I and beam II, and

$$\frac{dv}{dx}(0) = \frac{dv}{dx}(l_1). \tag{7}$$

It is to be noted that, in a real construction, boundary condition (7) is fulfilled only approximately.

The solution of system (1) is similar to the solution given of S. Timoshenko [**18**] and is of the following form:

$$\begin{aligned} v(x) &= \frac{PK_0}{EJ_2}\left(x^3 - \frac{3}{2}l_1x^2 + \frac{1}{2}l_1^2x\right), \\ w(y) &= \frac{P}{EJ_2}\left(-\frac{1}{6}y^3 - K_1y^2 + K_2y + K_3\right), \end{aligned} \tag{8}$$

where

$$K_0 = \frac{l_2l_1^{-1}}{12J_1J_2^{-1} + l_1l_2^{-1}}, \qquad K_1 = 3l_1K_0\frac{J_1}{J_2},$$

$$K_2 = \frac{1}{2}l_1^2K_0, \qquad K_3 = \frac{1}{3}l_2^2l_1K_0\left(3\frac{J_1}{J_2} + \frac{l_1}{l_2}\right).$$

3. The solution of the elasto-plastic problem

Let the load, P, be so large that the stresses in beam II satisfy the inequality $\|\sigma_y\| \geq \sigma_t$, where σ_t is the elongation yield point [**5**]. To examine the elasto-plastic bending we use the mathematical model developed by A. A. Ilyushin [**6**] by considering only the ideal plasticity case (σ_t constant) and neglecting Bauschinger's effect [**5**]. Plastic zones may appear and develop in different parts of beam II: a) on the lower edge, b) on the upper edge and c) on both the upper and lower edges and for the plastic zone $\sigma_y = \sigma_t$. We consider the Case (a) shown in Fig. 2 (the plastic zones are shaded).

Let x_0 and y_0 be the width and length of the plastic zone (see Fig. 2). Let $P = P_1$ be a load for which two points of plasticity, A and B, exist in beam II. For

$P \geq P_1$ these points go into two plastic zones. For $P = P_2 \geq P_1$ the plastic joint appears on the line AB. If $P \geq P_2$ the structure is not destroyed, but, in this case, plastic instability in the structure is possible. Later we shall consider only loads for which $P_1 \leq P \leq P_2$. The solution for $P \geq P_2$ may be found in the author's paper [**10**].

The bending moment $M(y)$ in the cross-section y of beam II is defined by the expression

$$M(y) = M_0 - Py. \tag{9}$$

Plasticity starts to develop in the cross-section $y = -l_2$, where the moment M attains its maximum. We express the moment $M(y)$ through the deformation $\varepsilon_y = x\varkappa$, where $\varkappa$ is the beam curvature [**6**]. Then the stresses σ_y are given by

$$\begin{aligned} \sigma_y &= E\varepsilon_y && \text{for} \quad \varepsilon_y \leq \varepsilon_t, \\ \sigma_y &= \sigma_t = E\varepsilon_t && \text{for} \quad \varepsilon_y \geq \varepsilon_t. \end{aligned} \tag{10}$$

Here, ε_t is the elongation deformation yield limit [**5**]. Therefore

$$\begin{aligned} M(\varkappa) &= M_k - \frac{E}{\varepsilon_t^3} 3\varkappa^2 && \text{for} \quad M_t \leq M \leq M_k, \\ M(\varkappa) &= EJ_2\varkappa && \text{for} \quad 0 \leq M \leq M_t, \end{aligned} \tag{11}$$

where

$$M_t = EJ_2 \frac{2\varepsilon_t}{h_2}, \qquad M_k = EJ_2 \frac{3\varepsilon_t}{h_2},$$

M_t and M_k are the values of the bending moment $M(y)$ for $x_0 = -h_2/2$ and $x_0 = 0$, respectively. Formula (11) determines the smooth function $M(\varkappa)$.

We substitute the expression for the curvature $\varkappa$ [**6**],

$$\varkappa = \frac{d^2 w}{dy^2},$$

where $w(y)$ is the deflection of beam II, into formula (11). Integrating twice the equation for $w(y)$, using the boundary conditions (2) and the continuity of $w(y)$ and $w'(y)$ in the cross-section $y = -l_2 + y_0$ we get

$$\begin{aligned} w(y) &= \frac{1}{6} a_1 y^3 + \frac{1}{2} a_2 y^2 + b_1 y + b_2 && \text{for} \quad -l_2 + y_0 \leq y \leq 0, \\ w(y) &= \frac{4}{3a_3^2} (a_3 y + a_4)^{\frac{3}{2}} + b_3 y + b_4 && \text{for} \quad -l_2 \leq y \leq -l_2 + y_0. \end{aligned} \tag{12}$$

Here

$$a_1 = -\frac{P}{EJ_2}, \qquad a_2 = \frac{M_0}{EJ_2}, \qquad a_3 = \frac{3P}{E\varepsilon_t^3}, \qquad a_4 = \frac{-3M_0}{\varepsilon_t^3} + \frac{3h_2^2}{4\varepsilon_t^2},$$

$$b_1 = \frac{2}{a_3}\left(s^{\frac{1}{2}} - t^{\frac{1}{2}}\right) - \frac{1}{2} a_1 y_1^2 - a_2 y_1,$$

$$b_2 = \frac{4}{3a_3^2}\left(s^{\frac{3}{2}} - t^{\frac{3}{2}}\right) - \frac{2y_1}{a_3} s^{\frac{1}{2}} - \frac{2l_2}{a_3} t^{\frac{1}{2}} + \frac{a_1}{3} y_1^3 + \frac{a_2}{2} y_1^2,$$

$$b_3 = -\frac{2}{a_3} t^{\frac{1}{2}}, \qquad b_4 = \frac{4}{3a_3^2} t^{\frac{3}{2}} - \frac{2l_2}{a_3} t^{\frac{1}{2}},$$

$$y_1 = -l_2 + y_0, \qquad s = a_3 y_1 + a_4, \qquad t = a_4 - a_3 l_2.$$

TABLE 1. The values of y_0 and M_0 vs. the load P.

P	y_0	M_0	Comments
1856.880	0.00000763	−642.750	$P = P_1$
1900.000	0.00916009	−657.700	–
2000.000	0.02885708	−692.390	–
2100.000	0.04654484	−727.360	–
2200.000	0.06241098	−762.800	–
2300.000	0.07658007	−798.970	–
2400.000	0.08913924	−836.170	–
2500.000	0.10012967	−874.780	–
2600.000	0.10567431	−915.360	–
2700.000	0.11719043	−958.690	–
2800.000	0.12288006	−1006.040	–
2815.320	...	...	$P = P_*$
2910.000	0.12644817	−1064.640	–
2992.000	0.12529217	−1119.230	–
2992.990	0.12550089	−1119.230	$P = P_2$

The value of y_0 may be found from the condition

$$M(y_1) = M_t, \tag{13}$$

and expressed in the form

$$y_0 = \frac{M_0}{P} - \frac{E\varepsilon_t h_2^2}{6P} + l_2. \tag{14}$$

To complete the solution of the elasto-plastic problem for beam II we have to get M_0 from condition (6), where $w''(0)$ is defined from the first condition in (12). As a result we have the following fourth-degree equation for M_0:

$$A_4 M_0^4 + A_3 M_0^3 + A_2 M_0^2 + A_1 M_0 + A_0 = 0, \tag{15}$$

where the coefficients A_4, A_3, A_2, A_1 and A_0 are constants depending upon E, ε_t, h_2, J_1, J_2 and P. Thus for the elastic problem we have

$$M_0 = -\frac{6Pl_2^2 J_1}{l_1 J_2 + 12 l_2 J_1}. \tag{16}$$

Comparing the roots of equation (15) with (16) one can select the proper root.

In the following example, we set $E = 1.15 \cdot 10^7 \mathrm{N/cm}^2$, $\varepsilon_t = 1.565 \cdot 10^3$, $h_1 = h_2 = 0.5$ cm, $l_1 = l_2 = 0.75$ cm. In Table 1, one can find the load P and the corresponding y_0 and M_0 for the elastic and the elasto-plastic problems. The values P_1, P_2, P_* may be obtained from the formulae

$$P_1 = \frac{M_t}{l_2}\frac{l_1 J_2 + 12 l_2 J_1}{l_1 J_2 + 6 l_2 J_1}, \qquad P_2 = \frac{M_k - M_0}{l_2}, \qquad P_* = P_1 \frac{M_k}{M_t}. \tag{17}$$

TABLE 2. The maximal deflections under loading and unloading.

P	$w_m^0(0)$	$v_m^0(l_1/4)$	$w_m^1(0)$	$v_m^1(l_1/4)$
1856.880	0.00067073	0.000023582	0.000000014	0.0000000027
1900.000	0.00068628	0.000024125	0.000015569	0.0000005453
2000.000	0.00072266	0.000025403	0.000051943	0.0000018229
2100.000	0.00075970	0.000026683	0.000088984	0.0000031038
2200.000	0.00079799	0.000027985	0.000127284	0.0000044051
2300.000	0.00083814	0.000029310	0.000167436	0.0000057305
2400.000	0.00088093	0.000030673	0.000210219	0.0000070928
2500.000	0.00092739	0.000032093	0.000256681	0.0000085133
2600.000	0.00097888	0.000033581	0.000308163	0.0000100014
2700.000	0.00103752	0.000035171	0.000366810	0.0000115909
2800.000	0.00110682	0.000036908	0.000436111	0.0000133284
2815.320	...	...		
2910.000	0.00121440	0.000040349	0.000543685	0.0000167689
2992.000	0.00130135	0.000041058	0.000630641	0.0000174780
2992.990	...	...		

Now we consider the deformation of the horizontal elastic bar I. If we solve the first equation of system (1) with the boundary conditions (3), (5) and (7), we obtain

$$v(x) = \frac{2}{l_1^2}\frac{dw}{dy}(0)\left(x^3 - \frac{3}{2}l_1x^2 + \frac{1}{2}l_1^2x\right), \tag{18}$$

where $w'_y(0)$ is equal to the coefficient b_1 in formulae (12). Numerical results are presented in Table 2. Let $w_m^0(0)$ denote the maximal deflection of beam II which occurs at $y = 0$ and $v_m^0(x)$ the maximal deflection of bar I, which occurs at $x = l_1/4$ and $x = 3l_1/4$. According to solution (18), $v(x)$ is determined as an implicit function of P.

4. The solution of the problem with the elastic unloading

Finally, we consider the elastic unloading of the structure after the loading. For the elastic problem, $P \leq P_1$ and the deformations $w(y)$ and $v(x)$ are equal to zero after the unloading. For the elasto-plastic problem, $w(y)$ and $v(x)$ decrease with the unloading but not to zero. We consider the unloading as a bending under the opposite load P_3 [**6**]. In Table 2, the maximal deflections for the unloading are denoted by v_m^1 and w_m^1.

From this solution it is possible to find the stress-strain state of the entire structure and to evaluate the maximum deformations and the deflections of the entire plate.

It is to be noted that this solution may be used only as a qualitative, but not quantitative, approximation. To obtain a more accurate solution we must take into account the transverse shear deformation [**2, 13, 19**] and the Bauschinger's effect [**5**].

References

1. S. M. Bauer, S. B. Filippov, A. L. Smirnov and P. E. Tovstik, *Asymptotic methods in mechanics with application to thin shells and plates*, Asymptotic Methods in Mechanics (A. L. Smirnov and R. Vaillancourt, eds.), CRM Proceedings and Lecture Notes, this volume, Amer. Math. Soc., Providence, R.I., 1993, pp. 3–140.
2. L. H. Donnell, *Beams, plates and shells*, McGraw-Hill, New York, 1976 (774 pp.).
3. G. R. Heppler and J. S. Hansen, *Timoshenko beam finite elements using trigonometric basis functions*, AIAA J. **26** (1988), no. 11, 1378–1386.
4. C. D. Hill, G. E. Blandford and S. T. Wang, *Post-buckling analysis of steel space trusses*, J. Struct. Eng. **4** (1989), 900–919.
5. R. Hill, *The mathematical theory of plasticity*, Oxford Univ. Press, Fair Lawn, 1950 (384 pp.).
6. A. A. Ilyushin, *Plasticity*, Gostekhizdat, Moscow, 1948 (376 pp.). (Russian)
7. A. D. Kovalenko, *Thermoelasticity*, Wolters Noorhoff Publishing, Groningen, 1969 (251 pp.).
8. P. Lardeur and J. L. Batoz, *Composite plate analysis using a new discrete shear triangular finite element*, Internat. J. Numer. Methods Engrg. **2** (1989), 343–359.
9. Sun Liangxin and Shi Li, *Thermoelasticity analysis of laminated plates and shells using hybrid-stress elements*, J. Nanjing Aeronaut. Inst. **2** (1991), 38–47.
10. G. V. Pavilainen, *Elastoplastic deformation of structures*, Vestnik Leningrad. Univ. Mat. Mekh. Astronom. (1992), no. 1, 67–73; English transl., Len. Univ. Mech. Bull. (1992), no. 1, 16–23.
11. V. B. Petrov and A. S. Saimanin, *Numerical solution of stationary and non-stationary heat conduction problems for plate-beam systems in space*, Raschety na prochnost **29** (1989), 79–87. (Russian)
12. Chandra Ram, Krishna Prem, and D. N. Trikha, *Elastic-plastic analysis of steel space structures*, J. Struct. Eng **4** (1990), 939–955.
13. G. Shi and S. N. Atluri, *Elasto-plastic large deformation analysis of space-frames: a plastic-hinge and stress-based explicit derivation of tangent stiffnesses*, Int. J. Numer. Methods Engrg. **26** (1988), no. 3, 589–615.
14. K. P. Soldatos, *On certain refined theories for plate bending*, Trans. ASME Ser. E, J. Appl. Mech. **55** (1988), no. 4, 994–995.
15. W. Sosnowski, *Simplified elastic-plastic analysis of stiffened plates and space frames*, Int. J. Space Struct. **3** (1988), no. 1, 1–14.
16. A. Suri, *Theoretical and experimental buckling analysis of stabilized webs*, Modell. Simul. and Contr. **18** (1988), no. 3, 11–41.
17. T. R. Tauchert and N. N. Huang, *Thermal buckling of symmetric angle-ply laminated plates*, Compos. Struct. 4: Proc. 4th Int. Conf. Paisley, 27th-29th July **1** (1987), 424–435.
18. S. Timoshenko, *Theory of elasticity*, McGraw-Hill, New York, London, 1934 (575 pp.).
19. ——, *Strength of materials*, Van Nostrand, New York, 1956 (468 pp.).

ABSTRACT. The elasto-plastic deformations of a thin rectangular plate reinforced by ribs in an uniform thermal field are analyzed. The influence of the thermal load on the vertical displacement of the plate and the horizontal displacement of the ribs are studied for elastic unloading (cooling) which comes after loading (heating). The numerical elasto-plastic solution is compared with the corresponding elastic analytical solution.

INSTITUTE OF MATHEMATICS AND MECHANICS, ST. PETERSBURG STATE UNIVERSITY, 2 BIBLIOTECHNAYA SQ., STARY PETERHOF, 198904 ST. PETERSBURG, RUSSIA

E-mail address: als@hq.math.lgu.spb.su

Centre de Recherches Mathématiques
CRM Proceedings and Lecture Notes
Volume **3**, 1993

Elastic Wave Propagation through Elastic Shells

Yury P. Shcheviev

1. Introduction

In theoretical acoustics, the propagation of an elastic wave through a system of plane layers is studied in detail; but the propagation mechanism of sound energy through a system of elastic shells has not yet been analyzed in such detail. However numerous experimental studies of the sound-proofing properties of multilayer sound-proofing casings revealed peculiarities of frequency characteristics, which cannot be explained on the basis of the well-developed theory of elastic wave propagation through a system of plane layers. In this paper, we consider the propagation of sound through two concentric elastic shells (casings) placed on a rigid foundation (see Fig. 1).

The sound field inside a thin elastic shell is analyzed in [**2**, **4**, **7**, **8**]. The subscripts 1 and 2 will refer to the external and internal shells, respectively.

The vibrations of the shells of a casing, generated by a source of sound, depend on the azimuthal (latitude) and polar (meridian) angles, θ and ϕ, respectively as shown in Fig. 1. The origin of the spherical coordinate system is situated at the geometrical center of the hemispheres. We assume that the source of sound, P, is located inside the shells on the axis of symmetry, i.e. on the OZ-axis of the spherical coordinate system (r, θ, φ) at distance ρ from the origin, where θ and φ denote the latitude and meridian angles, respectively.

Due to the position of the source on the OZ-axis the problem becomes axisymmetric and hence the system of differential equations for the shell vibrations can be simplified.

In the case of boundaries with ideal properties (and only in this case) the influence of the foundation on the shell vibrations and sound radiation may be examined by applying the method of electrical images. As a result, the initial problem becomes a problem of sound wave diffraction from two coherent point sources of equal capacity placed in region III (see Fig. 2), on the two concentric elastic spheres (see [**6**]).

Each of the spheres consists of two hemispheres connected in such a way that the tangential components of the displacement on the contours $\theta = \pi/2$ can be neglected, if $r = R_1$ and $r = R_2$.

2. The momentless theory of thin elastic shells

We consider the frequencies and shell thickness for which the condition $h \ll \lambda$ is

---

1991 *Mathematics Subject Classification.* Primary: 73D25; Secondary: 73K15.
Key words and phrases. Wave diffraction, sound propagation.
This is the final form of the paper.

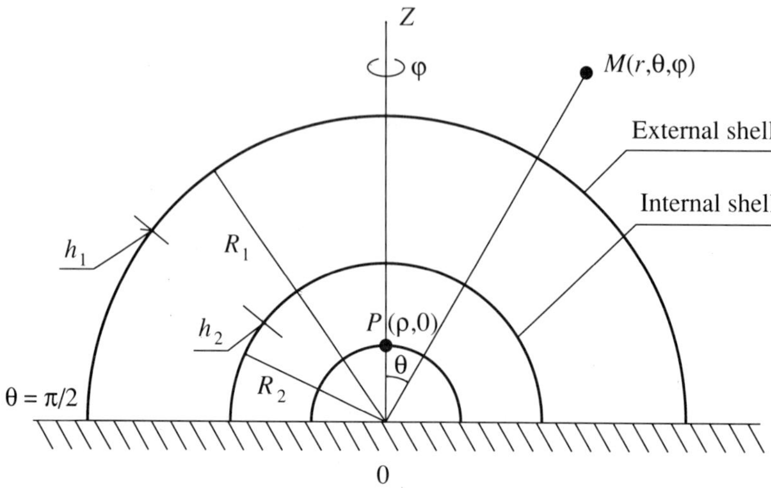

FIGURE 1. Casing of two concentric shells.

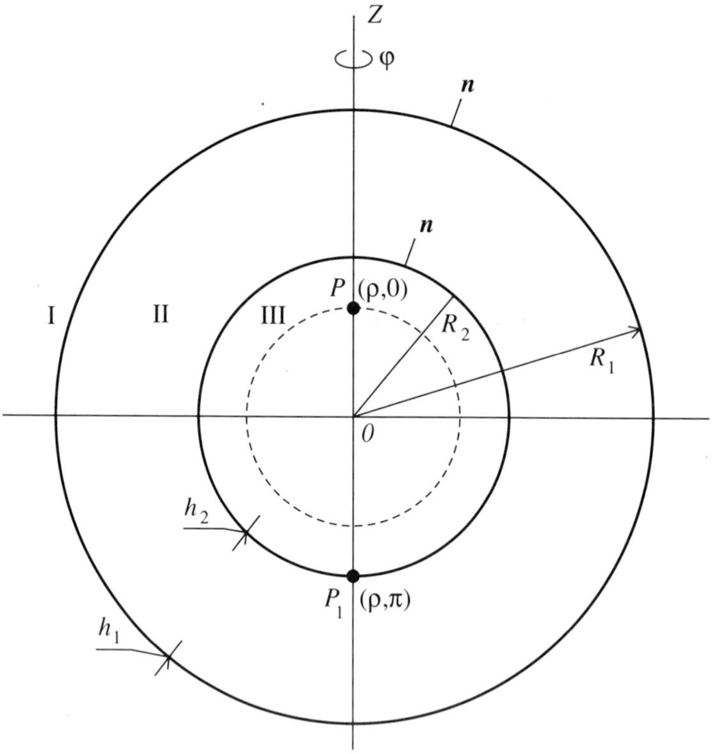

FIGURE 2. Location of point sources of elastic wave.

satisfied (where h is the shell thickness and λ is the sound wavelength). In this case, it is possible to use the momentless theory of thin elastic shells, which assumes that the shell displacement vector is constant along the shell thickness and the bending stresses are negligible in comparison with the stresses caused by the forces on the neutral surface (see Section 2.6 of the Survey [**1**]). Therefore, under Kirchhoff–Love type hypotheses, we may write the following system of differential equations for the forced vibrations of spherical shells, which we express in terms of displacement [**3**] (the subscripts $i = 1$ and 2 refer to the external and interval spheres, respectively):

$$\begin{gathered}\frac{\partial}{\partial\theta}\left[\frac{1}{\sin\theta}\frac{\partial}{\partial\theta}(u_{i\theta}\sin\theta)\right] - 2\frac{\partial u_{ir}}{\partial\theta} + (1-\sigma_i)\left(u_{i\theta} + \frac{\partial u_{ir}}{\partial\theta}\right) \\ + \frac{1-\sigma_1^2}{E_i}R_i\rho_{0i}\omega^2 u_{i\theta} = 0, \\ \frac{1+\sigma_i}{\sin\theta}\frac{\partial}{\partial\theta}(u_{i\theta}\sin\theta) - 2(1+\sigma_i)u_{ir} + \frac{1-\sigma_1^2}{E_i}R_i^2\rho_{0i}\omega^2 u_{ir} = -\frac{1-\sigma_1^2}{E_i h_i}R_i^2\Delta p_i,\end{gathered} \tag{1}$$

where $u_{i\theta}$ and u_{ir} are the tangential and radial components of the shell displacements, respectively, σ_i is Poisson's ratio, E_i is Young's modulus, h_i is the shell thickness, R_i is the radius of the shell neutral surface, Δp_i is the pressure difference on both sides of each of the two shells, ρ_{0i} is the shell material density and ω is the vibration angular frequency of the sound source.

Because of axial symmetry, $u_{i\theta} = 0$ when $\theta = 0$ and π and $r = R_i$ $(i = 1, 2)$.

According to the boundary conditions for the two hemispheres, each being freely supported by the other, $u_{i\theta} = 0$, and $\theta = \frac{\pi}{2}$, $r = R_i$.

Equations (1) are to be solved under the above restrictions on the tangential components, $u_{i\theta}$, of the displacement vector. The sound field in each one of the regions I, II and III (see Fig. 2), has to satisfy Helmholtz's equation

$$\nabla^2 p_i(r) + k^2 p_i(r) = \begin{cases} \delta(r-\rho) + \delta(r-\rho_1), & r \in III, \\ \qquad 0, & r \in II, \quad r \in I, \end{cases}$$

where p_i is the sound pressure in the i-th region, $k = 2\pi/\lambda$ is the wave number, ρ_1 is the radius vector of the imaginary source P_1 $(P_1(\rho_1, \pi))$ and $\delta(r)$ is Dirac's delta function.

The usual way to solve this problem is by direct substitution of the expressions for the acoustic fields into the equation of the shell motion [**5, 10, 11**]. For this, it is necessary to use the boundary conditions at $r = R_i$, that is, the continuity of the radial velocity of the medium particles under sound propagation through the shell from one region to another and the equality of the vibration velocities of the medium particles and shell. This method of solution gives rise to a large system of algebraic equations, which is difficult to solve analytically.

A more suitable procedure is to determine the mechanical impedances of the shells from the equations of the shell motion and then simplify the corresponding system of algebraic equations by splitting it into two smaller systems. For these systems, the boundary conditions are given in the following form [**11**]:

$$\begin{aligned} &\frac{1}{i\omega\rho_a}\frac{\partial p_1}{\partial r} = \frac{1}{i\omega\rho_a}\frac{\partial p_2}{\partial r}, \qquad p_1 - p_2 = \frac{\partial u_{1r}}{\partial t}z_1, \qquad \text{if } r = R_1, \\ &\frac{1}{i\omega\rho_a}\frac{\partial p_2}{\partial r} = \frac{1}{i\omega\rho_a}\frac{\partial p_3}{\partial r}, \qquad p_2 - p_3 = \frac{\partial u_{2r}}{\partial t}z_2, \qquad \text{if } r = R_2, \end{aligned} \tag{2}$$

where $z_{1,2}$ are the mechanical impedances of the shells and ρ_a is the air density. These conditions are typical of piston-like shell vibrations and they determine the sound generation in the spectrum of sound waves.

By expanding the shell pressure and the shell displacements into angular functions satisfying the restrictions for shell vibrations and substituting them into equation (1), we can express the mechanical shell impedances of each vibration mode as follows:

$$z_{in} = i\omega M_i \left\{ 1 - \frac{n(n+1)\left(1-\sigma_i^2\right) - 2(1+\sigma_i) - 2\left(1-\sigma_i^2\right)}{N_i\left[n(n+1) - N_i - (1-\sigma_i)\right]} \right\}, \tag{3}$$

where $n = 2k$, $k = 0, 1, 2, \ldots$, and $i = 1, 2$. It is to be noted that for odd vibration modes ($n = 2k + 1$), $z_{ni} \to \infty$, i.e. a shell freely supported on a rigid foundation acts as an absolutely rigid body. In equation (3), $M_i = \rho_i h_i$ is the running shell mass and $N_i = (1 - \sigma_i^2) R_i^2 \rho_i \omega^2 / E_i$.

3. Sound propagation through two elastic shells

Now we consider the sound propagation through two elastic shells with different mechanical properties.

By taking (2) into account, we represent the sound field in region III (see Fig. 2) as:

$$\begin{aligned} p_3 &= \sum_{n=0}^{\infty} A_n P_n(\cos\theta) J_n(kr) \\ &\quad + ik \sum_{n=0}^{\infty} (2n+1)\left[1 + (-1)^n\right] P_n(\cos\theta) \begin{cases} J_n(kr) H_n(k\rho), & r < \rho, \\ J_n(kr) H_n(kr), & r > \rho, \end{cases} \end{aligned} \tag{4}$$

where J_n and H_n are Bessel's and Hankel's functions of order n, respectively, and P_n are Legendre's polynomials of degree n.

The sound pressure in region II is represented as:

$$p_2 = \sum_{n=0}^{\infty} \left[B_n J_n(kr) + C_n Y_n(kr)\right] P_n(\cos\theta),$$

where $Y_n(kr)$ are Neumann's function of order n.

In region I, the general solution of the wave equation which satisfies the radiation condition with the time factor $\exp(-i\omega t)$, is [**11**]

$$p_1 = \sum_{n=0}^{\infty} D_n P_n(\cos\theta) H_n(kr). \tag{5}$$

The unknown coefficients A_n, B_n, C_n, and D_n are determined from the following boundary conditions:

$$\begin{cases} \dfrac{\partial p_{1n}}{\partial r} = \dfrac{\partial p_{2n}}{\partial r}, \quad \dfrac{p_{1n} - p_{2n}}{z_{1n}} = \dfrac{1}{i\omega\rho_a}\dfrac{\partial p_{1n}}{\partial r}, & \text{if } r = R_1, \\ \dfrac{\partial p_{2n}}{\partial r} = \dfrac{\partial p_{3n}}{\partial r}, \quad \dfrac{p_{2n} - p_{3n}}{z_{2n}} = \dfrac{1}{i\omega\rho_a}\dfrac{\partial p_{3n}}{\partial r}, & \text{if } r = R_2, \end{cases} \tag{6}$$

where the subscript n to a function indicates that it is evaluated at the n-th vibration mode.

Since the impedances z_{in}, for n odd, go to infinity, the system of boundary conditions is written as:

$$(7)\qquad \begin{cases} \dfrac{\partial p_{1n}}{\partial r} = \dfrac{\partial p_{2n}}{\partial r}, \quad \dfrac{p_{1n} - p_{2n}}{z_{1n}} = \dfrac{1}{i\omega\rho_a}\dfrac{\partial p_{1n}}{\partial r}, & \text{if } r = R_1, \\ \dfrac{\partial p_{2n}}{\partial r} = \dfrac{\partial p_{3n}}{\partial r}, \quad \dfrac{p_{2n} - p_{3n}}{z_{2n}} = \dfrac{1}{i\omega\rho_a}\dfrac{\partial p_{3n}}{\partial r}, & \text{if } r = R_2, \end{cases}$$

for $n = 2k$, $k = 0, 1, 2, \dots$, and

$$(8)\qquad \begin{cases} \dfrac{\partial p_{1n}}{\partial r} = \dfrac{\partial p_{2n}}{\partial r}, \quad \dfrac{\partial p_{1n}}{\partial r} = 0, & \text{if } r = R_1, \\ \dfrac{\partial p_{2n}}{\partial r} = \dfrac{\partial p_{3n}}{\partial r}, \quad \dfrac{\partial p_{3n}}{\partial r} = 0, & \text{if } r = R_2, \end{cases}$$

for $n = 2k + 1$.

The substitution of relations (4)–(5) into the boundary conditions (8) produces a homogeneous system of algebraic equations which has only the trivial solution for all the unknown coefficients. For the sound field in region I we assume that $D_n = 0$, for $n = 2k + 1$, $k = 0, 1, 2 \dots$

Substituting the expressions (4)–(5) into the boundary conditions (7) with $n = 2k$, $k = 0, 1, 2, \dots$, we obtain the following system of algebraic equations for the unknowns A_n, B_n, C_n, D_n:

$$(9)\qquad \begin{aligned} &D_n H_n'(kR_1) - B_n J_n'(kR_1) - C_n H_n'(kR_1) = 0, \\ &D_n\Big[H_n(kR_1) + \frac{k}{i\omega\rho_a} H_n'(kR_1) z_{1n}\Big] - B_n J_n(kR_1) - C_n H_n(kR_1) = 0, \\ &B_n J_n'(kR_2) + C_n H_n'(kR_2) - A_n J_n'(kR_2) = 2ik(2n+1) J_n(kR_2) H_n'(kR_2), \\ &B_n J_n(kR_2) + C_n H_n(kR_2) + A_n\Big[\frac{k}{i\omega\rho_a} J_n'(kR_2) z_{2n} - J_n(kR_2)\Big] \\ &\qquad = 2ik(2n+1) J_n(k\rho)\Big[H_n(kR_2) - \frac{k}{i\omega\rho_a} H_n'(kR_2) z_{2n}\Big]. \end{aligned}$$

In particular, for the coefficient D_n we get

$$(10)\qquad D_n = \frac{D_1}{D},$$

where

$$(11)\qquad \begin{aligned} D = J_n'(kR_2) H_n'(kR_1) \frac{\Delta}{(i\rho_a c)^2} \Big\{ &\Big[z_{2n} + z_S^{(1)}\Big] i\rho_a c \frac{\Delta_1}{\Delta} + \Big[z_{1n} + z_S^{(2)}\Big] i\rho_a c \frac{\Delta_2}{\Delta} \\ &+ \Big[z_{1n} + z_S^{(2)}\Big]\Big[z_{2n} + z_S^{(1)}\Big] + i\rho_a c\Big(i\rho_a c \frac{\Delta_3}{\Delta}\Big)\Big\}, \end{aligned}$$

$$(12)\qquad D_1 = -\frac{2k(2n+1) J_n(k\rho)}{(kR_1)^2 (kR_2)^2},$$

and

$$(13)\qquad \begin{aligned} H_n(kR_2) J_n'(kR_2) - H_n'(kR_2) J_n(kR_2) &= \frac{1}{i(kR_2)^2}, \\ J_n(kR_1) Y_n'(kR_1) - Y_n'(kR_1) J_n(kR_1) &= -\frac{1}{i(kR_1)^2}. \end{aligned}$$

Here c is sound speed in air, z_{2n} and z_{1n} are the mechanical impedance of the internal and external shells, respectively, and

$$z_S^{(1)} = -i\rho_a c\frac{J_n(kR_2)}{J_n'(kR_1)}, \qquad z_S^{(2)} = i\rho_a c\frac{H_n(kR_1)}{H_n'(kR_1)} \qquad \text{and} \qquad z_S^{(3)} = i\rho_a c\frac{\Delta_1}{\Delta}$$

are the radiation resistances of, respectively, the internal shell into the internal region III, the external shell into the external region I and the external shell into region II, provided the surface of the internal shell is absolutely rigid. Here

$$\begin{aligned}
\Delta \; &= J_n'(kR_1)Y_n'(kR_2) - J_n'(kR_2)Y_n'(kR_1),\\
\Delta_1 &= J_n'(kR_2)Y_n(kR_1) - J_n(kR_1)Y_n'(kR_2),\\
\Delta_2 &= J_n'(kR_1)Y_n(kR_2) - J_n(kR_2)Y_n'(kR_1),\\
\Delta_3 &= J_n'(kR_2)Y_n(kR_1) - J_n(kR_1)Y_n'(kR_2).
\end{aligned}$$

Finally we can represent the sound transparency $\tau = p_1/p_0$ through the coefficients D_n, where p_0 and p_1 are the sound pressures with and without the casing, respectively.

4. Sound source placed at the origin of the coordinate system

Now we consider the case where the sound source is placed at the origin of the coordinate system ($\rho = 0$). In this case, with the notation

$$\begin{aligned}
W = \left[z_{20} + z_s^{(1)}\right] i\rho_a c\Delta_1 + \left[z_{10} + z_s^{(2)}\right] i\rho_a c\Delta_2 \\
+ \left[z_{10} + z_s^{(2)}\right]\left[z_{20} + z_s^{(1)}\right]\Delta + i\rho_a(i\rho_a c)\Delta_3,
\end{aligned}$$

the sound pressure, p_1, is

$$p_1 = \frac{2ik\rho_a c(-i\rho_a c)H_0(kr)}{\Delta_4 W}, \tag{14}$$

where

$$\begin{aligned}
\Delta \; &= J_1(kR_1)Y_1(kR_2) - J_1(kR_2)Y_1(kR_1),\\
\Delta_1 &= J_0(kR_1)Y_1(kR_2) - J_1(kR_2)Y_0(kR_1),\\
\Delta_2 &= J_0(kR_2)Y_1(kR_1) - J_1(kR_1)Y_0(kR_2),\\
\Delta_3 &= J_0(kR_2)Y_0(kR_1) - J_0(kR_1)Y_0(kR_2),\\
\Delta_4 &= (kR_1)^2(kR_2)^2 J_1(kR_2)H_1(kR_1),
\end{aligned}$$

$$z_s^{(1)} = i\rho_a c\,\frac{J_0(kR_2)}{J_1(kR_2)}, \qquad z_s^{(2)} = i\rho_a c\,\frac{H_0(kR_1)}{H_1(kR_1)},$$

and z_{10} and z_{20} are determined by formula (3) for appropriate values of the parameters h_i, E_i, σ_i, R_i and ρ_i.

In the case $z_{20} = 0$, i.e. for the resonance of the internal shell, the pressure of the passing sound wave is determined by the properties of the external shell. In the case $z_{10} = z_{20} = 0$, the sound energy propagates freely, as if both shells were absent. Because the resonance frequency ω_0 substantially depends upon the radius of the sphere ($\omega \sim 1/R$) [**13**], the equality of mechanical impedances can be easily avoided.

We consider two extreme cases: $R_1 \approx R_2$ and $R_1 \gg R_2$.

In the first case, the radii of the shells are practically equal, i.e. $R_1 - R_2 = \varepsilon \ll 1$. Since

$$\Delta = O(\varepsilon), \qquad \Delta_3 = O(\varepsilon), \qquad \Delta_1 = \Delta_2 = -\frac{1}{(k\bar{R})^2} + O(\varepsilon),$$

where $\bar{R} \approx R_i$, the pressure p_1 is

$$p_1 = \frac{2ik\rho_a c}{(k\bar{R})^2} \frac{H_0(kr)}{J_1(k\bar{R})H_1(k\bar{R})(z_{10} + z_{20}) + \frac{\rho_a c}{(k\bar{R}_1)^2}} + O(\varepsilon). \tag{15}$$

Thus, if the difference between the radii of the shells is small, the entire mechanical impedance is equal to the sum of the impedances of the shells. It is evident that for n shells for which $R_i \approx R_2$, sound-proofing is multiplied by n.

In the second case, $R_1 \gg R_2$, a detailed analysis with an estimate for each term in (15) is difficult, because this case practically does not differ from the general case described by formula (14). Nevertheless it is clear, in this case, that the resonance frequency of the external shell, of radius R_1, is significantly lower than the resonance frequency of the internal shell, of radius R_2. Hence, the two equalities, $z_{10} = 0$ and $z_{20} = 0$, cannot hold simultaneously. When $z_{20} = 0$, the impedance of the external shell is controlled by the mass, and if the shell thickness h_1 is large enough, the external shell has high sound-proofing efficiency.

As the frequency of the source is reduced, the mechanical impedance, z_{10}, of the external shell goes to 0, while the impedance of the internal shell increases to the same order as $1/\omega$. In this case, sound-proofing is provided only by the internal shell. If the frequency is less than the resonance frequencies of both shells, the mechanical shell resistance goes up sharply and the efficiency of the sound attenuation may be infinitely high.

For real structures, the most probable case is $\omega > (\omega_0)_2$, where $(\omega_0)_2$ is the resonance frequency of the internal shell. In this case, the impedances of both shells are controlled by their mass. In general, in this case the sound-proofing effect is realized by the external shell, since $\omega \gg (\omega_0)_1$, where $(\omega_0)_1$ is the resonance frequency of the external shell.

To determine the resonance frequencies of the vibrations of the medium in the region between the shells, it is necessary to solve the system of transcendental equations:

$$\begin{cases} \Delta \ = 0, \\ \Delta_1 = 0, \\ \Delta_2 = 0, \\ \Delta_3 = 0. \end{cases}$$

However, it is easier to use numerical calculations directly with formula (14), and vary, if necessary, the values of kR_1 and kR_2. The modulus, $|\tau|$, of the coefficient of the sound transparency under the conditions $kR_1 \ll 1$ and $kR_2 \ll 1$ (which, for example, are always satisfied for steel and air) is

$$|\tau| \approx \frac{(\rho_a c)^2}{|\cos kR_2|\sqrt{A^2 + B^2}}, \tag{16}$$

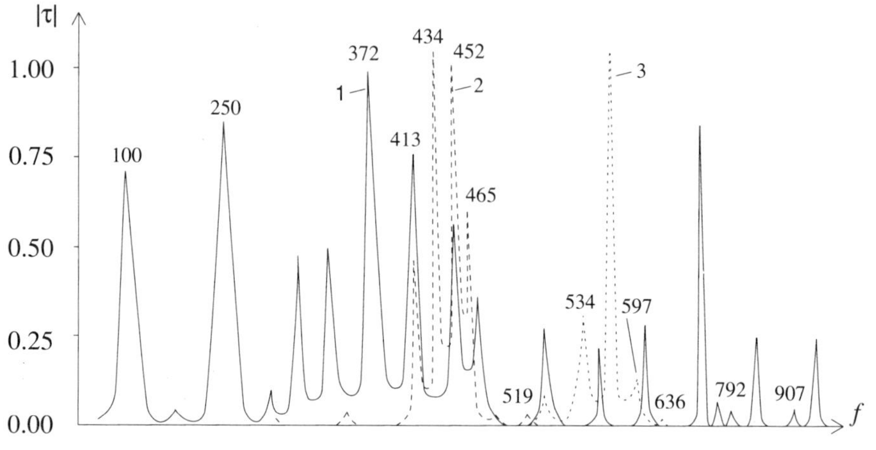

FIGURE 3. Effect of frequency, f, on coefficient of sound transparency of a two-shell casing.

where

$$
\begin{aligned}
A &= \left[\sin k(R_2 - R_1)\Re 1 - \cos k(R_2 - R_1)\Re 2\right],\\
B &= \left[\sin k(R_2 - R_1)\Im 1 - \cos k(R_2 - R_1)\Im 2\right],\\
\Re 1 &= \left(\omega a_1 - \frac{b_1}{R_1^2}\right)\left(\frac{b_2}{r_2^2\omega}\right) + \rho_a c\tan(kR_2)\left(\frac{b_1}{R_1^2\omega} - a_1\omega\right) + (\rho_a c)^2,\\
\Re 2 &= \rho_a c\left[\omega(a_1 + a_2) - \frac{1}{\omega}\left(\frac{b_2}{R_2^2} + \frac{b_1}{R_1^2}\right)\right] + (\rho_a c)^2\tan(kR_2),\\
\Im 1 &= \rho_a c\left(\frac{b_2}{R_2^2} - a_2\omega\right) - (\rho_a c)^2\tan(kR_2),\\
\Im 2 &= (\rho_a c)^2,
\end{aligned}
$$

and

$$
a_i = \rho_{0i}h_i, \qquad b_i = \frac{2h_iE_i}{1-\sigma_i}, \qquad i = 1, 2. \tag{17}
$$

Resonance of the inner region III occurs if $\cos kR_2 = 0$. The additional condition of the asymptotic representation of the spherical functions (see [**9**]) provides the equality $kR_2 = \pi\left(\frac{1}{2} + m\right)$, where $m = 8, 9, 10, \ldots$.

The method used here to evaluate the coefficient of sound transparency is suitable, in principle, for calculating the efficiency of a casing consisting of a large number of shells.

Studies have shown that the installation of a second shell under certain conditions may increase the sound-proofing efficiency over the entire frequency interval. In the precise solution of the problem of sound transmission through the system of shells obtained above, the interaction of both shells has been taken into account, and this yielded complicated conditions for shells vibrations. This interaction is controlled by the gap between the shells and is connected with the change in the radiation resistances of both shells working together in the system, and these differ from the radiation resistance of a single shell.

5. Numerical results

To illustrate formula (16) numerical results of the sound-proofing effect of spherical hemispheres are shown in Fig. 3.

The effect of the frequency on the coefficient of sound transparency is illustrated for different values of the geometric parameters R_i and h_i:

a) by curve 1 for $R_1 = 1.5$ m, $R_2 = 0.35$ m, $h_1 = h_2 = 0.0015$ m,
b) by curve 2 for $R_1 = 1.5$ m, $R_2 = 0.35$ m, $h_1 = 0.0015$ m, $h_2 = 0.010$ m,
c) by curve 3 for $R_1 = 0.6$ m, $R_2 = 0.35$ m, $h_1 = 0.015$ m, $h_2 = 0.010$ m.

A numerical experiment showed, as expected, the resonant character of the effect of frequency on the coefficient of sound transparency. At low frequencies, the sound transmitted is controlled by the resonance of the internal volume limited by the shells. At high frequencies, the influence of the thickness of each shell is significant.

We can optimize the geometric dimensions of the shells to provide maximum sound-proofing at prescribed frequencies. By properly choosing the distance between the two shells so that there is no resonance from region II to region III under the resonance of region III, it is clear that the installation of the second shell sharply increases the sound-proofing efficiency over a wide frequency interval. It is easy to determine the optimal ratio of the shells' dimensions, for which the resonances of the regions III and III $\cup$ II are separated from each other.

Indeed at resonance of region III the following equality is valid

$$kR_2 = \pi\left(\frac{1}{2} + m\right),$$

where m is the resonance number in region III. For the same frequency, if $\cos kR_1 = \pm 1$, then $kR_1 = \pi n$. Here n is the antiresonance number in the region II $\cup$ III.

The numbers m and n are connected only by the inequality $n \geq m + 1$, since $R_2 > R_1$. Hence we get the following relations [**9**]:

$$\frac{R_1}{R_2} = \frac{n}{m + \frac{1}{2}}, \qquad m, n = 8, 9, 10, \ldots, \qquad n \geq m + 1, \tag{18}$$

which do not depend upon the frequency. Thus, at the resonance frequencies of region III, if conditions (18) are fulfilled, there are antiresonances of region III $\cup$ II and the sound energy will be "locked up" in the region between the shells.

For the ratios of the external to the internal shells' dimensions (18), we consider two cases:

(1) Resonance in the region III $\cup$ II, i.e. $kR_1 = \pi\left(\frac{1}{2} + n_1\right)$, where n_1 is the resonance number of the region II $\cup$ III. In this case,

$$kR_2 = \pi\left(m + \frac{1}{2}\right)\frac{1 + 2n_1}{2n},$$

then it follows that resonance of region III does not occur for any value of n_1 and n, because $(1 + 2n_1)/(2n)$ cannot be an odd integer. Since $n_1 \to \infty$ as $n \to \infty$, the resonance density becomes so large that the number one in the numerator cannot be neglected.

(2) Antiresonance in region III, i.e. $kR_2 = \pi m_1$, where m_1 is the antiresonance number in region III. In this case, the relation

$$kR_1 = \pi n \frac{2m_1}{2m+1}$$

indicates that there is no antiresonance in the region II ∪ III. However, this does not lead to the decay of the construction efficiency. If region III is "locked up", i.e. is in antiresonance, the sound-proofing of the casing is practically equal to zero.

6. Conclusion

The vibration regime of the shells is determined by the resonances of the interconnected internal volumes and the mechanical resistances of the shells. The interaction of the shells is regulated by the gap between the shells and connected with the changes in the radiation resistances of each of the shells.

References

1. S. M. Bauer, S. B. Filippov, A. L. Smirnov and P. E. Tovstik, *Asymptotic methods in mechanics with application to thin shells and plates*, Asymptotic Methods in Mechanics (A. L. Smirnov and R. Vaillancourt, eds.), CRM Proceedings and Lecture Notes, this volume, Amer. Math. Soc., Providence, R.I., 1993, pp. 3–140.
2. R. D. Doolittle and H. Uberall, *Sound scattering by elastic cylindrical shells*, JASA **39** (1966), no. 2, 272–275.
3. A. L. Gol′denveizer, *Theory of elastic thin shells* (G. Herrmann, ed.), Pergamon Press, New York, Oxford, London, Paris, 1961 (658 pp.).
4. R. D. Goodman and R. Stern, *Transmission of sound by elastic spherical shells*, JASA **34** (1962), no. 3, 338–344.
5. V. T. Grinchenko and Ch. V. Vovk, *Wave problems of sound dispersion on elastic shells*, "Naukova Dumka", Kiev, 1986 (240 pp.). (Russian)
6. H. Hönl, A. W. Maue, and K. Westphahl, *Theorie der Beugung*, Handbuch der Physik, Band 25/1, Springer-Verlag, Berlin, 1961, pp. 218–253. (German)
7. R. Hickling, *Analysis of echoes from a solid elastic sphere in water*, JASA **34** (1962), no. 10, 1582–1592.
8. ———, *Analysis of echoes from a hollow metallic sphere in water*, JASA **36** (1964), no. 6, 1124–1137.
9. S. N. Rzhevkin, *A course of lectures on the theory of sound*, Moscow University Press, Moscow, 1970 (335 pp.). (Russian)
10. E. L. Shenderov, *Sound radiation and dispersion*, Sudostroenie, Leningrad, 1989 (302 pp.). (Russian)
11. Yu. P. Shcheviev, *Some diagnosis problems of acoustic methods*, Leningrad University Press, Leningrad, 1977 (150 pp.). (Russian)
12. ———, *Analysis and synthesis of nonuniform acoustic media*, Leningrad University Press, Leningrad, 1984 (215 pp.). (Russian)
13. E. Skuchik, *The foundation of acoustics*, Mir, Moscow, 1976 (542 pp.). (Russian)

ABSTRACT. The diffraction problem of elastic waves on a system of thin elastic shells is considered. The mechanism of sound propagation through a two-layer system is studied. The sound-proofing effect of a two-shell casing is estimated. To illustrate the formulae obtained, numerical results for the sound-proofing effect of spherical hemispheres are considered in detail.

DEPARTMENT OF ACOUSTICS, INSTITUTE OF CINEMA ENGINEERING, 13 PRAVDA ST., 196126, ST. PETERSBURG, RUSSIA

E-mail address: als@hq.math.lgu.spb.su

Centre de Recherches Mathématiques
CRM Proceedings and Lecture Notes
Volume 3, 1993

Dynamic Stability and Forced Vibrations of a Horizontal Rotor with a Cracked Shaft

Yury B. Shneerson

1. Introduction

In recent years, the analysis of the dynamics of rotors with transverse cracks has attracted a great deal of attention (see [**2, 3, 5, 8, 9**]), because it yields a very simple method for an early diagnosis of a crack. The analysis of the mathematical model of this system is complicated by two factors. Firstly, a cracked shaft has no elastic symmetry and secondly, the asymmetry is changing with time as the crack opens and closes (see Figs. 1, 2, 3).

These factors prevent a reduction of the problem to differential equations with constant coefficients. Moreover, the system of differential equations to be analyzed has coefficients which are not sinusoidal but consist of several harmonics.

In this paper, dynamically unstable zones and amplitudes of forced vibrations of a stationary rotating cracked shaft are analyzed. The asymptotic multi-scale method [**4**] is applied to the system of differential equations with variable coefficients. This method has been used in previous papers [**6, 7**] for the investigation of asymmetric rotors mounted in asymmetric elastic supports with damping.

2. Equations of motion

We consider a weightless horizontal shaft of length l, mounted in two rigid supports, and rotating with angular velocity ω. The concentrated mass, m, is positioned on the shaft at equal distance from the supports (see Fig. 1). We assume that the shaft has a small transverse crack which alternately opens and closes during the rotation (see Figs. 2 and 3).

We introduce rotating coordinates (ξ, η, ζ), attached to the shaft, and fixed coordinates (x, y, z). The ξ-axis is directed from the center of the cracked cross-section towards the crack. The η-axis is orthogonal to the ξ-axis and lies in the same cross-section. The ζ-axis is parallel to the nondeformed shaft line. The x- and y-axes are horizontal and vertical, respectively, and the z-axis coincides with the ζ-axis.

We suppose that the dynamic displacements of the shaft are much smaller than its static deflection. Hence, we may assume the shaft stiffness coefficients c_ξ and c_η, which are connected with the (ξ, ζ)- and (η, ζ)-plane deformations, respectively, to be harmonic functions of time. Therefore, the coefficients may be written in the

---

1991 *Mathematics Subject Classification.* Primary: 73H10; Secondary: 34E05, 70K28.
Key words and phrases. Asymptotic methods, dynamic stability, parametric resonance.
This is the final form of the paper

1065-8580/93 \$1.00 + \$.25 per page

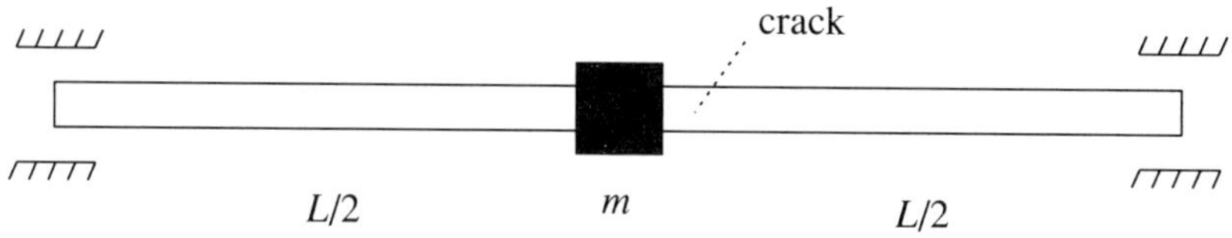

FIGURE 1. Mathematical model of a cracked shaft, with concentrated mass m.

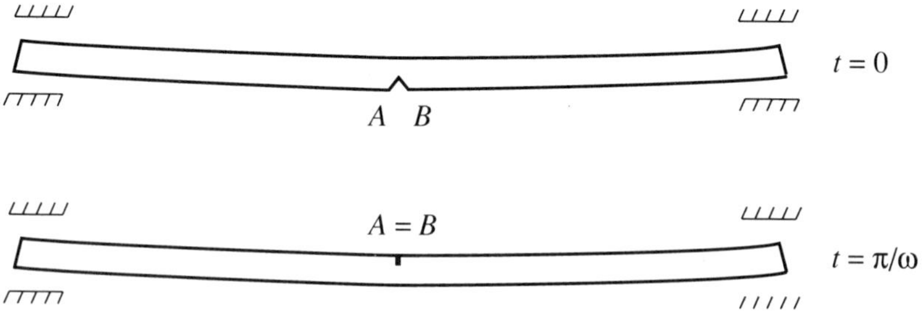

FIGURE 2. Behaviour of crack edges during rotation.

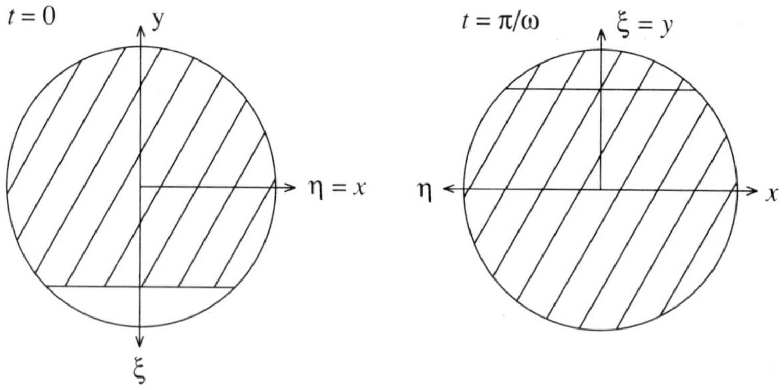

FIGURE 3. Shaft cross-section containing crack during rotation.

form [**8**]:

$$c_\xi = \frac{c_0}{2}(1 - \cos \omega t) + \frac{c_1}{2}(1 + \cos \omega t),$$
$$c_\eta = \frac{c_0}{2}(1 - \cos \omega t) + \frac{c_2}{2}(1 + \cos \omega t), \tag{1}$$

where $c_0 > c_1 > c_2$ are constant.

When the crack is closed ($t = k\pi/\omega$, where $k = 1, 3, 5, \dots$), c_ξ and c_η attain their maximum values which are equal to c_0. Therefore we can consider c_0 as the static stiffness coefficient of the uncracked beam with hinged supports. This coefficient may be evaluated using the one-dimensional beam theory (Kirchhoff hypotheses) or any of the more complex beam models (see the Survey [**1**], Sections 2.6 and 4.5).

When the crack is open ($t = k\pi/\omega$, where $k = 0, 2, 4, \dots$), c_ξ and c_η attain their minimum values c_1 and c_2, respectively. The values of c_1 and c_2 depend on the size and form of the crack. To find these values one has to apply the three-dimentional elasticity theory and numerical methods (e.g. finite element methods). Sometimes, it is easier to obtain these values from experiment.

In this paper, we assume that the effect of the crack size on the values c_1 and c_2 is known theoretically and/or experimentally. We consider a shaft with a small crack and suppose that c_0, c_1 and c_2 are known. Since the crack is small, we assume that the relative differences between these coefficients are small parameters.

To derive the equation of motion, we write the formulae for the kinetic and potential energies of the system. The kinetic energy may be written directly by means of the non-rotating coordinates. The potential energy is derived initially in the rotating (ξ, η, ζ)-coordinates, and then converted to the non-rotating coordinates:

$$\xi = x\cos\omega t + y\sin\omega t,$$
$$\eta = -x\sin\omega t + y\cos\omega t.$$

Now we apply Lagrange's equations which lead to the equation of motion (similar to that of [**6**]). In complex form it is

$$(2)\quad \frac{d^2w}{dt^2} + \alpha\omega_0\frac{dw}{dt} + \omega_0^2 w + \gamma\omega_0^2\left(e^{i\Omega\omega_0 t} + e^{-i\Omega\omega_0 t}\right)w - \delta\omega_0^2\left(e^{i\Omega\omega_0 t} - 2e^{2i\Omega\omega_0 t} + e^{3i\Omega\omega_0 t}\right)\overline{w} = -ig,$$

where

$$w = u + iv, \qquad \Omega = \frac{\omega}{\omega_0}, \qquad \omega_0 = \sqrt{\frac{c}{m}}, \qquad \alpha = \frac{r}{\sqrt{mc}},$$
$$c = \frac{2c_0 + c_1 + c_2}{4}, \qquad \gamma = \frac{2c_0 - c_1 - c_2}{8c}, \qquad \delta = \frac{c_1 - c_2}{8c}.$$

Here u and v are the displacements of the concentrated mass in the positive x- and y-directions, respectively, and g is the acceleration due to gravity.

3. Asymptotic analysis

We analyze equation (2) by applying the asymptotic multi-scale method [**4**]. We assume the crack size and the damping to be small, i.e.

$$\delta \ll 1, \qquad \gamma \ll 1, \qquad \alpha \ll 1.$$

We use the small parameter

$$\mu = \max\{\alpha, \delta, \gamma\}$$

to write

$$\gamma = \mu\Gamma, \qquad \delta = \mu\Delta, \qquad \alpha = \mu R,$$

where Γ, Δ and R are values of the asymptotic order $O(\mu^0)$. Thus γ, δ and α are of the order $O(\mu^1)$. For greater convenience, we take an arbitrary value Ω_0 and consider values of Ω in the neighbourhood of Ω_0. Therefore, we can assume that

$$\Omega = \Omega_0(1 + \mu\beta)$$

and introduce the new variable

$$\tau = (1 + \mu\beta)\omega_0 t.$$

If we keep only the terms of the asymptotic orders $O(\mu^0)$ and $O(\mu^1)$, then (2) becomes:

$$\frac{d^2w}{d\tau^2} + \mu R\frac{dw}{d\tau} + (1 - 2\mu\beta)w + \mu\Gamma\left(e^{i\Omega\tau} + e^{-i\Omega\tau}\right)w - \mu\Delta\left(e^{i\Omega\tau} - 2e^{2i\Omega\tau} + e^{3i\Omega\tau}\right)\overline{w} = -ig\,\frac{1 - 2\mu\beta}{\omega_0^2}. \tag{3}$$

The unknown variable w may be expanded into the asymptotic series

$$w = w_0 + \mu w_1 + \mu^2 w_2 + \cdots \tag{4}$$

Using the multi-scale method, we can consider each w_i as a function of the parameters τ_0, τ_1, τ_2 etc., where $\tau_0 = \tau$, $\tau_1 = \mu\tau$, $\tau_2 = \mu^2\tau$, etc. Applying this method, we substitute expression (4) into equation (3), separate all terms of the same order and denote $D_i = \frac{\partial}{\partial\tau_i}$. Taking into account only the terms of the orders $O(\mu^0)$ and $O(\mu^1)$, we obtain the system of two equations:

$$D_0^2 w_0 + w_0 = -\frac{ig}{\omega_0^2}, \tag{5}$$

$$D_0^2 w_1 + w_1 + 2D_0D_1w_0 + 2\beta w_0 + RD_0w_0 - \Delta\overline{w}_0\left(e^{i\Omega_0\tau_0} - 2e^{2i\Omega_0\tau_0} + e^{3i\Omega_0\tau_0}\right) + \gamma w_0\left(e^{i\Omega_0\tau_0} + e^{-i\Omega 0\tau_0}\right) = -2\,\frac{i\beta g}{\omega_0^2}, \tag{6}$$

where w_0 may be found from (5):

$$w_0 = -\frac{ig}{\omega_0^2} + A(\tau_1)\,e^{i\tau_0} + B(\tau_1)\,e^{-i\tau_0}, \tag{7}$$

and the functions $A(\tau_1)$ and $B(\tau_1)$ are yet unknown.

When we substitute (7) into (6), we assume that the coefficients in the terms including $\exp(i\tau_0)$ and $\exp(-i\tau_0)$ are equal to zero. This assumption [**4**] is necessary as we seek an approximate solution of (3) in the form (4). Equating to zero the mentioned coefficients, we obtain a system of four ordinary differential equations with constant coefficients:

$$2\frac{d}{d\tau_1}\begin{bmatrix} A_1 \\ A_2 \\ B_1 \\ B_2 \end{bmatrix} + \begin{bmatrix} c_{11} & c_{12} & c_{13} & c_{14} \\ c_{21} & c_{22} & c_{23} & c_{24} \\ c_{31} & c_{32} & c_{33} & c_{34} \\ c_{41} & c_{42} & c_{43} & c_{44} \end{bmatrix}\begin{bmatrix} A_1 \\ A_2 \\ B_1 \\ B_2 \end{bmatrix} + \begin{bmatrix} d_1 \\ d_2 \\ d_3 \\ d_4 \end{bmatrix} = 0, \tag{8}$$

where $A = A_1 + iA_2$, $B = B_1 + iB_2$. The coefficient matrix $C = (c_{ij})$ and vector $\mathbf{d} = (d_i)$ may be easily evaluated. These are as follows.

If $\Omega_0 = 2$:

$$C = \begin{bmatrix} R & -2\beta + \Delta & 0 & \Gamma \\ 2\beta + \Delta & R & -\Gamma & \\ 0 & N & R & -2\beta \\ \Gamma & 0 & -2\beta & R \end{bmatrix}, \qquad \mathbf{d} = \begin{bmatrix} 0 \\ 0 \\ 0 \\ 0 \end{bmatrix}. \tag{9}$$

If $\Omega_0 = 1$:

$$(10)\quad C = \begin{bmatrix} R & -2\beta - 2\Delta & 0 & 0 \\ 2\beta - 2\Delta & R & 0 & 0 \\ 0 & 0 & R & 2\beta \\ 0 & 0 & -2\beta & R \end{bmatrix}, \qquad \mathbf{d} = \begin{bmatrix} 0 \\ 0 \\ -g(\Delta + \Gamma)/\omega_0^2 \\ 0 \end{bmatrix}.$$

If $\Omega_0 = \frac{2}{3}$:

$$(11)\quad C = \begin{bmatrix} R & -2\beta & 0 & 0 \\ 2\beta & R & 0 & 0 \\ 0 & 0 & R & 2\beta \\ 0 & 0 & -2\beta & R \end{bmatrix}, \qquad \mathbf{d} = \begin{bmatrix} 0 \\ 0 \\ R \\ 0 \end{bmatrix}.$$

If $\Omega_0 = \frac{1}{2}$:

$$(12)\quad 2C = \begin{bmatrix} R & -2\beta + \Delta & 0 & \Gamma \\ 2\beta + \Delta & R & -\Gamma & 0 \\ 0 & N & R & -2\beta \\ \Gamma & 0 & -2\beta & R \end{bmatrix}, \qquad \mathbf{d} = \begin{bmatrix} g\Delta/\omega_0^2 \\ 0 \\ 0 \\ 0 \end{bmatrix}.$$

If $\Omega_0 = \frac{1}{3}$:

$$(13)\quad C = \begin{bmatrix} R & -2\beta & 0 & 0 \\ 2\beta & R & 0 & = 0 \\ 0 & 0 & R & 2\beta \\ 0 & 0 & -2\beta & R \end{bmatrix}, \qquad \mathbf{d} = \begin{bmatrix} 0 \\ -g/\omega_0^2 \\ 0 \\ 0 \end{bmatrix}.$$

For the other cases we obtain:

$$(14)\quad C = \begin{bmatrix} R & -2\beta & 0 & 0 \\ 2\beta & R & 0 & 0 \\ 0 & 0 & -R & 2\beta \\ 0 & 0 & -2\beta & R \end{bmatrix}, \qquad \mathbf{d} = \begin{bmatrix} 0 \\ 0 \\ 0 \\ 0 \end{bmatrix}.$$

The solution of system (8) may be written in the form:

$$(15)\quad \begin{bmatrix} A_1 \\ A_2 \\ B_1 \\ B_2 \end{bmatrix} = \begin{bmatrix} A_1^{(0)} \\ A_2^{(0)} \\ B_1^{(0)} \\ B_2^{(0)} \end{bmatrix} + \sum_{j=1}^{4} e^{\lambda^{(j)}\tau_1} \begin{bmatrix} A_1^{(j)} \\ A_2^{(j)} \\ B_1^{(j)} \\ B_2^{(j)} \end{bmatrix},$$

where $\lambda^{(1)}$, $\lambda^{(2)}$, $\lambda^{(3)}$ and $\lambda^{(4)}$, are the roots of the characteristic equation corresponding to (8). If the real part of each $\lambda^{(j)}$ is negative, then the rotation with the corresponding constant angular velocity $\omega = \Omega_0\omega_0(1 + \mu\beta)$ is stable. If any of the $\lambda^{(j)}$ has a positive real part, then the rotation is unstable.

4. Dynamic stability analysis

We study the stability of the cracked shaft corresponding to system (8) by means of the Routh–Hurwitz criterion.

4.1. In the case $\Omega_0 = 2$ (i.e. $\omega \approx 2\omega_0$), the unstable interval is

$$2\omega_0(1 - \Psi_1) < \omega < 2\omega_0(1 + \Psi_1), \tag{16}$$

where

$$\Psi_1 = \sqrt{\delta^2 + 2\gamma^2 - 2\alpha^2 + \delta(\delta^2 + \gamma^2)^{1/2}}.$$

If damping is very large and/or the crack is so small that the expression under the square root is negative, then there are no unstable intervals near $2\omega_0$.

4.2. In the case $\Omega_0 = 1$, (i.e. $\omega \approx \omega_0$), the unstable interval is

$$\omega_0(1 - \Psi_2) < \omega < \omega_0(1 + \Psi_2), \tag{17}$$

where

$$\Psi_2 = \sqrt{\delta^2 - \frac{\alpha^2}{4}}.$$

If $\alpha > 2\delta$, there are no unstable intervals near ω_0.

4.3. In the case $\Omega_0 = \frac{2}{3}$, (i.e. $\omega \approx \frac{2}{3}\omega_0$), the unstable interval is

$$\frac{2\omega_0}{3}(1 - \Psi_3) < \omega < \frac{2\omega_0}{3}(1 + \Psi_3), \tag{18}$$

where

$$\Psi_3 = \frac{1}{2}\sqrt{\delta^2 - \alpha^2}.$$

If $\alpha > \delta$, there are no unstable intervals near $2\omega_0/3$.

4.4. In all the other cases, there are no unstable intervals.

Therefore, the rotation may get unstable only if the angular velocity is close to $2\omega_0$, ω_0 or $2\omega_0/3$ and the physical parameters have such values that the corresponding quantities Ψ_1, Ψ_2, Ψ_3 are real.

5. Forced vibration analysis

We can also analyze the forced vibrations induced by the interaction of the crack with gravity, i.e. the crack may induce forced vibrations (rotor precession). These vibrations do not depend on the rotor unbalance.

The amplitudes and angular velocity of these vibrations are very important for a crack diagnosis. By considering the coefficients d_i in (9)–(14) we see that such vibrations may occur only if ω is not far from one of the values ω_0, $\omega_0/2$ or $\omega_0/3$. It should also be mentioned that it is reasonable to consider forced vibrations only in stable zones.

Applying expressions (7) and (15), we can write the formulae for the weight-induced forced vibrations of a cracked shaft as follows:

$$w_0 = -\frac{ig}{\omega_0^2} + A^{(0)}e^{i\omega t/\Omega_0} + B^{(0)}e^{-i\omega t/\Omega_0}, \tag{19}$$

where

$$A^{(0)} = A_1^{(0)} + iA_2^{(0)}, \qquad B^{(0)} = B_1^{(0)} + iB_2^{(0)}.$$

Separating the real and imaginary parts of (19), we have

$$u = \left(A_1^{(0)} + B_1^{(0)}\right)\cos\frac{\omega t}{\Omega_0} + \left(B_2^{(0)} - A_2^{(0)}\right)\sin\frac{\omega t}{\Omega_0}, \tag{20}$$

$$v = -\frac{g}{\omega_0^2} + \left(A_2^{(0)} + B_2^{(0)}\right)\cos\frac{\omega t}{\Omega_0} + \left(A_1^{(0)} - B_1^{(0)}\right)\sin\frac{\omega t}{\Omega_0}. \tag{21}$$

Hence, the possible angular velocities of the analyzed forced vibrations are:

$$\begin{array}{lll} \omega, & \text{for } \omega \text{ close to } \omega_0, & (\Omega_0 = 1), \\ 2\omega, & \text{for } \omega \text{ close to } \dfrac{\omega_0}{2}, & (\Omega_0 = \frac{1}{2}), \\ 3\omega, & \text{for } \omega \text{ close to } \dfrac{\omega_0}{3}, & (\Omega_0 = \frac{1}{3}). \end{array}$$

The amplitudes have the following values.

If $\omega \approx \omega_0$:

$$\begin{aligned} A_1^{(0)} &= -\frac{g\alpha(\delta+\gamma)}{\omega_0^2\,(4\Psi^2 - 4\delta^2 + \alpha^2)}, \\ A_2^{(0)} &= \frac{2g(\Psi-\delta)(\delta+\gamma)}{\omega_0^2\,(4\Psi^2 - 4\delta^2 + \alpha^2)}, \\ B_1^{(0)} &= \frac{g\alpha\gamma}{\omega_0^2\,(4\Psi^2 + \alpha^2)}, \\ B_2^{(0)} &= \frac{2\beta g\gamma}{\omega_0^2\,(4\Psi^2 + \alpha^2)}. \end{aligned} \tag{22}$$

If $\omega \approx \omega_0/2$:

$$\begin{aligned} A_1^{(0)} &= \frac{2g\delta\alpha}{\omega_0^2\,(4\Psi^2 + \alpha^2)}, \\ A_2^{(0)} &= -\frac{4g\delta\Psi}{\omega_0^2\,(4\Psi^2 + \alpha^2)}, \\ B_1^{(0)} &= 0, \\ B_2^{(0)} &= 0. \end{aligned} \tag{23}$$

If $\omega \approx \omega_0/3$:

$$\begin{aligned} A_1^{(0)} &= -\frac{g\delta\alpha}{\omega_0^2\,(4\Psi^2 + \alpha^2)}, \\ A_2^{(0)} &= \frac{2g\delta\Psi}{\omega_0^2\,(4\Psi^2 + \alpha^2)}, \\ B_1^{(0)} &= 0, \\ B_2^{(0)} &= 0. \end{aligned} \tag{24}$$

We see from the expressions (22), (23) and (24) that the weight-induced forced vibration of a cracked shaft is a forward precession (for $\omega \approx \omega_0/2$ and $\omega \approx \omega_0/3$) or the sum of forward and backward precessions (for $\omega \approx \omega_0$). In the first case the motion of the shaft axle is circular, but, in the second case it is elliptical. It is interesting to remark that precession takes place in spite of the fixed direction of the external force (which is the gravity force). The precession amplitude depends upon the parameter δ connected with the crack size. This fact is very important for a crack diagnosis.

6. Conclusion

The dynamics of a horizontal cracked shaft exhibits three types of behaviour: the crack may cause

(a) instability of rotation,
(b) a stable circular or elliptical motion of the curved shaft axle that may be called precession or forced vibration,
(c) a stable shaft rotation that does not essentially differ from the rotation of the same shaft before the crack occurred.

Case (a) may occur only if, firstly, the angular velocity of rotation ω is not far from one of the values $2\omega_0$, ω_0, $2\omega_0/3$ and, secondly, the crack size parameter δ is large enough as compared with the damping parameter α.

Case (b) may take place for ω near ω_0, $\omega_0/2$ or $\omega_0/3$ (for $\omega \approx \omega_0$ the precession may occur only if stability conditions are valid).

Case (c) takes place for angular velocities which are far from $2\omega_0$, ω_0, $2\omega_0/3$, $\omega_0/2$ and $\omega_0/3$.

The results of this paper show that it is possible to detect the development of a transverse crack by means of an experimental analysis of the shaft dynamic behaviour. The most significant fact for the crack diagnosis is the development of large vibration components of angular velocity 3ω (for ω close to $\omega_0/3$) or 2ω (for ω close to $\omega_0/2$). If the amplitude increases with time for ω close to $2\omega_0/3$ it may also mean that there is a crack. The angular speed intervals containing the values ω_0 and $2\omega_0$ are not so significant for a crack diagnosis because in these cases vibrations may be caused by other factors.

References

1. S. M. Bauer, S. B. Filippov, A. L. Smirnov and P. E. Tovstik, *Asymptotic methods in mechanics with application to thin shells and plates*, Asymptotic Methods in Mechanics (A. L. Smirnov and R. Vaillancourt, eds.), CRM Proceedings and Lecture Notes, this volume, Amer. Math. Soc., Providence, R.I., 1993, pp. 3–140.
2. R. Gasch, M. Person and B. Weitz, *Dynamic behaviour of the Laval rotor with a cracked hollow shaft – A comparison of crack models*, Int. Conf. on Vibrations in Rotating Machinery, (13–15 Sept. 1988), Inst. of Mech. Eng., London, 1988, pp. 463–472.
3. I. W. Mayes and W. G. R. Davies, *Analysis of the response of a multi-rotor-bearing system containing a transverse crack in a rotor*, ASME J. of Vibr., Acoust., Stress and Reliab. in Des. **106** (1984), 139–145.
4. A. H. Nayfeh, *Perturbation methods*, Wiley, New York, 1973 (426 pp.).
5. H. D. Nelson and C. Nataraj, *The dynamics of a rotor system with a cracked shaft*, ASME J. of Vibr., Acoust., Stress and Reliab. in Des. **108** (1986), 189–196.
6. Yu. B. Shneerson, *Dynamic stability of an anisotropic flexible shaft in anisotropic elastic damper supports*, Vestnik Leningrad. Univ. Mat. Mekh. Astronom. (1991), no. 8, 109–114; English transl., Len. Univ. Mech. Bull. (1991), no. 2, 44–49.
7. ——, *Dynamic stability of an asymmetric rotor on a thin shaft rotating in anisotropic elastic supports with damping*, Vestnik Leningrad. Univ. Mat. Mekh. Astronom. (1992), no. 8, 92–98; English transl., Len. Univ. Mech. Bull. (1992), no. 2, 32–38.
8. A. Tamura, Y. Iwata and H. Sato, *Unstable vibration of a rotor with a transverse crack*, Int. Conf. on Vibrations in Rotating Machinery (13–15 Sept. 1988), Inst. of Mech. Eng., London, 1988, pp. 647–653.

9. B. C. Wen and Y. B. Wang, *Theoretical research calculation and experiments of cracked shaft dynamic response.*, Int. Conf. on Vibrations in Rotating Machinery (13–15 Sept. 1988), Inst. of Mech. Eng., London, 1988, pp. 473–478.

ABSTRACT. The dynamic behaviour of a rotating shaft with a transverse crack is analyzed. The problem is very important for an early diagnosis of a crack. It leads to differential equations with periodic coefficients, which are asymptotically solved in the paper. The explicit expressions are derived for the amplitudes and frequencies of the vibrations and for the instability zones.

INSTITUTE OF MATHEMATICS AND MECHANICS, ST. PETERSBURG STATE UNIVERSITY, 2 BIBLIOTECHNAYA SQ., STARY PETERHOF, 198904 ST. PETERSBURG, RUSSIA

E-mail address: als@hq.math.lgu.spb.su

Centre de Recherches Mathématiques
CRM Proceedings and Lecture Notes
Volume **3**, 1993

Edge Effect under Large Axisymmetric Deformations of Shells of Revolution

Petr E. Tovstik

1. Introduction

Large axisymmetric deformations of thin elastic shells of revolution with physical and geometrical nonlinearities are studied. The shell behaviour is described by nonlinear equations of Kirchhoff–Love type. The nonlinear expressions for the stress resultants and the stress couples for the deformations of the middle surface are those proposed by K. F. Chernyh [**2**]. The system of ordinary differential equations describing the problem is of the sixth order and contains a small geometric parameter, i.e. the relative shell thickness, which makes it possible to use the asymptotic integration method. As a result, the entire shell stress-strain state separates into the membranous state and the edge effect. The purpose of this paper is to study the effect of the loading order on the form of the asymptotic solution. It is necessary to emphasize that we consider not only small deformations but also deformations of the order of 1. Large deformations may exist without destroying the structure only for rubber-like materials. Such large deformations may only be stretching, otherwise buckling of the axisymmetric state occurs.

2. Equations of equilibrium and elasticity relations

The middle surface of the undeformed shell is described by the formulae

$$r_0 = r_0(s_0), \qquad \varphi_0 = \varphi_0(s_0), \qquad x_{30} = x_{30}(s_0), \qquad s_1 \le s_0 \le s_2, \tag{1}$$

where s_0 is the length of the generatrix, r_0 is the distance between a point on the middle surface and the axis of rotation, φ_0 is the angle between the normal to the shell and the axis of rotation and x_{30} is the vertical coordinate (see Fig. 1).

After deformation, the same variables are denoted by s, r, φ, x_3. The following formulae are valid [**2**]:

$$\begin{gathered} r_0' = \cos\varphi_0, \qquad x_{30}' = -\sin\varphi_0, \qquad r' = \lambda_s \cos\varphi, \qquad {}' = \frac{d}{ds_0}, \\ x_3' = -\lambda_s \sin\varphi, \qquad \lambda_s = s', \qquad \lambda_\theta = \frac{r}{r_0}, \qquad h = h_0\lambda_\xi, \\ \varkappa_s = \frac{\lambda_\xi}{\lambda_s}\varphi' - \varphi_0', \qquad \varkappa_\theta = \frac{1}{r_0}\left(\frac{\lambda_\xi}{\lambda_\theta}\sin\varphi - \sin\varphi_0\right), \end{gathered} \tag{2}$$

1991 *Mathematics Subject Classification.* Primary: 73K15; Secondary: 73C50.
Key words and phrases. Shell of revolution, large deformation, asymptotic method.
This is the final form of the paper.

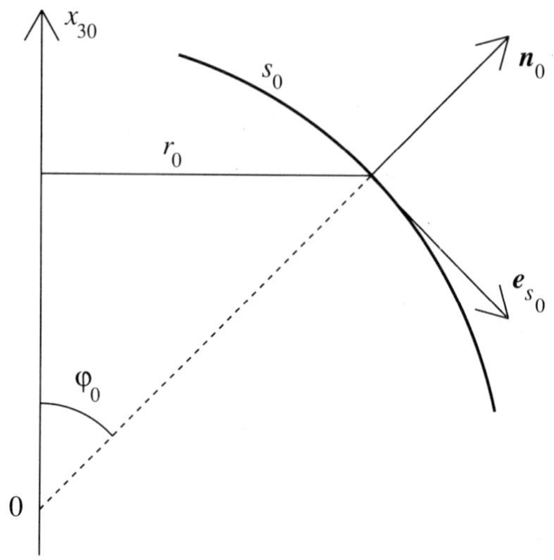

FIGURE 1. A shell of revolution.

where λ_s, λ_θ, λ_ξ are the extension ratios (for incompressible material $\lambda_s\lambda_\theta\lambda_\xi = 1$), $\varkappa_s$, $\varkappa_\theta$ are the variations of the principal curvatures of the middle surface, and h_0, h are the shell thickness before and after deformation, respectively. The equations of the projections of a shell element on the axes of equilibrium, after deformation, are given in [**2**] in the form

$$\begin{aligned}(r_0T_s)' - T_\theta \cos\varphi + r_0\varphi' N_s + r_0\lambda_s\lambda_\theta q_s &= 0,\\ (r_0N_s)' - T_\theta \sin\varphi - r_0\varphi' T_s + r_0\lambda_s\lambda_\theta q_n &= 0,\\ (r_0M_s)' - M_\theta \cos\varphi - r_0\lambda_s N_s &= 0,\end{aligned} \tag{3}$$

where the tangential stress-resultants, T_s and T_θ, the bending moments, M_s and M_θ, and the shearing stress-resultant, N_s, correspond to the unit length of the middle surface before deformation. The loading intensities, q_s and q_n, correspond to unit area after deformation.

The shell material is supposed to be isotropic and homogeneous. The unknown functions in equations (3) have to satisfy the elasticity relations [**2**], which can be written in the form

$$\begin{aligned}T_s &= S_s + \frac{1}{\lambda_s}\varphi' M_s, & S_s &= Gh_0t_1,\\ T_\theta &= S_\theta + \frac{M_\theta}{r_0\lambda_\theta}\sin\varphi, & S_\theta &= Gh_0t_2,\\ M_s &= \frac{1}{12}Gh_0^3(a_{11}\varkappa_s + a_{12}\varkappa_\theta), & M_\theta &= \frac{1}{12}Gh_0^3(a_{21}\varkappa_s + a_{22}\varkappa_\theta),\end{aligned} \tag{4}$$

where t_i and a_{ij} are given dimensionless functions of the extension ratios λ_s and λ_θ, depending on the form of the elasticity potential and G is the shear modulus for small deformations.

3. Equations and boundary conditions in dimensionless form

We introduce new (starred) dimensionless variables by the following formulae:

$$\begin{aligned}(r_0, r, s_0, s, x_{30}, x_3) &= R(r_0^*, r^*, s_0^*, s^*, x_{30}^*, x_3^*),\\ h_0 = h_* H(s_0^*), &\qquad (\varkappa_s, \varkappa_\theta) = \frac{1}{R}(\varkappa_s^*, \varkappa_\theta^*),\\ (T_s, T_\theta, S_s, S_\theta, N_s) &= G h_* (T_s^*, T_\theta^*, S_s^*, S_\theta^*, N_s^*),\\ (M_s, M_\theta) = G h_* R(M_s^*, M_\theta^*), &\qquad (q_s, q_n) = \frac{G h_*}{R}(q_s^*, q_\theta^*),\end{aligned} \tag{5}$$

and transform system (1)–(4) into dimensionless form. Here R is the characteristic linear size of the middle surface and h_* is the characteristic shell thickness. From now on we delete the $*$. Then equations (1)–(3) do not change form, but (2.4) becomes

$$\begin{aligned} T_s &= S_s + \lambda_s^{-1}\varphi' M_s, & S_s &= H t_1,\\ T_\theta &= S_\theta + \frac{M_\theta}{r_0 \lambda_\theta}\sin\varphi, & S_\theta &= H t_2,\\ M_s &= \mu^2 H^3(a_{11}\varkappa_s + a_{12}\varkappa_\theta), & M_\theta &= \mu^2 H^3(a_{21}\varkappa_s + a_{22}\varkappa_\theta),\end{aligned} \tag{6}$$

where $\mu > 0$ is a small parameter given by

$$\mu^2 = \frac{h_*^2}{12R^2}. \tag{7}$$

Now we impose an additional assumption on the functions t_i and a_{ij} in formulae (6) to restrict the elastic potential of the material. Let z be one of these functions. We introduce the variables ε_1 and ε_2 as follows

$$\lambda_s = 1 + \varepsilon_1, \qquad \lambda_\theta = 1 + \varepsilon_2,$$

and expand z in the form

$$z(\lambda_s, \lambda_\theta) = z_0 + \sum_{i+j=1}^{\infty} z_{ij}\varepsilon_1^i \varepsilon_2^j. \tag{8}$$

Let us suppose that the orders of all the coefficients z_{ij} are not greater than 1. For small deformations, i.e.

$$\varepsilon \ll 1, \qquad \varepsilon = \max\{|\varepsilon_1|, |\varepsilon_2|\} \tag{9}$$

this assumption allows us to use, in (6), the approximate formulae

$$\begin{aligned} t_i(\lambda_s, \lambda_\theta) &= 2\frac{\varepsilon_i + \nu\varepsilon_j}{1-\nu},\\ a_{ij}(\lambda_s, \lambda_\theta) &= a_{ij}(1,1) = a_{ij}^0,\end{aligned} \tag{10}$$

and to apply the linear shell theory. On the edges $s_0 = s_1$ and $s_0 = s_2$ we consider six types of boundary conditions

$$\begin{aligned} x_3 - x_{30} = b_1 &\quad \text{or} \quad T_s \sin\varphi - N_s\cos\varphi = b_2,\\ \lambda_\theta - 1 = b_3 &\quad \text{or} \quad T_s\cos\varphi + N_s \sin\varphi = b_4,\\ \varphi - \varphi_0 = b_5 &\quad \text{or} \quad M_s = \mu^2 b_6,\end{aligned} \tag{11}$$

where b_i are given constants.

We choose one condition from each line of (11) according to the edge support conditions. If the edge is clamped in the corresponding direction we use the left-hand conditions. Since at the top of a cupola $r_0 = 0$, instead of (11) we use the limitation conditions for the solutions.

4. The momentless problem

To estimate the loading order we introduce the dimensionless load, $q > 0$, as follows

$$(q_s, q_n, b_j) = q\left(q_s^0, q_n^0, b_j^0\right), \tag{12}$$

where q_s^0 and q_n^0 are given functions of s_0, and b_j^0 are given constants. Let us suppose that the orders of the variables q_s^0, q_n^0, b_j^0 are not greater than 1 and the orders of some of them are equal to 1. In special cases some of the variables q_s^0, q_n^0, b_j^0 may be equal to zero.

Firstly, we study the large deformations ($q \sim 1$). System (1)–(3), (6) contains a small parameter μ in the derivatives and is singularly perturbed. If $\mu \neq 0$, the system is of order six and if $\mu = 0$ it is of order four.

If $\mu = 0$ and $M_s = M_\theta = N_s = 0$, we obtain the nonlinear momentless system [**2**]:

$$\begin{aligned} (r_0 T_s^0)' &= T_\theta^0 \cos\varphi^0 - r_0\lambda_s^0\lambda_\theta^0 q_s, \\ r_0 T_s^0(\varphi^0)' &= -T_\theta^0 \sin\varphi^0 + r_0\lambda_s^0\lambda_\theta^0 q_n, \\ r_0(\lambda_\theta^0)' &= \lambda_s^0\cos\varphi^0 - \lambda_\theta^0\cos\varphi_0, \\ (x_3^0)' &= -\lambda_s^0\sin\varphi^0. \end{aligned} \tag{13}$$

The functions λ_s^0 and T_θ^0 on the right-hand sides of (13) may be found from (6):

$$T_s^0 = Hf_1(\lambda_s^0, \lambda_\theta^0), \qquad T_\theta^0 = Hf_2(\lambda_s^0, \lambda_\theta^0). \tag{14}$$

If $s_0 = s_j$, the boundary conditions (11) have the form

$$\begin{aligned} x_3^0 - x_{30} &= b_1 \qquad \text{or} \qquad T_s^0\sin\varphi^0 = b_2, \\ \lambda_\theta^0 - 1 &= b_3 \qquad \text{or} \qquad T_s^0\cos\varphi^0 = b_4. \end{aligned} \tag{15}$$

The nonlinear boundary problem (13)–(15) corresponds to Problem 5.1.2 in Section 5.1 of the Survey [**1**]. The existence and uniqueness of the solution of this problem depend upon: (a) the form of the shell, (b) the loading character and the loading order, (c) the elasticity relations and (d) the boundary conditions. We are interested only in solutions for which $T_s^0 \geq 0$ and $T_\theta^0 \geq 0$ for all s_0. If compressing stress-resultants ($T_s^0 < 0$ or $T_\theta^0 < 0$) are developing in the shell, it looses stability. Indeed, according to (5), for $\mu > 0$ shell buckling occurs if :

$$T_s^0 \sim \mu \qquad \text{or} \qquad T_\theta^0 \sim \mu, \tag{16}$$

and this buckling is accompanied by numerous small concavities. (See the book [**4**] by E. I. Grigolyuk and V. V. Kabanov and Section 2.7 of the Survey [**1**]).

Stability may also be lost if $T_s^0 > 0$ and $T_\theta^0 > 0$. In this case, the shell thickness becomes infinitely small and we say that the solution does not exist.

5. The edge effect

Let the solution of the momentless boundary problem (4.2)–(4.4) be smooth. We study the full boundary problem (2.2), (2.3), (3.2), (3.7) for $\mu > 0$. The solution algorithm for this problem as $\mu \to 0$ is described by A. B. Vasil′eva and V. F. Butuzov [**5**] and in Sections 5.1 and 5.2 of the Survey [**1**].

We denote by z any one of the unknown functions. If $q \sim 1$, we represent z as a sum of three asymptotic series

$$z = z^0 + z^{1e} + z^{2e} = \mu^{\alpha_0(z)} \sum_{j=0}^{\infty} \mu^j z_j^0(s_0) + \sum_{i=1}^{2} \mu^{\alpha_{ie}(z)} \sum_{j=0}^{\infty} \mu^j z_j^{ie}(\xi_i), \tag{17}$$

where

$$\xi_i = \frac{s_0 - s_i}{\mu}.$$

The first series, z^0, is the main (internal) solution. The last two series, z^{1e} and z^{2e}, are the edge effect solutions which decrease away from the edges $s = s_1$ and $s = s_2$, respectively, i.e.

$$\begin{aligned} z_j^{1e} &\to 0 \quad \text{for} \quad \xi_1 \to \infty, \\ z_j^{2e} &\to 0 \quad \text{for} \quad \xi_2 \to -\infty. \end{aligned} \tag{18}$$

According to (17) the index of variation, t, (see A. L. Gol′denweizer's book [**3**] and Section 1.7 of the Survey [**1**]) of the edge effect solutions is equal to 1. The intensity indices of the main solution and the edge effect solutions are denoted in (17) by $\alpha_0(z)$ and $\alpha_{ie}(z)$, respectively. The function z_0^0 coincides with the momentless solution which is assumed to exist. The main terms z_0^{1e} and z_0^{2e} of the edge effect solutions may be found from the nonlinear equations (described below) and the other terms, z_j^i, for $j \geq 1$, satisfy linear differential equations. We are going to evaluate only the functions z_0^{2e} and z_1^0.

For all the functions obtained in Section 3 of this paper, the intensity index $\alpha_0 = 0$. The intensity indices of the functions M_s, M_θ, N_s are equal to 2 ($\alpha_0 = 2$).

We study the edge effect near the edge $s_0 = s_2$. Later we shall omit the superscript and the subscript 2 and denote the functions z^{2e} and α_{2e} as z^e and α_e, respectively. The intensity of the edge effect depends upon the third boundary conditions in (11). Firstly, suppose the condition $\varphi - \varphi_0 = b_5$ is fulfilled for $s_0 = s_2$. Then for φ^e we obtain

$$\alpha_e(\varphi) = 0, \qquad \varphi_0^e = \beta_0 = b_5 + \varphi_0 - \varphi_0^0 \quad \text{for} \quad s_0 = s_2. \tag{19}$$

If $t = 1$, we find the intensity indices for the remaining functions. For the functions T_s, T_θ, S_s, S_θ, N_s and λ_s, the intensity index $\alpha_e = 0$, and for the functions M_s, M_θ, λ_θ, r and x_3, the intensity index $\alpha_e = 1$. Since $(\varphi^k)' \sim \mu^{-1}$, it follows from these estimates that, for the edge effect solution, both terms for T_s and T_θ on the right-hand sides of (6) are of the same order.

To find the solution z_0^e we apply the main features of the method described in [2]. The first two equations in (3) yield

$$\begin{aligned} &\frac{dT_{s0}^e}{d\xi_2} + N_{s0}^e \frac{d\varphi_0^e}{d\xi_2} = 0, \\ &\frac{dN_{s0}^e}{d\xi_2} - T_{s0} \frac{d\varphi_0^e}{d\xi_2} = 0, \\ &T_{s0} = T_{s0}^0 + T_{s0}^e. \end{aligned} \tag{20}$$

If we use (19), from (20) we get that

$$T_{s0}^e = T_{s0}^0(\cos\beta - 1), \qquad N_{s0}^e = T_{s0}^0 \sin\beta, \qquad \beta = \varphi_0^e. \tag{21}$$

Now we use the variable β instead of ξ_2 ($0 \le \beta \le \beta_0$) and obtain

$$\begin{aligned} &\frac{d}{d\beta}\left[(M_{s0}^e)^2\right] = 2aT_{s0}^0\lambda_{s0}\sin\beta, \\ &\frac{d}{d\beta}(\lambda_{\theta 0}^e) = \frac{a}{r_0 M_{s0}^e}\left(\lambda_{s0}\cos(\beta + \varphi_0^0) - \lambda_{s0}^0\cos\varphi_0^0\right), \\ &\frac{d}{d\beta}(x_{30}^e) = \frac{a}{M_{s0}^e}\left(\lambda_{s0}^0\sin\varphi_0^0 - \lambda_{s0}\sin(\beta + \varphi_0^0)\right), \\ &\frac{d\xi_2}{d\beta} = \frac{a}{M_{s0}^e}, \end{aligned} \tag{22}$$

where

$$\begin{aligned} &a = a(\lambda_{s0}, \lambda_{\theta 0}^0) = \frac{\lambda_{\xi 0}}{\lambda_{s0}} H^3 a_{11}(\lambda_{s0}, \lambda_{\theta 0}^0), \\ &\lambda_{s0} = \lambda_{s0}^0 + \lambda_{s0}^e, \qquad \lambda_{\xi 0} = \lambda_{\xi 0}^0 + \lambda_{\xi 0}^e. \end{aligned} \tag{23}$$

The constants T_{s0}^0, λ_{s0}^0, $\lambda_{\theta 0}^0$, $\lambda_{\xi 0}^0$ and H are the values of the corresponding functions at $s_0 = s_2$. The value of λ_{s0} in (23) may be found from (6) and it is

$$Ht_1(\lambda_{s0}, \lambda_{\theta 0}^0) = T_{s0}^0 \cos\beta - \frac{(M_{s0}^k)^2}{a\lambda_{s0}}. \tag{24}$$

According to (18), all the functions z_0^e are equal to zero if $\beta = 0$. This is why the first three equations in (5.6) are integrated with zero initial conditions. From the last equation in (5.6) we get

$$\xi_2(\beta) = -\int_\beta^{\beta_0} \frac{a}{M_{s0}^e}\, d\beta \tag{25}$$

and $\xi_2 \to -\infty$ as $\beta \to 0$.

We denote by z^0 any of the unknown functions in (4.2). To find the first term of the main solution z_1^0 we substitute the following expansion in (4.2):

$$z^0 = z_0^0 + \mu z_1^0, \tag{26}$$

and keep all terms of order μ (after an expansion in a power series in μ). As a result we get a linear boundary problem for z_1^0.

We now study the boundary conditions for z_1^0. Firstly we consider a clamped edge ($x_3 - x_{30} = b_1, \lambda_\theta - 1 = b_3$). Then we have

$$x_{31}^0 = -x_{30}^e(\beta_0), \qquad \lambda_{\theta 1}^0 = -\lambda_{\theta 0}^e(\beta_0) \qquad \text{for} \qquad s_0 = s_2, \tag{27}$$

where x_{30}^e and $\lambda_{\theta 0}^e$ may be found from (22).

It is necessary to use a special algorithm to solve the linear boundary value problem for the functions z_1^0 numerically. However, it is possible to use the algorithm as for the nonlinear momentless boundary problem (13) to find the sum (26) and correct the boundary conditions. In this case, the boundary conditions become

$$x_3^0 - x_{30} = b_1 - \mu x_{30}^e(\beta_0), \qquad \lambda_\theta^0 - 1 = b_3 - \mu \lambda_{\theta 0}^e(\beta_0). \tag{28}$$

If the stress boundary conditions (15) are given, then the corrected conditions have the form

$$\begin{aligned} T_s^0 \sin\varphi^0 &= b_2 - \mu(A\sin\varphi_0^0 + B\cos\varphi_0^0), \\ T_s^0 \cos\varphi^0 &= b_4 - \mu(A\cos\varphi_0^0 - B\sin\varphi_0^0), \end{aligned} \tag{29}$$

where

$$A = \int_0^{\beta_0} (F_1\cos\beta + F_2\sin\beta)\frac{a}{r_0 M_{s0}^e}\,d\beta,$$

$$B = \int_0^{\beta_0} (F_1\sin\beta - F_2\cos\beta)\frac{a}{r_0 M_{s0}^e}\,d\beta,$$

and

$$\begin{aligned} F_1 &= T_\theta(\lambda_{s0}, \lambda_{\theta 0}^0)\cos(\beta + \varphi_0^0) - T_\theta(\lambda_{s0}^0, \lambda_{\theta 0}^0)\cos\varphi_0^0 - r_0\lambda_{\theta 0}^0\lambda_{s0}^e q_s, \\ F_2 &= T_\theta(\lambda_{s0}, \lambda_{\theta 0}^0)\sin(\beta + \varphi_0^0) - T_\theta(\lambda_{s0}^0, \lambda_{\theta 0}^0)\sin\varphi_0^0 - r_0\lambda_{\theta 0}^0\lambda_{s0}^e q_n. \end{aligned} \tag{30}$$

Here λ_{s0} is the same as in (23).

6. The case of a freely supported edge

Now we consider the case where the condition $M_s = \mu^2 b_6$ is given as $s_0 = s_2$. In this case the intensity of the edge effect solution is less than the one in Section 5 (the intensity indices of all the functions are increased by 1). The edge effect solution may be found from the second-order linear differential equation

$$\begin{aligned} \varphi^e &= \mu\varphi_0^e + O(\mu^2), \\ a^0\frac{d^2\varphi_0^e}{\xi_2^2} &= \lambda_{s0}^0 T_{s0}^0 \varphi_0^e, \qquad a^0 = a(\lambda_{s0}^0, \lambda_{\theta 0}^0), \\ \varphi_0^e &= \frac{b_6 - M_{s0}^0}{a^0} = b_6^*, \qquad \text{for} \qquad s_0 = s_2, \end{aligned} \tag{31}$$

from which we get

$$\varphi_0^e = \frac{b_6^*}{p}e^{p(s_0 - s_2)/\mu}, \qquad p^2 = \frac{\lambda_{s0}^0 T_{s0}^0}{a_0}. \tag{32}$$

In this case, the momentless solution is sufficiently exact since $z^0 = z_0^0 + O(\mu^2)$.

7. The linear approximation

Let now the loading order be less than the one in the previous Sections 4–6 (i.e. $q \ll 1$), and set

$$q = \mu^{t_q}, \qquad t_q > 0. \tag{33}$$

In this case the approximate formulae (10) are valid and with some error the linearized system of equations (3) may be used to solve the problem. To estimate the order of this error for different values of t_q, we represent all the unknown functions as the sum, $z = z^0 + z^{1e} + z^{2e}$, of the momentless solution, z^0, and the two edge effect solutions, z^{1e} and z^{2e}, and study the neighbourhood of the edge $s_0 = s_2$ while neglecting the interference of the edge effect solutions. Then

$$z = z^0 + z^e. \tag{34}$$

Let the edge $s_0 = s_2$ be clamped. We take into account the projections of the displacement on the tangent u and on the normal w to the generatrix before deformation and the rotation angle, ψ, of the tangent for which

$$\begin{aligned} u &= (r - r_0)\cos\varphi_0 - (x_3 - x_{30})\sin\varphi_0, \\ w &= (r - r_0)\sin\varphi_0 + (x_3 - x_{30})\cos\varphi_0, \\ \psi &= \varphi - \varphi_0. \end{aligned} \tag{35}$$

Therefore the boundary conditions for $s_0 = s_2$ are as follows:

$$\begin{aligned} u &= b_1^* = \frac{b_3}{r_0}\cos\varphi_0 - b_1\sin\varphi_0, \\ w &= b_3^* = \frac{b_3}{r_0}\sin\varphi_0 + b_1\cos\varphi_0, \\ \psi &= b_5. \end{aligned} \tag{36}$$

Later we shall find only the second term of the momentless solution, z^0, and the first term of the edge effect solution, z^e. We suppose that all the momentless functions are of the order q

$$z^0 \sim q, \qquad z^0 = \{T_s^0, T_\theta^0, \varepsilon_s^0, \varepsilon_\theta^0, u^0, w^0, \psi^0\}. \tag{37}$$

Then, after linearizing (2) and (3), we find the momentless solution by quadratures:

$$\begin{aligned} T_s^0 &= \frac{1}{r_0\sin\varphi_0}\left[P + \int_{s_0}^{s_2}(q_s\sin\varphi_0 - q_n\cos\varphi_0)r_0\,ds_0\right], \\ T_\theta^0 &= \frac{1}{\sin\varphi_0}(r_0q_n - r_0\varphi_0'T_s^0), \\ u_0 &= \frac{1}{\sin\varphi_0}\left[C + \int_{s_0}^{s_2}(r_0\varphi_0'\varepsilon_\theta^0 - \varepsilon_s^0\sin\varphi_0)\,ds_0\right], \\ w_0 &= \frac{1}{\sin\varphi_0}(r_0\varepsilon_\theta^0 - u_0\cos\varphi_0), \end{aligned} \tag{38}$$

where ε_θ^0 and ε_s^0 are obtained from (10). Here P and C are arbitrary constants which may be found from the boundary conditions. We satisfy only one boundary condition at each edge. If the edge is clamped, we use the first condition in (7.4).

It follows from (38) that if the value of $\sin\varphi_0$ is equal or close to zero, then assumption (37) may be invalid. Therefore in this section we do not consider shallow shells for which the value of $\sin\varphi_0$ is small and shells similar to tori for which the value of $\sin\varphi_0$ is equal to zero at some specific points. For such shells, assumption (37) is not fulfilled and it is necessary to use special methods. At the same time, shells similar to cupolae, whose curvature at the top is not equal to zero, are not excluded from our consideration since the 0/0 indeterminacies in (38) may be resolved. We remark that no special consideration is required for shallow shells and shells similar to tori if $q \sim 1$ (see Sections 4 to 6).

According to this assumption, the relative error in z^0 is of the order $q = \mu^{t_q}$ and is caused by the linearization of equations (3).

We find now the edge effect solution z^e. The second equation in (3) becomes

$$\mu^2 a_{11}^0 \frac{d^4 w^e}{ds_0^4} - T_s^0 \frac{d^2 w^e}{ds_0^2} + \frac{2(1+\nu)w^e}{R_2^2} = 0, \qquad R_2 = \frac{r_0}{\sin\varphi_0}, \tag{39}$$

whose main terms coincide with the so-called nonlinear edge effect equation [**4**]. The other unknown functions may be expressed in terms of w^e as follows:

$$\begin{gathered} \psi^e = -\frac{dw^e}{ds_0}, \qquad M_s^e = -\mu^2 a_{11}^0 \frac{d^2 w^e}{ds_0^2}, \qquad N_s^e = -\mu^2 a_{11}^0 \frac{d^3 w^e}{ds_0^3}, \\ \varepsilon_\theta^e = \frac{w^e}{R_2}, \qquad \varepsilon_s^e = -\nu\varepsilon_\theta^e, \qquad T_\theta^e = 2(1+\nu)\varepsilon_\theta^e, \\ u^e = \int_{-\infty}^{s_0} (\varepsilon_s^e - \varphi_0' w^e)\, ds_0, \qquad T_s^e = \frac{\cos\varphi_0}{r_0} \int_{-\infty}^{s_0} T_\theta^e\, ds_0. \end{gathered} \tag{40}$$

We consider the functions $(a_{11}^0, T_s^0, R_2, \varphi_0, \varphi_0')$ at $s_0 = s_2$ and suppose that $H(s_2) = 1$. Then equation (39) has constant coefficients. If $T_s^0 \geq 0$, equation (39) has two solutions for each edge, which decrease exponentially away from the corresponding edge. The index of variation t of the edge effect solution depends upon the loading order q whose influence appears in the second term on the left-hand side of (39) ($T_s^0 \sim q$).

If $t_q \geq 1$, all the solutions of equation (39) have the same index of variation $t = 1/2$. If $t_q > 1$, the second term in (39) may be neglected to an error of the order of μ^Δ ($\Delta = t_q - 1$) and we get the simple edge effect solution (see [**3**] and Section 2.3 of the Survey [**1**]). If $t_q = 1$, all the terms on the left-hand side of (39) are of the same order.

If $0 < t_q < 1$, equation (39) has two edge effect solutions with different indices of variation in a neighbourhood of each edge (see Fig. 2), namely,

$$t_1 = 1 - \frac{t_q}{2}, \qquad t_2 = \frac{t_q}{2}. \tag{41}$$

If $t_q = 0$, two of the edge effect solutions are transformed into momentless solutions and the other two have the intensity indices $t = 1$ (see Section 5).

If $t_q < 1$, the solution of equation (39) has the form

$$w^e(s_0) = C_1 e^{p_1(s_0 - s_2)} + C_2 e^{p_2(s_0 - s_2)}, \qquad p_k \sim \mu^{t_k}, \tag{42}$$

where C_i are arbitrary constants which can be found from the boundary conditions

$$w^e(s_2) = b_3^* - w^0(s_2), \qquad (w^e)'(s_2) = -b_5 - (w^0)'(s_2), \tag{43}$$

and $C_1 \sim 1$, $C_2 \sim q$. The intensity of the edge effect for the remaining unknown functions can be found from (40).

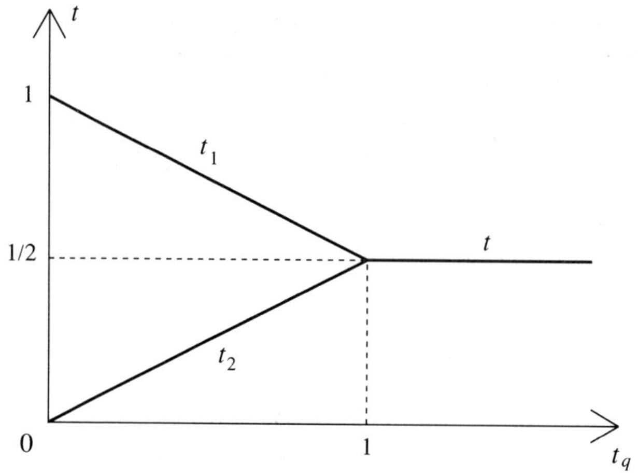

FIGURE 2. The indices of variation, t, of the edge effect solutions vs. the intensity order, t_q, of the loading.

If, in the power series expansion, we consider the next terms in μ, we find the error η of solution (34), in which we keep only the main terms,

$$\eta \sim \mu^{\Delta}, \qquad \Delta = \min\left\{\frac{1}{2}, \frac{t_q}{2}\right\}. \tag{44}$$

Therefore, if t_q is close to zero, the error η is essential. If the next terms are kept in (34), this error decreases but at the expense of lengthy computations.

Now, when $t_q > 0$, we use the algorithm of Sections 4 and 5, which was first proposed for $t_q = 0$, and we keep only the first terms in the series (17). The error, η_1, in this solution is of the order

$$\eta_1 \sim \mu^{\Delta_1}, \qquad \Delta_1 = 1 - t_q. \tag{45}$$

It follows from this formula that the algorithm of Sections 4 and 5 may be used only if $t_q < 1$.

If $t_q < 2/3$, comparing (34) and (35) we get (see Fig. 3) that the order η_1 is much less than η and the algorithm described in Sections 4 and 5 provides the smallest error.

For $t_q = 2/3$, we may take $\Delta = \Delta_1 = 1/3$.

8. A hemisphere under internal pressure

As an example, we study the expansion of a hemisphere of radius R under internal pressure. Taking into account (5), formulae (1) become

$$\varphi_0 = s_0, \qquad r_0 = \sin\varphi_0, \qquad x_{30} = \cos\varphi_0, \qquad 0 \le \varphi_0 \le \frac{\pi}{2}. \tag{46}$$

The edge $\varphi_0 = \pi/2$ is considered to be clamped, i.e.

$$x_3 = 0, \qquad \lambda_\theta = 1, \qquad \varphi = \frac{\pi}{2} \qquad \text{for} \qquad \varphi_0 = \frac{\pi}{2}. \tag{47}$$

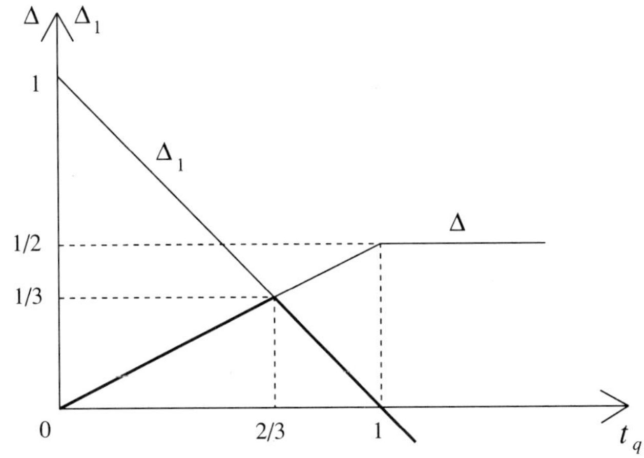

FIGURE 3. The orders of the errors, Δ and Δ_1, of the approximate solutions vs. the loading intensity order, t_q.

Let the shell thickness be constant ($H = 1$). The material is incompressible ($\nu = 1/2$) and in (3) we assume that $q_s = 0$ and $q_n = q$. The elasticity relations take the following form [**2**]:

$$t_1 = \frac{2}{n\lambda_s}(\lambda_s^n - \lambda_\xi^n), \qquad t_2 = \frac{2}{n\lambda_\theta}(\lambda_\theta^n - \lambda_\xi^n),$$
$$\lambda_\xi = \frac{1}{\lambda_s\lambda_\theta}, \qquad a_{11} = 4\lambda_\xi^{n+1}\lambda_s^2. \tag{48}$$

The formulae for the other a_{ij} in (6) are not written here because they are not used in these approximations. Formulae (48) correspond to the elasticity potential,

$$\Phi = 2\frac{G}{n^2}(\lambda_1^n + \lambda_2^n + \lambda_3^n - 3), \tag{49}$$

and may be obtained by using some hypotheses about the distribution of the strains and stresses in the shell thickness [**2**].

Firstly, from the linear equations of Section 7 we find the solution for $q \ll 1$. If we consider the relative variation of the shell height, Δx_3, and the bending moment, M_s, at the shell edge, $\varphi_0 = \pi/2$, as the functions which characterize the shell deformations, then we get

$$T_s^0 = T_\theta^0 = \frac{q}{2}, \qquad \varepsilon_s^0 = \varepsilon_\theta^0 = w^0 = \frac{q}{12},$$
$$w^e = \frac{w^0}{p_1 - p_2}\left[p_2 e^{p_1(\varphi_0 - \pi/2)} - p_1 e^{p_2(\varphi_0-\pi/2)}\right],$$
$$M_s\left(\frac{\pi}{2}\right) = -\frac{\sqrt{3}}{6}\mu q, \qquad \Delta x_3 = \frac{x_3(0) - x_3^0(0)}{x_3^0(0)} = w^0 + u^e\left(\frac{\pi}{2}\right),$$
$$u^e\left(\frac{\pi}{2}\right) = w^0\sqrt{\frac{3q}{8} + 3^{3/2}\mu}, \tag{50}$$

where p_1 and p_2 are the roots of the equation

$$4\mu^2 p^4 - T_s^0 p^2 + 3 = 0, \qquad \Re p_1 > 0, \qquad \Re p_2 > 0. \tag{51}$$

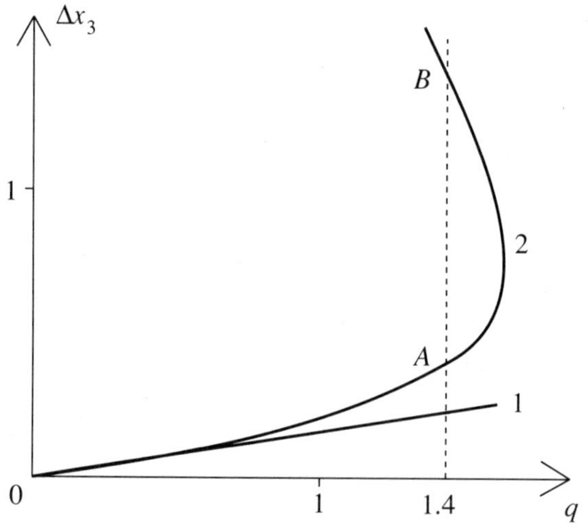

FIGURE 4. The hemisphere displacement, Δx_3, vs. internal pressure, q, according to the linear (1) and nonlinear (2) theories, respectively.

The first term in the formula for Δx_3 corresponds to the momentless solution and the second term is the correction to first order caused by the edge effect.

If $q \sim 1$, firstly we seek the solution of system (13) which is bounded as $\varphi_0 = 0$ and satisfy the condition $\lambda_\theta^0 = 1$ as $\varphi_0 = \pi/2$. Then, near the edge $\varphi_0 = \pi/2$, we get the edge effect solution from system (22). At last, we solve system (13) again and obtain its solution which satisfies the condition $\lambda_\theta^0 = 1 - \mu\lambda_{\theta 0}^e(\beta_0)$ as $\varphi_0 = \pi/2$ (see (27)). In both cases we solve system (13) by means of the shooting method. This means that we seek an initial value $\lambda_\theta^0(0)$ such that the condition for λ_θ^0 is fulfilled if $\varphi_0 = \pi/2$. For $\varphi_0 = 0$, we also assume that

$$\lambda_s^0 = \lambda_\theta^0, \qquad T_s^0 = T_\theta^0 = 2\frac{(\lambda_\theta^0)^n - (\lambda_\xi^0)^n}{n\lambda_\theta^0}, \qquad x_3^0 = 0. \tag{52}$$

The variation to first approximation, on the shell height, is

$$\Delta x_3^0 = 1 - x_{30}^0\left(\frac{\pi}{2}\right), \tag{53}$$

and, to second approximation (by taking into account the edge effect solution),

$$\Delta x_3 = 1 - x_3^0\left(\frac{\pi}{2}\right) - \mu x_3^e(\beta_0). \tag{54}$$

For the numerical analysis we take $n = 2$ in (48). The loading parameter q satisfies the inequalities $0.1 \le q \le 1.57$. The relative thickness is $h_0/R = 0.05$.

The function $\Delta x_3(q)$, which is found from formulae (50) (in the linear theory) and (54) (in the nonlinear theory), is plotted in Fig. 4. A comparison of these results shows good agreement if q is small and gets worse if q increases.

If $q \ge 1.58$, the solution to the nonlinear boundary value problem does not exist and if $q \le 1.57$ two solutions exist (they are shown in Fig. 4 for $1.3 \le q \le 1.57$).

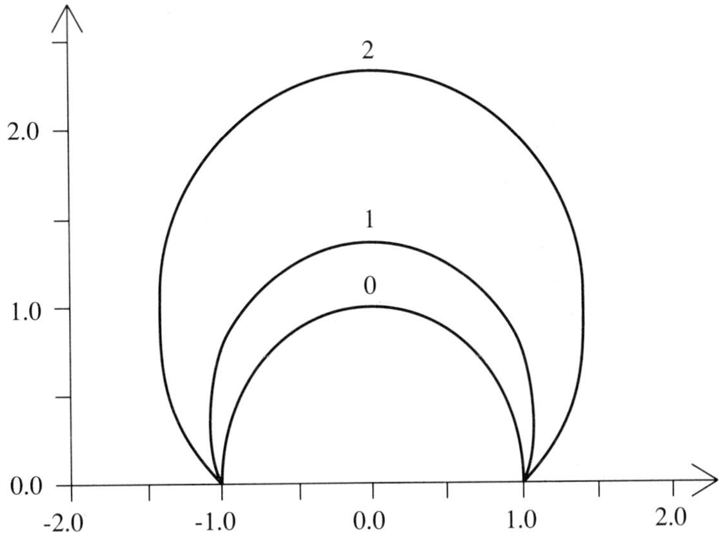

FIGURE 5. The initial form of a shell (0) and a deformation pair, (1) and (2), under loading $q = 1.4$.

In Fig. 5, the initial form of the shell and two shell stress-strain states are plotted for the load $q = 1.4$.

If we compare the whole sphere we get

(55)
$$T_s = T_\theta = \text{const}, \qquad \lambda_s = \lambda_\theta = \text{const},$$
$$q = \frac{4}{n}\left(\lambda_\theta^{n-3} - \lambda_\theta^{-2n-3}\right) = q(\lambda_\theta).$$

For $n < 3$ and $q < q_*$, equation (55) has two solutions, $\lambda_\theta^{(1)}$ and $\lambda_\theta^{(2)}$, but for $q > q_*$ it has no solutions. Here

(56)
$$q_* = \max_{\lambda_\theta}\{q(\lambda_\theta)\} = \frac{12}{2n+3}\left(\frac{3-n}{2n+3}\right)^{(3-n)/(3n)}.$$

When $n = 2$, we have $q_* = 1.24$ and this value is essentially less than $q = 1.57$. This difference is caused by the supporting effect of the clamped edge (the hemisphere resists greater pressure than the entire sphere). If $n > 3$, equation (55) has only one solution for all q (see Fig. 6).

Let n_* be such that the nonlinear boundary value problem for the hemisphere has only one solution for $n > n_*$ and all q, and no solutions for $n < n_*$ and sufficiently large q. An additional analysis reveals that $n_* \simeq 3$.

9. Conclusion

For $\mu \to 0$, the degeneracy of the entire boundary value problem of the sixth order into the membranous problem is singular. The full solution consists of the membranous solution and the edge effect. The character of this degeneracy depends on the loading order.

In the case of a large loading for which the dimensionless parameter q (see (12)) has order 1, the membranous system (13) is of the order 4. The solution of

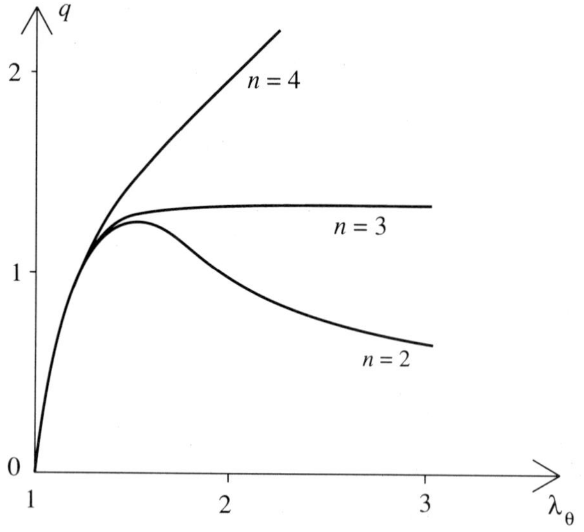

FIGURE 6. The function $q(\lambda_\theta)$ for different values of n.

this system satisfies, at each edge, two boundary conditions (15). The remaining boundary conditions (one at each edge) are satisfied by the edge effect solutions (see Section 5).

In the case of relatively small loading ($q \sim \mu$) the degeneracy changes character. The membranous system is of the second order and the edge effect solution may be found from equation (39) of the fourth order. Therefore the membranous solution satisfies one boundary condition but two boundary conditions are satisfied by the edge effect solutions. In this case the solution may be found by quadratures (see Section 7).

In the intermediate cases ($\mu \ll q \ll 1$) any of the above algorithms may be used. The first of them has smaller error than the second one for $q < \mu^{2/3}$ (see Fig. 3).

References

1. S. M. Bauer, S. B. Filippov, A. L. Smirnov and P. E. Tovstik, *Asymptotic methods in mechanics with application to thin shells and plates*, Asymptotic Methods in Mechanics (A. L. Smirnov and R. Vaillancourt, eds.), CRM Proceedings and Lecture Notes, this volume, Amer. Math. Soc., Providence, R.I., 1993, pp. 3–140.
2. K. F. Chernykh, *Nonlinear elasticity theory in engineering designs*, Mashinostroenie, Moscow, 1986 (360 pp.). (Russian)
3. A. L. Gol'denveizer, *Theory of elastic thin shells* (G. Herrmann, ed.), Pergamon Press, New York, Oxford, London, Paris, 1961 (658 pp.).
4. E. I. Grigolyuk and V. V. Kabanov, *Stability of shells*, "Nauka", Moscow, 1978 (359 pp.). (Russian)
5. A. B. Vasil'eva and V. F. Butuzov, *Asymptotic expansions of solutions of singular perturbed equations*, Nauka, Moscow, 1973 (272 pp.). (Russian)

ABSTRACT. Large axisymmetric deformations of thin elastic shells of revolution with physical and geometrical nonlinearities are studied. The asymptotic method which is connected with a small parameter (the relative shell thickness) is used. The complete stress-strain state of the shell is separated into the membranous state and the edge effect. The dependence of the asymptotic solution on the order of the loading is studied. The nonlinear solution is compared with its linear approximation. As an example, the deformations of a hemisphere with a clamped edge under internal pressure is analyzed.

INSTITUTE OF MATHEMATICS AND MECHANICS, ST. PETERSBURG STATE UNIVERSITY, 2 BIBLIOTECHNAYA SQ., STARY PETERHOF, 198904 ST. PETERSBURG, RUSSIA

E-mail address: als@hq.math.lgu.spb.su

Centre de Recherches Mathématiques
CRM Proceedings and Lecture Notes
Volume **3**, 1993

Turning Points and Caustics in Linear Problems of Thin Shell Free Vibrations and Buckling

Petr E. Tovstik

The equations for thin elastic shells contain a natural small parameter, $\mu > 0$, connected with the relative shell thickness, h/R. Thus in solving these equations it is convenient to use asymptotic integration methods. Some problems of shell vibrations and buckling may be reduced to linear ordinary differential equations. To solve the others we have to consider partial differential equations. If the coefficients of these equations are variable, then the construction of the asymptotic expansions of their solutions may be complicated by the turning points or caustics. These cases are discussed in this report.

1. Ordinary differential equations

1.1 Axisymmetric vibrations. The axisymmetric free vibration problem for shells of revolution may be reduced to the equation

$$\mu^4 \sum_{k=1}^{6} a_k(x) \frac{d^k w}{dx^k} + \sum_{k=0}^{2} b_k(x) \frac{d^k w}{dx^k} = 0, \qquad a_6 = 1, \qquad b_2 = k_2^2(x) - \Lambda, \tag{1}$$

where Λ is the unknown frequency parameter (see formula (2.6.15) in Subsection 2.6.4 of the Survey [**1**]) and $k_2(x)$ is the shell curvature. When

$$\min k_2^2 \leq \Lambda \leq \max k_2^2,$$

equation (1) has turning points at the roots $x = x_*$ of $b_2(x) = 0$.

If

$$b_2'(x_*) = \frac{1}{2} b_1(x_*) \neq 0,$$

the asymptotic expansion of the solutions of (1) has the form [**9**]

$$w^{(i)} \simeq \sum_{k=0}^{\infty} \varepsilon^k C_k(x) v_k^{(i)}(\eta) + \varepsilon \delta_{i5} w^*(x,\mu), \qquad i = 1,2,3,4,5, \qquad \varepsilon = \mu^{4/5}, \qquad \eta = \varepsilon^{-1} \left(\frac{5}{4} \int_{x_*}^{x} b_2^{1/4}\, dx \right)^{4/5}, \tag{2}$$

1991 *Mathematics Subject Classification.* Primary: 34E20; Secondary: 35B40, 73D30, 73H05, 73K15.

Key words and phrases. Shell, vibration, buckling, singular perturbation, asymptotic integration method, turning point, caustic.

Presented at the First European Congress of Mathematics, in Paris in July 1992.

This is the final form of the paper.

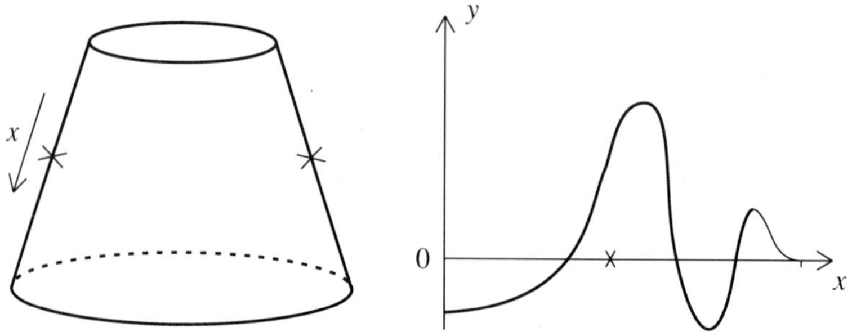

FIGURE 1. Axisymmetric vibration mode of a conical shell of revolution with turning point.

(3) $$w^{(6)} \simeq \sum_{k=0}^{\infty} \mu^{4k} w_k^{(6)}(x), \qquad \sum_{k=0}^{2} b_k(x) \frac{d^k w_0^{(6)}}{dx^k} = 0,$$

where the functions C_k, $w_k^{(6)}$, w^* and η are regular at $x = x_*$. The standard functions $v_k^{(i)}$ satisfy the conditions

(4) $$\frac{d^5 v_0}{d\eta^5} - \eta \frac{dv_0}{d\eta} - v_0 = 0, \qquad v_{k+1} = \int v_k \, d\eta.$$

Unfortunately, the expansions in (2) are valid only when $\varepsilon\eta \to 0$ or $x \to x_*$. If $|x - x_*| \sim 1$, the expansions in (2) for $i = 4, 5$ are not valid and it is necessary to piece them with simpler expansions of the form

(5) $$w(x, \mu) \simeq \sum_{k=0}^{\infty} D_k(x) \mu^k \exp\left(\mu^{-1} \int b_2^{1/4} \, dx\right)$$

and (3) for $x < x_*$ and $x > x_*$. Uniform asymptotic expansions for the solutions of (1) are unknown.

Expansion (2) has been used for the solution of boundary value problems [**2, 3, 9**]. One of the modes of axisymmetric vibrations of a conical shell of revolution is shown in Fig. 1. The crosses in Figs. 1–5 indicate turning points.

The case of two turning points of this type has been studied in [**16**] for a spheroidal shell.

Moreover, for a spheroidal shell a double turning point, x_*, can also exist at which

$$b_2(x_*) = b_2'(x_*) = 0, \qquad b_2''(x_*) \neq 0.$$

Asymptotic solutions can be obtained in a neighbourhood of x_* by means of the standard equation [**10**]

(6) $$\frac{d^6 v}{d\eta^6} = \eta^2 \frac{d^2 v}{d\eta^2} + 4\eta \frac{dv}{d\eta} + \gamma v.$$

1.2. Nonaxisymmetric vibrations. In the case of nonaxisymmetric vibrations, besides μ, there is one more parameter, m, which is equal to the wave number in the circular direction.

If m is small (fixed), formulae (2) keep their forms [**12**] and the number of solutions of the form (3) is equal to 3.

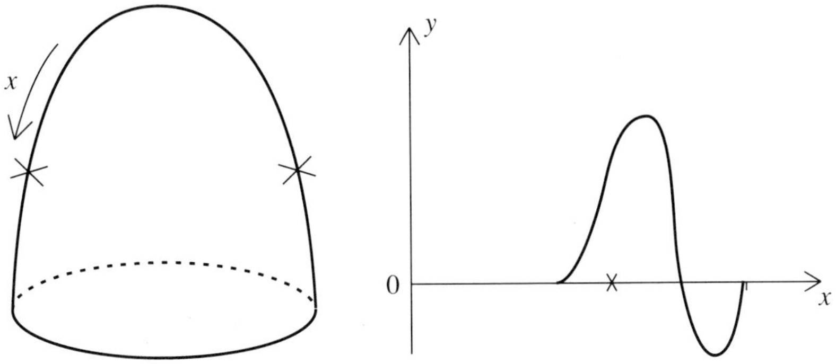

FIGURE 2. Vibration modes of a paraboloid.

If m is large ($m \sim \mu^{-1}$) the problem can be reduced to the following system [**3**]

$$\mu \frac{dy}{dx} = A(x,\mu)y, \qquad A(x,\mu) = \sum_{k=0}^{\infty} A_k(x)\mu^k, \tag{7}$$

where A is an 8×8 matrix. The buckling problem for the momentless axisymmetric initial stress state of a convex shell of revolution can also be reduced to system (7) [**11, 15**].

At the turning points $x = x_*$, the equation

$$\det(A_0(x) - \lambda E) = 0 \tag{8}$$

has multiple roots. Away from the turning points, the solutions of (7) have the expansions

$$y^{(i)}(x,\mu) \simeq \sum_{k=0}^{\infty} y_k^{(i)}(x)\mu^k \exp\left(\mu^{-1}\int \lambda_i\, dx\right). \tag{9}$$

Near a simple turning point, (where $\lambda_1(x_*) = \lambda_2(x_*) = 0$) two solutions have the form

$$\begin{gathered} y^{(j)}(x,\mu) \simeq \sum_{k=0}^{\infty} a_k^{(j)}(x)\mu^k v(\eta) + \mu^{1/3}\sum_{k=0}^{\infty} b_k^{(j)}(x)\mu^k \frac{dv}{d\eta}, \\ \frac{d^2v}{d\eta^2} - \eta v = 0, \qquad \eta = \mu^{-2/3}\left(\frac{3}{2}\int_{x_*}^{x} i\lambda_j\, dx\right)^{2/3}, \qquad j = 1,2, \end{gathered} \tag{10}$$

where v are standard Airy functions.

The free vibrations modes and buckling of shells are shown in Figs. 2, 3 and 4.

In Fig. 3, and in Figs. 7 and 8 below, we denote by $\mathbf{q}$ the external pressure on the shell. In Fig. 4, M is a twisting moment.

Let us consider a shell of revolution the curvature of which changes sign at $x = x_*$ as shown in Fig. 5. For the vibrations and buckling of such a shell, four solutions have expansions (7), where $\lambda_1 = \lambda_2$ and $\lambda_3 = \lambda_4$ for all x and $\lambda_1(x_*) = \lambda_3(x_*)$. Near the point x_* the solutions have the form (10) [**7**].

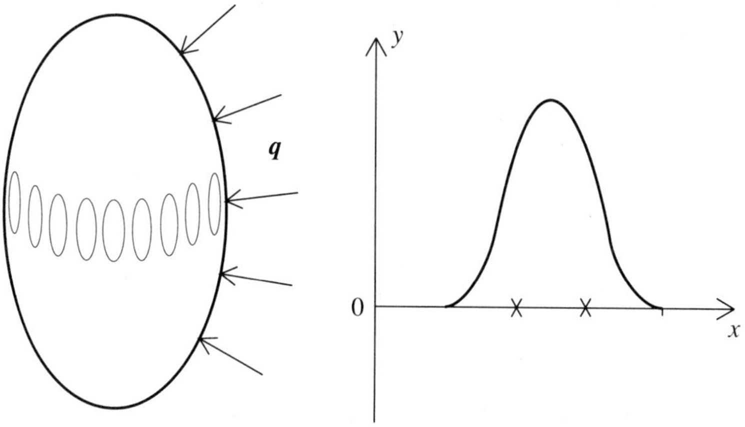

FIGURE 3. Buckling modes of a spheroidal shell under external pressure.

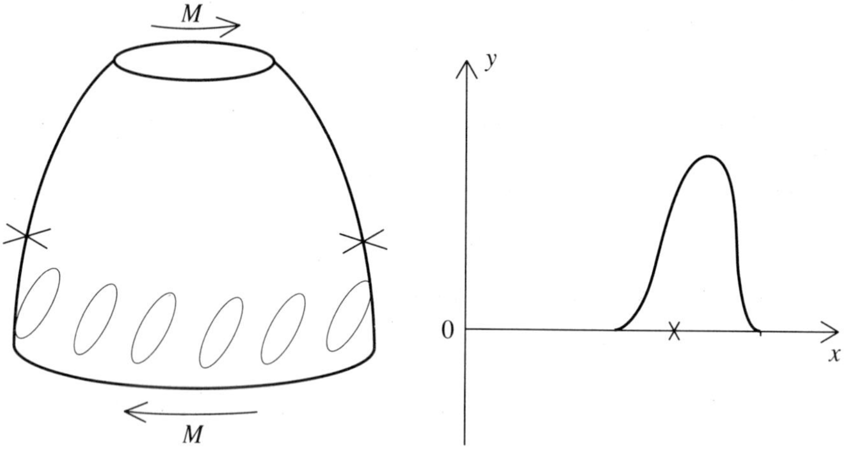

FIGURE 4. Buckling modes of a convex shell under torsion.

1.3. A pair of close turning points. Expansion (10) is useless if the distance between two turning points is small (i.e. $\left|x_*^{(2)} - x_*^{(1)}\right| = O\left(\mu^{1/2}\right)$, see Fig. 3). Let system (7) contain the unknown spectral parameter Λ. We seek the eigenvalues Λ for which this system has exponentially decreasing solutions away from the interval $\left[x_*^{(1)}, x_*^{(2)}\right]$.

Let $x_*^{(1)} = x_*^{(2)} = x_0$ for $\Lambda = \Lambda_0$. If equation (8) has the multiple root $\lambda = iq_0$, then its solution has the expansion [**14**]

$$(11)\qquad y(x,\mu) \simeq \sum_{k=0}^{\infty} \mu^{k/2} y_k(\xi) e^{(iq_0 x + a\xi^2)/\mu}, \qquad \xi = \frac{x - x_0}{\sqrt{\mu}}, \qquad \Re\, a < 0.$$

Here $y_k(\xi)$ are polynomials in ξ and $y_0(\xi) = H_m(\xi)$ is Hermite's polynomial of

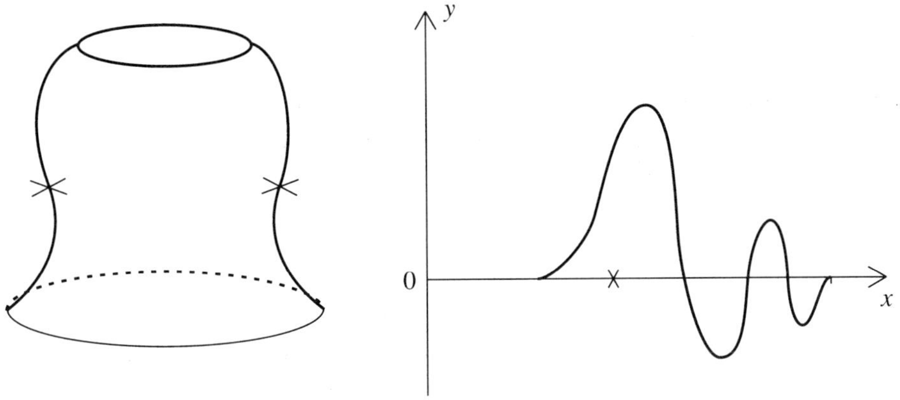

FIGURE 5. A shell of revolution the curvature of which changes sign.

degree m. The eigenvalues have the expansions

(12) $$\Lambda^{(m)} = \Lambda_0 + \mu r \left(m + \frac{1}{2}\right) + O\left(\mu^2\right), \qquad m = 0, 1, 2, \ldots$$

If $q_0 \neq 0$, these eigenvalues are asymptotically double.

Formulae (11) describe the buckling modes of a noncircular cylindrical shell under variable axial compression [**8**]. After separation of variables,

$$w(x, z) = y(x) \sin \frac{n\pi z}{l},$$

the two possible buckling modes corresponding to the same critical load are shown in Fig. 6. In Figs. 6 and 7, M_1 is a bending moment.

For a short shell, $q_0 = 0$ and the mode is similar to the one shown in Fig. 3.

2. Partial differential equations

2.1. Convex shells. The following system of partial differential equations of the fourth order:

(13) $$\mu^2 \Delta_0^2 w + \Lambda \Delta_1 w - \Delta_2 \Phi = 0, \qquad \mu^2 \Delta_0^2 \Phi + \Delta_2 w = 0,$$

may be used to solve some problems of free vibrations and buckling of shells. Here $Delta_1$ and Δ_2 are linear differential operators of the second order with variable coefficients, Δ_0 is the two-dimensional Laplace operator in curvilinear coordinates and Λ is the unknown eigenvalue (see Section 2.6 of the Survey [**1**]).

We seek solutions of system (13) which are exponentially decreasing away from the weakest point (x_0, y_0), in the form [**4, 13**]

(14) $$w(x, y, \mu) \simeq \sum_{k=0}^{\infty} \mu^k w_k(\xi, \eta) e^{iS(x,y)/\mu},$$

where

$$S(x, y) = p_0 x + q_0 y + V(x, y), \qquad \xi = \frac{x - x_0}{\sqrt{\mu}}, \qquad \eta = \frac{y - y_0}{\sqrt{\mu}},$$

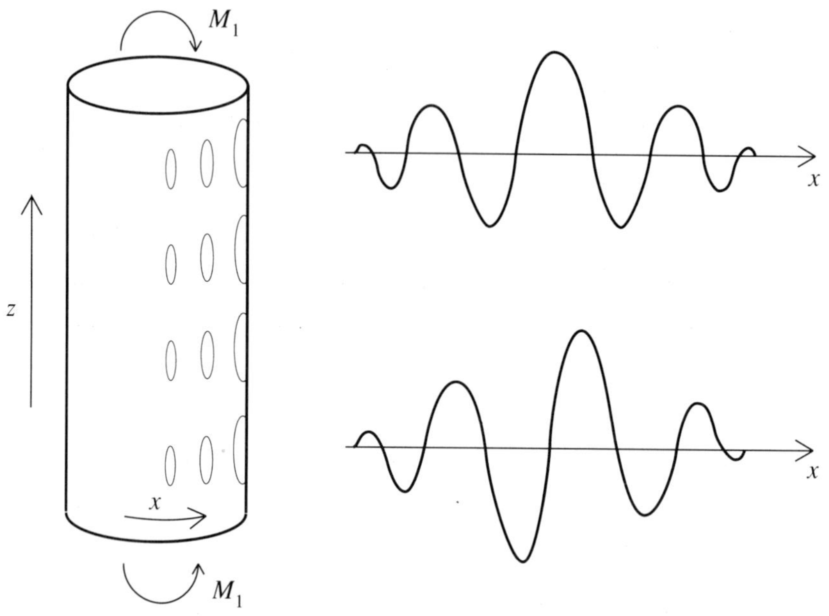

FIGURE 6. The two possible buckling modes corresponding to the same critical load.

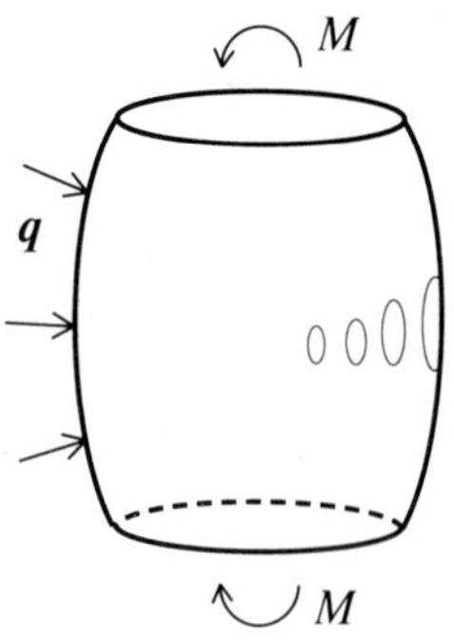

FIGURE 7. The buckling mode of a spheroidal shell under complex loading.

and $w_k(\xi,\eta)$ are polynomials and $V(x,y)$ is a homogeneous quadratic form in x and y with positive definite imaginary part.

The buckling mode of a spheroidal shell under complex loading is described by formula (14) (see [**6**] and Fig. 7).

2.2. Cylindrical and conical shells. The mode (14) is used mainly for convex shells. Let us study the localized modes of low frequency vibrations or buckling of cylindrical and conical shells. Since the variation of these modes in the direction of the generator is sufficiently less than in the circular direction, then one can obtain an asymptotic separation of variables.

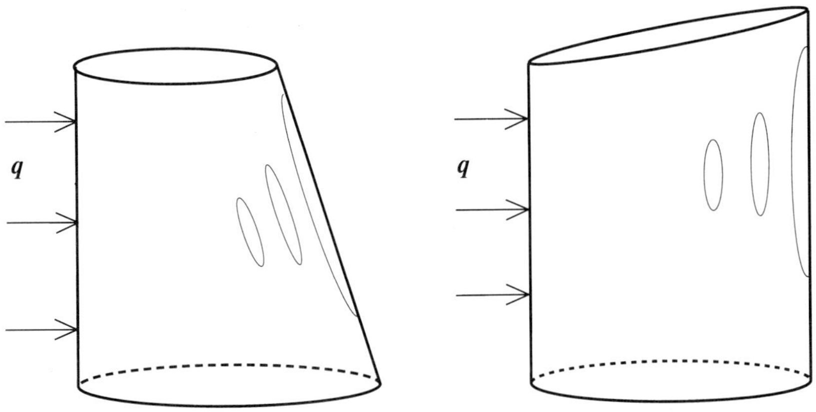

FIGURE 8. Vibration modes for two shells.

Let us rewrite system (13) in the form [**14**]

$$\mu^4\Delta_0^2 w + \mu^2\Lambda\Delta_1 w - \Delta_2\Phi = 0, \qquad \mu^4\Delta_0^2\Phi + \Delta_2 w = 0. \tag{15}$$

The solution of this system, localized near the weakest generator $y = y_0$, has the expansion [**14**]

$$w(x, y, \mu) \simeq \sum_{k=0}^{\infty} \mu^k w_k(x, \xi) e^{i\mu_0^q y + a\xi^2}, \qquad \xi = \frac{y - y_0}{\sqrt{\mu}}, \quad \Re a < 0, \tag{16}$$

where the $w_k(x, \xi)$ are polynomials in ξ and $w_0(x, \xi) = v_0(x)H_m(\xi)$. The function $v_0(x)$ satisfies an ordinary differential equation of the fourth order and the eigenvalues are asymptotically double.

Two shells with the free vibration modes (16) are shown in Fig. 8.

Under external lateral pressure these shells have similar buckling modes.

The buckling form localized in a neighbourhood of the asymptotic line of the shell of negative Gaussian curvature has been constructed in [**5**].

References

1. S. M. Bauer, S. B. Filippov, A. L. Smirnov and P. E. Tovstik, *Asymptotic methods in mechanics with application to thin shells and plates*, Asymptotic Methods in Mechanics (A. L. Smirnov and R. Vaillancourt, eds.), CRM Proceedings and Lecture Notes, this volume, Amer. Math. Soc., Providence, R.I., 1993, pp. 3–140.
2. S. B. Filippov, *Axisymmetric vibrations of scarfed shells of revolution*, Proceedings of the 12-th All-Union Conference on the Theory of Shells and Plates, Sudostroenie, Leningrad, 1975, pp. 230–233. (Russian)
3. A. L. Gol′denveizer, V. B. Lidsky and P. E. Tovstik, *Free vibrations of thin elastic shells*, "Nauka", Moscow, 1979 (383 pp.). (Russian)
4. V. P. Maslov, *Complex WKB method in nonlinear equations*, "Nauka", Moscow, 1977 (384 pp.). (Russian)
5. A. L. Maiboroda, *Buckling of revolution shells of a negative Gaussian curvature under asymmetric loads*, Vestnik Leningrad. Univ. Mat. Mekh. Astronom. (1989), no. 22, 61–65; English transl., Len. Univ. Mech. Bull. (1989), no. 4, 30–36.
6. G. I. Mikhasev, *Local loss of stability of a thin truncated ellipsoid acted upon by a combined load*, Vestnik Leningrad Univ. Mat. Mekh. Astronom. (1984), no. 19, 85–90; English transl., Len. Univ. Mech. Bull. (1984), no. 4, 63–71.

7. M. B. Petrov and P. E. Tovstik, *Low-frequency vibrations of shells of revolution with a curvature which changes sign*, Proceedings of the 12-th All-Union Conference on the Theory of Shells and Plates, vol. 3, Yerevan, 1980, pp. 118–124. (Russian)
8. G. V. Timofeeva, *Loss of stability by a cylindrical shell upon bending by a force and moments*, Vestnik Leningrad. Univ. Mat. Mekh. Astronom. (1984), no. 19, 99–101; English transl., Len. Univ. Mech. Bull. (1984), no. 4, 72–76.
9. P. E. Tovstik, *Asymptotic integration of the equations of free axisymmetric vibrations of a thin shell of revolution in the case of single turning point*, Izv. Akad. Nauk SSSR Mekh. Deform. Tverd. Tela **2** (1967), no. 4, 124–132; English transl., Mech. Solids **2** (1967), no. 4, 81–86.
10. ——, *On the symmetrical vibrations of the shell of revolution in the case of the turning point of the second order*, Vestnik Leningrad. Univ. Mat. Mekh. Astronom. (1967), no. 1, 118–124. (Russian)
11. ——, *Buckling of shells of revolution in the linear approximation*, Raschety prostranstvennykh konstruktsii, vol. 13, Stroiizdat, Moscow, 1970, pp. 118–138. (Russian)
12. ——, *Nonaxisymmetric vibrations of shells of revolution with small wave numbers along the parallels*, Issledovaniya po uprugosti i plastichnosti, vol. 8, Leningrad University Press, Leningrad, 1971, pp. 131–140. (Russian)
13. ——, *On the local buckling of shells*, Vestnik Leningrad. Univ. Mat. Mekh. Astronom. (1982), no. 13, 72–78. (Russian)
14. ——, *Some problems of the stability of cylindrical and conical shells*, Prikl. Mat. Mekh. **47** (1983), no. 5, 815–822; English transl., J. Appl. Math. Mech. **47** (1983), no. 5, 657–663.
15. ——, *On the forms of local buckling of thin elastic shells*, Trans. CSME **15** (1991), no. 3, 199–211.
16. P. E. Tovstik and S. B. Filippov, *Axisymmetric vibrations of a shell of revolution in the case of two points of rotation*, Izv. Akad. Nauk SSSR Mekh. Deform. Tverd. Tela **9** (1974), no. 5, 98–103; English transl., Mech. Solids **9** (1974), no. 5, 88–93.

ABSTRACT. Linear boundary value problems for the free vibrations and buckling of thin elastic shells are studied. Using a small parameter (the relative shell thickness), asymptotic solutions of some problems are constructed. Special attention is devoted to the cases of turning points and caustics, which often occur in buckling problems and in low-frequency vibration problems.

INSTITUTE OF MATHEMATICS AND MECHANICS, ST. PETERSBURG STATE UNIVERSITY, 2 BIBLIOTECHNAYA SQ., STARY PETERHOF, 198904 ST. PETERSBURG, RUSSIA

E-mail address: als@hq.math.lgu.spb.su

About the Authors

Maximilian Ya. Antimirov was born in 1937 in Grozny, Russia. He graduated from Kirgiz State University in Frunze (Kirgizstan) in theoretical physics in 1960. From 1960 to 1964, he was assistant and senior lecturer in the Department of Mathematics of the Grozny Petroleum Institute. In 1964–1965, he did his doctoral studies at the Computer Centre of the Latvian State University in Riga, Latvia, where he held a Senior Research Fellowship from 1965 to 1968. He obtained his Ph.D. (Candidate in physics and mathematics) in 1966. From 1968 to 1991 he was docent in the Department of Applied Mathematics of the Riga Technical University and is now professor in the same Department. In 1991, he obtained the degree of Doctor of Mathematics and Physics at the Physics Institute of the Academy of Sciences of Latvia and was promoted to full professorship in his University. He published more than 70 research papers.

Svetlana M. Bauer was born in Barnaul, Russia in 1954. She is Associate Professor in the Department of Hydroelasticity of the Faculty of Mathematics & Mechanics, St. Petersburg State University. She graduated in Applied Mathematics from St. Petersburg State University in 1975 and obtained her Ph.D. in Mechanics of Solids from the same university in 1980 for the thesis: *The stability of the thin elastic shells of revolution with imperfections of form.* She is the author of papers on asymptotic methods in dynamics of thin structures and elastic stability of thin elastic shells, published in *Optiko-mekhanicheskaya promyshlennost', Vestnik Leningradskogo Universiteta* and other journals.

Boris A. Ershov was born in Leningrad, Russia, in 1925. He is Professor in the Department of Hydroelasticity of the Faculty of Mathematics & Mechanics, St. Petersburg State University. He graduated in Applied Mathematics from St. Petersburg State University in 1949, and obtained his Ph.D. in Mechanics of Solids from St. Petersburg State University in 1953 and the degree of Doctor of Science in 1990 from the same university for the thesis: *Transient processes in the tied problems of hydroaeroelasticity.* He is the author of numerous papers on analytical methods in hydroaeromechanics and the theory of automatic control published in *Differential Equations, Vestnik Leningraskogo Universiteta* and other journals.

Sergei B. Filippov was born in Leningrad, Russia, in 1946. He is Senior Research Associate at the Laboratory of Applied Mechanics of the Institute of Mathematics & Mechanics, St. Petersburg State University. He graduated in Applied Mathematics from St. Petersburg State University in 1970 and obtained his Ph.D. in mechanics of solids from the same university in 1975 for the thesis: *Free vibrations of the thin shells of revolution.* He is the author of papers on vibrations and stability of thin shells published in *Mechanics of Solids*, *Vestnik Leningradskogo Universiteta* and other journals.

Andrei A. Kolyshkin was born in 1954 in Tallinn, Estonia. He graduated in applied mathematics from the Riga Technical University in Riga, Latvia, in 1976, and obtained his Ph.D. (Candidate in physics and mathematics) from Saint-Petersburg (then Leningrad) State University in 1981. In 1979–1980, he held a Research Fellowship in the Laboratory of Nondestructive Testing at the Riga Technical University. Since then he is with the Department of Applied Mathematics of the Riga Technical University, as assistant in 1980–1983, senior lecturer in 1983–1990 and associate professor (docent) from 1990. In 1987–1988, he visited the University of Ottawa under the auspices of the Canada-USSR Academic Exchange Program, and in 1991–1992 and 1993 as a visiting researcher. He published more than 30 research papers.

Aleksandr L. Maiboroda was born in Khmelnitskii, Ukraine in 1961. He is Research Associate at the Institute of Transportation Problems of the Russian Academy of Sciences. He graduated in Applied Mathematics from St. Petersburg State University in 1983 and obtained his Ph.D. in Mechanics of Solids from the same university in 1987 for the thesis: *The local buckling of the shells.* He is the author of papers on vibration and stability of elastic shells, and numerical and computer methods in mechanics of elastic structures, published in *Vestnik Leningradskogo Universiteta* and other journals.

Yuliya A. Mochalova was born in Leningrad, Russia, in 1965. She is a junior research associate at the Institute of Engineering Problems of the Russian Academy of Science. She graduated in Applied Mathematics from St. Petersburg in 1987. She is the author of papers on analytical methods in hydroelasticity, published in *Vestnik Leningradskogo Universiteta.*

Galina V. Pavilainen was born in Leningrad, Russia, in 1959. She is Research Associate at the Laboratory of Applied Mechanics of the Institute of Mathematics & Mechanics, St. Petersburg State University. She graduated in Applied Mathematics from St. Petersburg State University in 1981 and obtained her Ph.D. in Mechanics of Solids from the same university in 1983 for the thesis: *Elasto-plastic bending of transverse-isotropic plates.* She is the author of papers on analytical and numerical methods in nonlinear mechanics of thin structures and beams, published in *Vestnik Leningradskogo Universiteta* and other journals.

Ekaterina V. Polyakova was born in Leningrad in 1966. She is a Ph.D. student in the Department of Hydroelasticity of the Faculty of Mathematics & Mechanics, St. Petersburg State University. She graduated in Applied Mathematics from St. Petersburg State University in 1987. Her papers on analytical methods in hydroelasticity appeared in *Vestnik Leningradskogo Universiteta.*

Yury P. Shcheviev was born in Leningrad, Russia, in 1935. He is Professor and Head of the Department of Accoustics of the Faculty of Electrical Engineering of the St. Petersburg Institute of Cinema Engineering. He graduated in Engineering from the Leningrad Institute of Electrical Engineering in 1959, and obtained his Ph.D. in Engineering from the Krylov Central Research Institute (St. Petersburg) in 1967 and the degree of Doctor of Science in 1983 from the Polzunov

Turbine Institute (St. Petersburg) for the thesis: *Design and manufacturing theory of soundproofing constructions.* He is the author of more than 70 papers and four monographs (including *Some diagnosis problems of heterogeneous materials by acoustic methods* (1977) and *Analysis and synthesis of heterogeneous acoustic media* (1984)) on noise protection, synthesis of heterogeneous acoustic media and optimization methods of multi-layer protective system elements.

Yury B. Shneerson was born in Leningrad, Russia, in 1960. He is a Ph.D. student in the Department of Theoretical and Applied Mechanics of the Faculty of Mathematics & Mechanics, St. Petersburg State University. He graduated in Applied Mathematics from St. Petersburg State University in 1983. He is the author of papers on asymptotic and numerical methods in stability of shafts, published in *Vestnik Leningradskogo Universiteta.*

Andrei L. Smirnov was born in Leningrad, URSS, in 1956. He is Senior Research Associate at the Laboratory of Applied Mechanics of the Institute of Mathematics & Mechanics of the St. Petersburg State University and Associate Professor in the Department of Mathematics and Computer Sciences of the St. Petersburg Institute of Cinema Engineers, St. Petersburg State University. He graduated in Applied Mathematics from St. Petersburg State University in 1978 and obtained his Ph.D. in Mechanics of Solids from the same university in 1981 for the thesis: *Vibrations of the rotating shells of revolution.* He is the author of papers on asymptotic and numerical methods in mechanics of thin structures published in *Transactions of the ASME*, *Transactions of the CSME*, *Vestnik Leningradskogo Universiteta* and other journals.

Irina I. Strelkovskaya was born in Leningrad, Russia, in 1968. She is a Ph.D. student in the Department of Hydroelasticity of the Faculty of Mathematics & Mechanics, St. Petersburg State University. She graduated in Applied Mathematics from St. Petersburg State University in 1991. She is the author of papers on analytical methods in hydroelasticity, published in *Vestnik Leningradskogo Universiteta.*

Igor Yu. Teterin was born in Stalingrad, Russia, in 1952. He is Senior Research Associate at the Laboratory of Applied Mechanics of the Institute of Mathematics & Mechanics, St. Petersburg State University. He graduated in Applied Mathematics from St. Petersburg State University in 1974 and obtained his Ph.D. in solid mechanics from the same university in 1982 for the thesis: *Buckling of thin shells with defective forms.* He is the author of papers on buckling of thin structures published in *Vestnik Leningradskogo Universiteta* and other journals.

Petr E. Tovstik was born in Leningrad, Russia, in 1935. He is Professor and Head of the Department of Theoretical and Applied Mechanics of the Faculty of Mathematics & Mechanics, St. Petersburg State University. He graduated in Applied Mathematics from St. Petersburg State University in 1958, obtained his Ph.D. in Mechanics of Solids from the same university in 1963 and the degree of Doctor of Science in 1968 for the thesis: *Free vibrations and the stability of thin elastic shells.* He is the co-author, with A. L. Gol′denveiser and V. B. Lidsky, of a monograph entitled *Free vibrations of thin elastic shells* and numerous papers on asymptotic methods in mechanics of thin structures published in *Doklady of the Academy of Science of the USSR*, *Mechanics of Solids*, *Transactions of CSME*,

Differential Equations, *Soviet Applied Mathematics* and other journals. He received the First Prize for Research, Leningrad State University, 1970. He is a member of the National Committee on Theoretical and Applied Mechanics, and of the National Society of Acousticians. He is Vice-editor-in-chief of the journal *Vestnik Sankt-Petersburgskogo Universiteta, Ser. Math. Mekh. Astr.*

Rémi Vaillancourt was born in 1934 in Maniwaki, Québec, Canada. He took several bachelor's and master's degrees at the University of Ottawa, Ontario, Canada, and his Ph.D. under the direction of P. D. Lax and K. O. Friedrichs at the Courant Institute of Mathematical Sciences at New York University in 1969. He held an Office of Naval Research Fellowship at the University of Chicago where he worked under the direction of A. P. Calderón in 1969–1970. He is known for the Yamanugi–Nogi–Vaillancourt pseudo-difference operators and the Calderón–Vaillancourt theorem for non-classical pseudodifferential operators. Since 1970 he is with the department of Mathematics and jointly, since 1974, with the department of Computer Science at the University of Ottawa. He served as chairman of the former department in 1972–1976, as vice-president, president and past-president of the Canadian Mathematical Society in 1975–1981, and as chairman of the Canadian National Committee for the International Mathematical Union in 1979–1988. He is now president of the Canadian Applied Mathematics Society. He collaborated with late Hitoshi Kumano-go and Michihiro Nagase in translating and updating Kumano-go's monograph entitled *Pseudo-Differential Operators*, published by The MIT Press. He recently co-authored, with M. Ya. Antimirov and A. A. Kolyshkin, a monograph entitled *Applied Integral Transforms* in the CRM Monograph Series, published by AMS. His mathematical hobby is writing reviews for *Mathematical Reviews* and *Zentralblatt für Mathematik.*